Computational Intelligence and Mathematical Applications

Computational Intelligence and Mathematical Applications

Editors

Dr. Devendra Prasad
Dr. Suresh Chand Gupta
Dr. Anju Bhandari Gandhi
Dr. Stuti Mehla
Dr. Upasana Lakhina
Computer Science and Engineering,
Panipat Institute of Engineering and Technology, Samalkha, Panipat

CRC Press
Taylor & Francis Group
Boca Raton London New York

CRC Press is an imprint of the
Taylor & Francis Group, an **informa** business

First edition published 2024
by CRC Press
4 Park Square, Milton Park, Abingdon, Oxon, OX14 4RN

and by CRC Press
2385 NW Executive Center Drive, Suite 320, Boca Raton FL 33431

British Library Cataloguing-in-Publication Data
A catalogue record for this book is available from the British Library

ISBN: 978-1-032-87721-1 (pbk)
ISBN: 978-1-003-53411-2 (ebk)

DOI: 10.1201/9781003534112

Typeset in Times LT Std
by Aditiinfosystems

Contents

1. **Gold Price Prediction in Stock Exchange using LSTM and Bi-LSTM Multivariate Deep Learning Approaches** — 1
   Renu Popli, Isha Kansal, Rajeev Kumar, Vikas Khullar, Preeti Sharma and Devendra Parsad

2. **Application of Deep Neural Networks in the Detection of Surface Defects of Magnetic Tiles through the Analysis of Digital Images** — 6
   Jyoti Verma, Devendra Parsad, Isha Kansal, Renu Popli, Vikas Khullar and Rajeev Kumar

3. **Design of Micro-Machined Frequency and Pattern Reconfigurable Antennas** — 11
   Ashish Kumar, Devendra Parsad, Amar Partap Singh Pharwaha, Mandeep Jit Singh, Gurmeet Singh and Manpreet Singh

4. **Performance Evaluation of Junction-based Routing Protocols for Vehicular Ad hoc Networks (VANETs)** — 18
   Ishita Seth, Kalpna Guleria and Surya Narayan Panda

5. **AI Invasion in Banking: Impact of Digital Banking Adoption on IN-Branch Banking from Banker's Perspective** — 25
   Manju Bala and Sangeeta

6. **Random Forest Algorithm, Support Vector Machine for Regression Analysis** — 31
   Tushar Yadav, Nabha Varshey and Shanu khare

7. **Cloud Based Virtual Reality Training Platform for Occupational Safety** — 40
   Parishi Gupta, Kritika Gupta and Anup Lal Yadav

8. **Predicting Renewable Energy Demand using Machine Learning** — 47
   Anup Lal Yadav and Prabhneet Singh

9. **Volume Controller using Hand Gestures** — 53
   Anup Lal Yadav and Prabhneet Singh

10. **Blockchain and Data Security: Building Trust in a Digital World** — 60
    Dhananjay Batra, Shilpi Harnal, Paras Sachdeva, Rajeev Tiwari, Raghav Bali and Shuchi Upadhyay

11. **Fermatean Fuzzy Distance Measures with Pattern Recognition application** — 68
    Tamanna Sethi, Satish Kumarb, Megha Tanejaa and Rachna Khuranaa

12. **Intelligent Threat Mitigation in Cyber-Physical Systems** — 73
    Sandeep Singh Bindra and Alankrita Aggarwal

13. **Lane Detection Under Lousy Light Conditions** — 79
    Ritik Dwivedia and Kamal Kumar

14. **Data-Driven Precision Agriculture: Enhancing Crop Classification Accuracy and Agro-Environmental Analysis with Machine Learning** — 84
    Aaskaran Bishnoi, Gurwinder Singh and Ranjit Singh

15. **The Digital Dilemma—Exploring the Impact of Social Media on Mental Health** — 93
    Kumar Yashu, Akash Kumar, Vani Pandit, Kishan Raj, Anjali Gupta Somya and Sourabh Budhiraja

16. **IoT-Blockchain Integration for Optimizing the Performance of E-Health Systems: A Framework Design** — 99
    Deepak Singla, Sanjeev Kumar and Shipra Goel

17. **Issues Related to Accuracy and Performance during Classification Process in Sentiment Extraction from Image Description** 105
    Puneet Sharma and Sanjay Kumar Malik

18. **Human Activity Recognition from Sensor Data using Machine Learning** 111
    Prabneet Singh, Kshitij Kumar Pandey and Shubham Sharma

19. **Decoding Sentiments: A Comprehensive Review of Analytical Approaches and Associated Challenges** 118
    Richa Grover and Yakshi

20. **A Novel Outlook for Auto Coloring of Greyscale Images Using Python and OpenCV** 124
    Kulvinder Singh and Navjot Singh Talwandi

21. **Comparative Analysis of Machine Learning Algorithms used by Recommender Systems to Improve Marketing Strategies on E-commerce** 130
    Vikas Sethi and Rajneesh Kumar

22. **Evolution of Recommender System: Conventional to Latest Techniques, Application Domains and Evaluation Metrices** 136
    Vikas Sethi, Rajneesh Kumar, Suresh Chand Gupta, Sunny Kuhar and Sumit Kumar Rana

23. **Securing Cloud Data: A Comprehensive Analysis of Encryption Techniques and Recent Advancements** 143
    Juvi Bharti and Sarpreet Singh

24. **A IoT Based Student Registration Process in Education Sector** 149
    Gurpreet Singh, Mrinal Paliwal, Punit Soni and Rajwinder Kaur

25. **Unraveling the Nexus between Social Platforms and Psychological Well-Being** 155
    Kumar Yashu, Akash Kumar, Vani Pandit, Kishan Raj, Anjali Gupta Somya and Sourabh Budhiraja

26. **Rotten Apple Detection System Based on Deep Learning using Convolution Neural Network** 161
    Mrinal Paliwal, Gurpreet Singh, Punit Soni and Rajwinder Kaur

27. **Biometric System Integration in Public Health: Challenges and Solutions** 168
    Tajinder Kumar, Manoj Arora, Vikram Verma, Sachin Lalar and Shashi Bhushan

28. **Brain Tissue Segmentation Using Improvised 3D CNN Model on MRI Images** 174
    Saloni, Dhyanendra Jain, Anjani Gupta, Utkala Ravindran and Shruti Aggarwal

29. **Utilizing Transfer Learning Technique with Efficient NetV2 for Brain Tumor Classification** 180
    Deependra Rastogi, Heena Khera, Rajeev Tiwari and Shantanu Bindewari

30. **Believer: The Spiritual App—A New Era of Belief in the Digital Age** 185
    Megha, T. Mansi, K. Rajender, R. Kavita and S. Punit

31. **On Premises Vs Cloud Data Center: A Detailed Survey** 190
    Rabina Bagga, Kamali Gupta and Rahul Singla

32. **Survey and Observation of Task Scheduling in Cloud** 196
    Rabina Bagga and Kamali Gupta

33. **A Comprehensive Review of Soft Computing Approaches based on Different Applications** 202
    Harminder Kaur and Neeraj Raheja

34. **Ocular Disease Detection in Retina Fundus Images Using Deep Learning: Methods, Datasets, and Performance Metrics** 208
    Gurpreet Kaur and Amit Kumar Bindal

35. **Comparative Analysis of Classical and Quantum Machine Learning for Breast Cancer Classification** 216
    Manaswini De, Arunima Jaiswal and Tanya Dixit

36. **Comparative Analysis of Machine Learning and Deep Learning Algorithms for the Detection of Hate Speech** 225
    Arpana Jha, Arunima Jaiswal, Anshika Singh, Sampurnna Swain and Eshika Aggarwal

37. **Quantum Inspired Anomaly Detection for Industrial Machine Sensor Data** 231
    Tanisha Kindo, Arunima Jaiswal, Kirti Singh and Vanshika

38. **Challenges and Opportunities in IoT-based Smart Farming: A Survey Report** 236
    Isha Chopra, Amit Kumar Bindal and Zatin Gupta

39. **An Advanced Internet of Things-Based Health Monitoring System Assisted by Machine Learning** 241
    Venkateswaran Radhakrishnan, Husna Tabassum and Zatin Gupta

40. **Artificial Intelligence and Machine Learning in Women's Health: Unveiling PCOS with Image Analysis** 247
    Rishit Aggarwal, Shweta Jain, Anju Bhandari Gandhi, Drishti Dhingra and S.C. Gupta

41. **Hybrid FUCOM-TOPSIS based Matching of Users' Requirements with Life Insurance Policy Features** 253
    Asha Rani, Kavita Taneja and Harmunish Taneja

42. **Technological Marvels: Pioneering PCOS Detection through the Lens of Machine Learning and
    Deep Learning** 258
    Himanshi Jain, Shweta Jain, Anju Bhandari Gandhi, Rishit Aggarwal and S.C. Gupta

43. **Transforming Visits: Empowering Guests with QR-Based E-Ticketing** 265
    Deepti Dhingra, Mayank Jindal, Rahul Tomer and Jatin Rathee

44. **Prioritizing the Barriers of EVs adaptation in India using Fuzzy TOPSIS Methodology** 270
    Pardeep Kumar, Sunil Dhull, Anju Gandhi and Stuti Mehla

45. **Arduino Based Vehicle Accident Alert System Using GPS** 276
    Shanu Khare and Navpreet Kaur Walia

46. **Comparative Analysis of Heart Failure Detection Using Quantum Enhanced Machine Learning** 283
    Avni, Arunima Jaiswal, Amritaya Ray and Swati Gola

47. **Empirical Analysis of Depression Detection Using Bat and Firefly Algorithm** 289
    Kritika Shrivastava, Arunima Jaiswal, Suresh Kumar and Nitin Sachdeva

48. **Comparative Analysis of Optimisation Techniques for Depression Detection** 294
    Kritika Shrivastava, Arunima Jaiswal, Javed Miya and Nitin Sachdeva

49. **The Impact of Augmented Reality: Applications in Education** 300
    Khushi Rao, Daniel Kr Brahma, Jagriti Das, Alongbar Wary and Gaurav Indra

50. **Revolutionizing Attendance Management: Deep Learning and HOG-Based Smart System for
    Educational Institutions** 305
    Sunesh, Mamta, Priyanka and Aakanksha

Computational Intelligence and Mathematical Applications – Dr. Devendra Prasad et al. (eds)
© 2024 Taylor & Francis Group, London, ISBN 978-1-032-87721-1

List of Figures

1.1	*LSTM and Bi-LSTM training and validation loss metrics*	**3**
1.2	*LSTM and Bi-LSTM training and validation loss metrics*	**4**
1.3	*Truth versus predicted values for LSTM and Bi-LSTM separate*	**4**
3.1	*Various techniques to achieve reconfigurability in antennas*	**12**
3.2	*Micro-machined patch antenna*	**12**
3.3	*MEMS shunt capacitive switch*	**13**
3.4	*Reconfigurable micro-machined 1×2 patch antenna array*	**14**
3.5	*Frequency shift variations in 1×2 reconfigurable patch antenna array*	**14**
3.6	*Radiation patterns in switches ON and OFF conditions*	**14**
3.7	*Gain v/s frequency curve of the 1×2 antenna array structure*	**15**
3.8	*Reconfigurable micro-machined 2×2 patch antenna array*	**15**
3.9	*Frequency shift variations of the reconfigurable micro-machined 2×2 antenna array*	**16**
3.10	*Gain radiation pattern of reconfigurable micro-machined 2×2 antenna array in switches OFF and ON position*	**16**
3.11	*Gain v/s frequency curve of the 2x2 antenna design*	**16**
4.1	*VANET communication architecture*	**20**
4.2	*PDR to the speed of nodes*	**22**
4.3	*E2ED to the speed of nodes*	**22**
4.4	*Drop Ratio to the speed of nodes*	**23**
5.1	*'Economy creator' model*	**27**
5.2	*Representation of research analysis*	**28**
6.1	*Diagrammatic representation of methodology*	**34**
6.2	*Cancer dataset*	**36**
6.3	*Forest dataset*	**36**
6.4	*Wine database*	**36**
6.5	*Bike database*	**36**
6.6	*Evaluation of the regression method*	**38**
7.1	*VR Benefits in medical training*	**42**
7.2	*VR rendering framework*	**42**
7.3	*Operational flow*	**43**
7.4	*Real scenario of using VR*	**44**
8.1	*Energy management process on micro grid*	**49**
8.2	*Energy Consumption Forecasting Model*	**49**
8.3	*The architecture of neural networks*	**49**
8.4	*The planning of DNN and LSTM*	**50**
8.5	*Visualizing the resulting hourly dataset*	**50**
8.6	*Presenting model on the test dataset*	**51**

9.1	*Hand points*	**54**
9.2	*Use case diagram*	**56**
9.3	*Result I*	**57**
9.4	*Result II*	**57**
10.1	*(a) Centralized Networks, (b) Decentralized Network, (c) Distributed Ledger (adopted from Blockgeeks, 2016)*	**61**
10.2	*Biomedical security system with blockchain*	**62**
10.3	*Architecture of the blockchain-based open cyber threat intelligence system (BLOCIS)*	**63**
10.4	*Applications of blockchain technology*	**63**
10.5	*Architecture of blockchain-based IoT system for smart industry*	**64**
10.6	*The blockchain evoting system*	**65**
10.7	*Role of blockchain network in insurance company*	**66**
10.8	*Here is a table with some examples of fields in which blockchain is more helpful in the present than in the past, along with some relevant data*	**66**
11.1	*A timeline diagram to show flow of work*	**70**
12.1	*Block diagram of the proposed method*	**75**
13.1	*Proposed model for lane detection*	**80**
13.2	*Proposed neural network architecture for lane detection*	**81**
13.3	*Accuracy and loss of model*	**82**
13.4	*Output predicted by the model*	**83**
14.1	*Box plots of different crops*	**87**
14.2	*Histograms of parameter distributions*	**87**
14.3	*Heat map*	**87**
14.4	*Boxplot of parameter distributions*	**88**
14.5	*Price comparison for two crops*	**88**
14.6	*Correlation heatmap*	**89**
14.7	*Temperature variations*	**90**
14.8	*Rainfall trends*	**90**
15.1	*Indulgence of people into different activities over internet[3]*	**94**
15.2	*Positive and negative effects of social networks[4]*	**94**
15.3	*Positive and negative effects of social networks[8]*	**95**
15.4	*Average time spent during the depicted years.[10]*	**95**
15.5	*Distribution of social media platforms used[11]*	**95**
15.6	*Trends in social media use and mental health issues[13]*	**96**
15.7	*Relationship between social media usage and anxiety/depression[14]*	**96**
15.8	*The mental health and well-being profile of young adults using social media[15]*	**97**
16.1	*Alice signs her mails with her private key using a public key digital signature mechanism. By utilizing Alice's public key, Bob may confirm that Alice signed the message*	**100**
16.2	*Illustrate how to send/receive data in IPFS*	**101**
16.3	*Proposed system model*	**102**
16.4	*Contrasting various established methods*	**104**
17.1	*Key aspects of research in domain (ahmed et al. 2022)*	**106**
17.2	*Factors influencing accuracy during image classification in deep learning model*	**108**
19.1	*Methodology followed for Survey*	**119**
20.1	*Colourizing grey-scale image*	**125**
20.2	*Scrabbled shades turned into coloured picture*	**125**

20.3	*Experiment input and output I*	**127**
20.4	*Experiment input and output*	**127**
20.5	*Experiment input and output III*	**127**
21.1	*Machine learning techniques*	**131**
23.1	*Encryption model*	**143**
23.2	*Need of cloud encryption*	**144**
24.1	*Framework for students registration process using AIoT*	**152**
24.2	*Accuracy analysis of different semester*	**152**
24.3	*Miss ratio of different semester*	**153**
25.1	*Growth trajectory of social platform users over the past decade [3][4][5]*	**156**
25.2	*Correlation between self-esteem levels and social comparison tendencies among active social media users[7]*	**156**
25.3	*Impact of social media usage on psychological well-being over time [11]*	**156**
25.4	*Impact of social platform challenges on mental health[15]*	**157**
25.5	*Instances of cyberbullying reported in different demographics[16]*	**157**
25.6	*Schematic representation of the impact of ethical social platform design on user well-being[31]*	**158**
25.7	*Correlation between frequency of social platform usage and reported anxiety levels[34]*	**159**
26.1	*Sample images of fresh fruits and rotten fruits*	**162**
26.2	*Process flow of the proposed model*	**162**
26.3	*Architecture diagram of proposed model*	**164**
26.4	*The apple fruit disease detection system*	**165**
26.5	*Accuracy graph for the proposed system*	**166**
26.6	*Training and accuracy graph*	**166**
27.1	*Integration of biometric systems related to public health*	**169**
27.2	*Biometric Privacy and Security Considerations*	**170**
27.3	*Biometric Legal and ethical implications*	**171**
28.1	*Proposed 3D CNN model methodology*	**177**
29.1	*Dataset distribution and visualization (a) Dataset Label Distribution (b) Visualization of the random sample*	**181**
29.2	*Dataset distribution after distribution in train, test and validation*	**181**
29.3	*Image data augmentation*	**182**
29.4	*EfficientNetV2: Adding a combination of MBConvand Fused-MBConvblocks and transfer learning technique (a) Structure of MBConvand Fused-MBConvblocks [Tanet. al.], (b) EfficientNetV2 ith Transfer Learning Layer*	**182**
29.5	*Model performance and confusion matrix (a) Model Loss and Accuracy, (b) Confusion Matrix*	**183**
30.1	*Flow chart of spiritual app*	**187**
30.2	*Login/Registration*	**187**
30.3	*Home page*	**188**
30.4	*Audio section*	**188**
30.5	*Video section*	**188**
30.6	*Books*	**188**
31.1	*CC Infrastructure*	**191**
31.2	*Deployment Models of CC*	**191**
31.3	*Architecture of CC*	**192**
32.1	*Task scheduling processes*	**197**
33.1	*Components of model development*	**203**
33.2	*Fuzzy Inference System*	**203**
33.3	*Working of fuzzy system*	**203**

33.4   *Types of approaches for model development*   **204**

34.1   *Several diseases (i) Diabetic Retinopathy (DR) (ii) Diabetic Macular Edema (iii) Glaucoma (iv) Age-related Macular Degeneration (AMD) [4]*   **209**

34.2   *ODIR dataset Image [14]*   **211**

34.3   *Kaggle EyePacs dataset [15]*   **212**

34.4   *EFI dataset image [16]*   **212**

34.5   *OCTID [17]*   **212**

34.6   *Analysis of different methods: Accuracy Rate variables when all the occurrences are negative, whereas a container is constructive if various examples are positive [19]. In this, two variants of SVM methods, such as mi-SVM and MI-SVM. These methods are generalized as the powerful supervised SVM method*   **214**

34.7   *Performance based on different methods: Precision*   **214**

35.1   *Confusion matrix for train and test data set of one of the typical experimental cases*   **222**

36.1   *The above figure describes the generalized approach to our study of hate speech detection using the stated machine learning, deep learning, and ensemble techniques*   **226**

36.2   *LSTM algorithm which involves tokenization after preprocessing the data, creating an LSTM model with embedding and dense layers, compiling, training, and evaluating the model to produce multi-label classification*   **227**

36.3   *The above bar graph compares the accuracies of different ml, dl, and ensemble techniques for hate speech detection*   **229**

37.1   *Industrial machine anomaly detection workflow*   **232**

37.2   *Comparison of anomaly detection metrics*   **234**

37.3   *Performance Comparison - One-Class SVM vs. Quantum SVM*   **234**

38.1   *(a) and (b) Role of IoT devices in different domains of agriculture [8] [9]*   **237**

38.2   *IoT devices in different domains of agriculture [8]*   **237**

38.3   *Elements of crop monitoring [12]*   **238**

38.4   *Challenges in smart agriculture [19]*   **238**

38.5   *Model for the irrigation system [11]*   **238**

39.1   *System engineering methodology of a cutting-edge machine learning-based IoT-based health monitoring system*   **242**

39.2   *The Monitoring System of Health*   **243**

39.3   *Application of machine learning in the health monitoring system*   **244**

40.1   *Normal ovary vs polycystic ovary*   **248**

40.2   *Dataset A (First input is a polycystic ovarian ultrasound image and second input is a healthy fit ovarian ultrasound image)*   **249**

40.3   *Dataset B (First input is a healthy fit ovarian ultrasound image and Second input is a infected polycystic ovarian ultrasound image)*   **249**

40.4   *Workflow for training and evaluating our models*   **249**

40.5   *Simple architecture of PCONet*   **250**

40.6   *The first convolutional block of PCONeT*   **250**

40.7   *Outputs*   **251**

40.8   *Training and validation accuracy, precision, and loss*   **251**

41.1   *Ideal solution, anti-ideal Solution and closeness coefficient obtained from FUCOM-TOPSIS approach*   **256**

41.2   *Results of sensitivity analysis*   **256**

41.3   *Correlation coefficient between various test runs of the hybrid FUCOM-TOPSIS method*   **256**

42.1   *Normal and polycystic ovary*   **259**

42.2   *The complete view of research process*   **260**

42.3   *Implementation of image acquisition, pre-processing, and augmentation*   **261**

42.4	*Results of phase 0*	**261**
42.5	*Proposed architecture*	**263**
43.1	*Block diagram of the proposed model*	**266**
43.2	*Dashboard*	**268**
43.3	*Monument's info.*	**268**
43.4	*Payment*	**268**
44.1	*Shows the average CCi values of barriers of EVs*	**274**
45.1	*Entire amount of roads coincidences, examination of govt.of india from 2001 to 2010[2]*	**277**
45.2	*A vehicle accident alert systems block diagram*	**278**
45.3	*Architecture for automated smart accident detection*	**278**
45.4	*Usage circumstance diagrams for complete procedure of the classification*	**279**
45.5	*Implementation architecture*	**280**
46.1	*Quantum model building workflow*	**284**
46.2	*Quantum circuits generated by our qubit hyperparameter tuning code*	**285**
46.3	*Graphical analysis VQC model Accuracy*	**286**
46.4	*Graphical analysis VQC model metrics*	**286**
46.5	*Graphical analysis of QSVC model accuracy*	**287**
46.6	*Graphical analysis of QSVC model metrics*	**287**
46.7	*Graphical analysis of QML model Accuracies*	**287**
47.1	*Proposed methodology*	**290**
47.2	*Accuracies comparison of ML classifiers*	**291**
47.3	*ML classifiers BAT optimized accuracies*	**292**
47.4	*Firefly optimized accuracies*	**292**
47.5	*Comparison of bat & firefly optimized accuracies*	**292**
48.1	*Common symptoms of depression*	**294**
48.2	*Proposed methodology*	**295**
48.3	*Accuracies obtained by applying ML classifiers*	**297**
48.4	*Accuracies obtained by applying PSO optimization technique*	**297**
48.5	*Accuracies obtained by applying GWO optimization technique*	**298**
48.6	*Accuracies obtained by applying cuckoo optimization technique*	**298**
48.7	*Comparison of optimization techniques*	**298**
49.1	*Unity 3D Interface*	**301**
49.2	*Importing vuforia in unity*	**301**
49.3	*Creating image target using vuforia engine*	**302**
49.4	*3D image rendering*	**302**
49.5	*Sample images taken from application (a) Diaphragm (b) Human skeleton*	**303**
50.1	*Proposed attendance surveillance model*	**307**
50.2	*Automated communication results (a) E-mail results, (b) SMS result*	**309**

Computational Intelligence and Mathematical Applications – Dr. Devendra Prasad et al. (eds)
© 2024 Taylor & Francis Group, London, ISBN 978-1-032-87721-1

List of Tables

1.1	*Error analysis of LSTM, Bi-LSTM, and MLP (Multi-Layer Perceptron) algorithms using various evaluation metrics*	4
2.1	*Performance evaluation of various imaging algorithm*	8
2.2	*Comparing the system to other automated defect detection systems*	8
3.1	*Performance parameters of reconfigurable micro-machined antenna array with actuation of switches*	15
4.1	*Parameters considered for the simulation*	22
5.1	*Characteristics of participants*	27
6.1	*Dataset descriptions*	35
6.2	*Evaluation of regression method in wisconsin cancer dataset*	37
6.3	*Evaluation of regression method in forest fire dataset*	37
6.4	*Evaluation of regression method in wine quality dataset*	37
6.5	*Evaluation of regression method in bike sharing dataset*	37
11.1	*Estimated values of various FF compatibility measure related to the example*	71
12.1	*Literature survey*	75
14.1	*Crop summary*	86
14.2	*Insights from water quality metrics*	89
16.1	*Analysis of security requirements*	103
18.1	*Differentiation of Technologies for HAR*	114
19.1	*Research questions with objectives*	119
19.2	*Literature Review of Sentiment Analysis*	120
19.3	*Dataset used in sentiment analysis*	121
21.1	*Comparing different machine learning techniques in recommendation systems*	134
23.1	*Comparison of various encryption algorithms*	146
23.2	*Comparison of different proposed or modified encryption models*	146
24.1	*First round results of registration process*	152
24.2	*Second round results of registration process*	152
26.1	*Description of the number of images in the dataset*	163
26.2	*Different layers used in the proposed system*	164
26.3	*Recorded observations of simulating environment*	165
26.4	*Comparison of average loss and average accuracy in different epochs*	165
26.5	*Comparison of different method applied*	166
27.1	*Privacy and security considerations and impact on biometrics [6-7]*	170
27.2	*Legal and ethical implications and impact on biometrics [8]*	171
27.3	*Privacy-enhancing techniques and security measures and impact on biometrics [9]*	172
27.4	*Future directions and recommendations [3, 10-12]*	173
28.1	*Performance comparison of brain tissue segmentation methods*	177

28.2	*Comparison of segmentation performance with different hyper parameters*	177
28.3	*Comparison of segmentation performance with state-of-the-art methods on BRATS dataset*	177
28.4	*Computational cost comparison between proposed method and state-of-the-art methods*	178
28.5	*Summarized comparison between proposed model and state of art methods*	178
29.1	*Comparison*	184
30.1	*Review of Literature*	186
31.1	*Comparison between on premises and cloud*	193
32.1	*Related work on task scheduling based on identified algorithms*	199
34.1	*Analysis of existing methods*	210
34.2	*Data division from ODIR dataset*	211
34.3	*Kaggle EyePacs Data Detail*	212
34.4	*Result analysis based on different methods*	213
35.1	*Classical machine learning*	221
35.2	*Quantum machine learning*	221
36.1	*Results obtained by applying machine learning algorithms*	228
36.2	*Results obtained by applying deep learning algorithms*	228
36.3	*Results obtained by applying ensemble learning algorithms*	228
37.1	*Results obtained from unsupervised machine learning*	234
37.2	*Results obtained from Quantum machine learning*	234
38.1	*The difference in the technologies used in IoT-based farming and traditional farming*	237
39.1	*Displays the diagnostic criteria for standard health examinations*	242
39.2	*The table represents the pulse rate and states of the pulse rate of the patient*	244
41.1	*User inputs- demographic features*	255
41.2	*Priority score assigned to criteria with respect to the most preferred criteria C10*	255
41.3	*Result of hybrid FUCOM-TOPSIS method (Value of closeness coefficient of each life insurance policy)*	256
42.1	*List of acronyms*	259
42.2	*Results of phase 1*	262
42.3	*Results of phase 2*	262
42.4	*Results of phase 3*	262
42.5	*Results of phase 4*	264
43.1	*Technical requirements*	267
44.1	*List of barriers of EV adaptation*	271
44.2	*Criteria and alternative rating in linguistic terms (fuzzy rating)*	272
44.3	*Linguistic assessment of the criteria*	272
44.4	*Linguistic assessment of alternatives (Barriers)*	272
44.5	*Linguistic assessment of alternatives (Barriers of EV)*	273
44.6	*Aggregated closeness coefficients for alternative (Barriers of EVs)*	273
46.1	*Comprehensive metrics of VQC model*	286
46.2	*Comprehensive metrics of QSVC model*	286
46.3	*Accuracy achieved by QNN model*	286
49.1	*Analysis of the Data*	303
50.1	*Visual results of face identification under three imaging conditions*	308
50.2	*Accuracy results of proposed attendance surveillance model*	309

Preface

It is with great pleasure to present the proceedings of the International Conference on Computational Intelligence and Mathematical Applications (ICCIMA 2023), held on 21-22 December 2023, at Panipat Institute of Engineering and Technology, Panipat. This conference brought scholars, researchers, professionals, and intellectuals together from diverse fields to exchange ideas, share insights, and foster collaborations in Optimization, Computational Intelligence and Mathematical Applications.

The ICCIMA 2023 served as a platform for contributors to demonstrate their latest findings, discuss emerging trends, and explore innovations to the problems that different disciplines are currently experiencing. The conference's scope and depth of themes reflect our community's rich diversity of interests and levels of competence.

We extend our heartfelt gratitude to all the authors who contributed their valuable research papers to these proceedings. Their dedication to advancing knowledge and pushing the boundaries of their respective fields is commendable.

We would also like to express our appreciation to the members of the organizing committee, the program committee, and the reviewers for their tireless efforts in ensuring the high quality and relevance of the papers included in this volume.

Furthermore, we are thankful to AICTE and our management for their generous support, without which this conference would not have been possible.

Lastly, we hope that the discussions, insights, and collaborations sparked during the conference will continue to inspire future research endeavors and contribute to the advancement of science and technology.

Thank you for your participation and contribution to the success of the conference.

Team ICCIMA 2023

Acknowledgement

We would like to extend our sincere gratitude to all those who contributed to the success of the International Conference on Computational Intelligence and Mathematical Applications (ICCIMA 2023), held on 21-22 December 2023, at Panipat Institute of Engineering and Technology, Panipat. This conference would not have been possible without the collective effort and support of numerous individuals and organizations.

First and foremost, we express our appreciation to the authors who submitted their research papers and presentations. Their contributions have enriched the content of these proceedings and have significantly contributed to the advancement of knowledge in our field.

We are deeply thankful to Our Management, Director Prof. Dr. J.S. Saini, our editors Prof. Dr. Devendra Prasad, Prof. Dr. S. C. Gupta, Prof. Dr. Anju Bhandari Gandhi, Dr. Stuti Mehla, and Dr. Upasana Lakhina, and the organizing committee for their dedication, hard work, and meticulous planning, which ensured the smooth execution of the conference. Their commitment to excellence has set a high standard for future events.

We also extend our gratitude to the program committee and reviewers for their expertise, time, and constructive feedback in evaluating the submitted papers. Their rigorous assessment has ensured the quality and relevance of the papers included in this volume.

We are also grateful to the keynote speakers, session chairs, and panelists for sharing their insights, expertise, and perspectives, which have enriched the intellectual discourse of the conference.

Lastly, we extend our thanks to all the participants, attendees, volunteers, and staff members who contributed their time, energy, and enthusiasm to make the ICCIMA a vibrant and memorable event.

Thanks once again to each one who has played a part in making this conference a success. Your contributions are deeply appreciated.

Team ICCIMA 2023

About the Editors

Dr. Devendra Prasad is currently working as professor and Head in the Department of CSE(AIML) with Panipat Institute of Engineering and Technology (PIET), Samalkha Panipat, Haryana. He received the B. Tech, M.Tech, and a Ph.D. degree in CSE from Kumaon University Nainital, Kurukshetra University Kurukshetra, Haryana and MM(DU), Mullana, Ambala, Haryana respectively. Dr. Prasad has supervised 8 PhD Thesis and 28 MTech Thesis. He has published more than 60 research papers in different reputed international journals and conferences. He has published 22 National/International Design/Utility patents out of which 5 patents are granted. Dr. Prasad has written one Book and 4 Book Chapters with International Publishers. He has received Best paper Award from ICNGC2S-10), NITTTR Chandigarh, India. Dr. Prasad is a Member of IEEE, ACM, ACSE, and CSTA. His research interests are Blockchain Networks, Wireless Sensor Networks, Internet of Things (IoT) and Machine Learning.

Dr. S. C. Gupta is working as Head of CSE departments in PIET. During his 25 years of journey as teacher, he excelled in academics along with successful execution of administrative responsibilities, holding key positions at various prestigious institutes. As an active researcher he promoted young and fresh minds to envoke them in research. He guided numbers of M. Tech/ M.phil/MCA research scholars. He has published more than 50 national; international research articles/papers. He organized and cordinated various national/international events such as conferences, seminars, workshops, STCs in collaboration with MHRD Govt. of India. He is working as Nodal Coordinator for Virtual Labs programs of IIT, Delhi in PIET. He Contributed in B. Tech./M. Tech. curriculum design/revision program of KUK. His area of interest are software reusability, component based software development and reenginnering. He is an active member of various professional societies.

Dr. Anju Bhandari Gandhi has completed her B.Tech, M.Tech and Ph.D in Computer Science and Engineering and is currently working as a Professor Incharge-R&D at Panipat Institute of Engineering and Technology, Panipat, Haryana, India. She has 18 years of experience and published 65 research papers in reputed and Scopus indexed journals. She worked as convener of number of technical events like international conferences, seminars, FDPs, workshops and trainings. She has received number of awards and grants for research and innovation, establishing AICTE IDEA Lab. Her research area includes Artificial Intelligence, Machine learning and Data Science.

Dr. Stuti Mehla is working as an associate Professor in CSE departments in PIET. She has completed her B.tech, M.tech and P.hd degree in CSE from Kurukshetra university and Maharishi Markandeshwar University Mullana respectively in 2010,2012 and 2022. She is GATE Qualified and has 10 years of experience and published more than 30 research articles in reputed international journals and presented research papers in national and international conference. She is active member of IEEE and ACM. Her research interest is related to artificial intelligence, machine learning,data science and recommendation system.

Dr. Upasana Lakhina is working as an Assistant Professor in Panipat Institute of Engineering and Technology. She has achieved her B.tech and M.tech from kurukshetra university, Kurukshetra in 2014 and 2016 (Gold Medallist) and her Ph.D from University Teknologi PETRONAS, Malaysia (QS World Ranking 304). She is GATE qualified, and NET exempted and has done research collaboration with different researchers from Malaysia, France, Denmark and Vietnam. She is an active IEEE member and organized various conference and has a teaching experience of 7 years. She has also served as Judge in international Robotic Competition and also received best reviewer award. She started her research journey in 2015 as a scholar and published more than 20 articles in Journals and conferences. Her research area includes machine learning, optimization, microgrids, renewable energy and cloud computing.

Computational Intelligence and Mathematical Applications – Dr. Devendra Prasad et al. (eds)
© 2024 Taylor & Francis Group, London, ISBN 978-1-032-87721-1

Gold Price Prediction in Stock Exchange using LSTM and Bi-LSTM Multivariate Deep Learning Approaches

Renu Popli[1], Isha Kansal[2], Rajeev Kumar[3], Vikas Khullar[4], Preeti Sharma[5]
Chitkara University Institute of Engineering and Technology, Chitkara University, Punjab, India

Devendra Parsad[6]
Department of Artificial Intelligence and Machine Learning, Panipat Institute of Engineering and Technology (PIET), Panipat, Haryana, India

Abstract: As the economy of the entire world is dependent on gold prices, predictive study of gold prices is an absolute necessity. Given the global economic interdependence on gold prices, doing a predictive analysis of gold prices is imperative. Forecasting the price of gold in the stock exchange is a significant undertaking that carries substantial implications for investors and financial institutions. This study explores the field of gold price prediction through the utilization of Long Short-Term Memory (LSTM) and Bidirectional LSTM (Bi-LSTM), Multi Layer Perceptron (MLP). This paper presents a comprehensive examination and empirical investigation that reveals valuable insights into these models' predicted precision and pragmatic utility. The study begins by gathering historical data from the national stock exchange, encompassing crucial variables such as economic indicators, interest rates, currency exchange rates, and news emotion. These components possess the potential to exert a substantial influence on gold prices. A complete range of assessment metrics is utilized to assess the efficacy of the model, including Mean Squared Error (MSE), Root Mean Squared Error (RMSE), R2 Score, Normalised Mutual Information Coefficient (NMIC), Mean Absolute Percentage Error (MAPE), and Mean Absolute Error (MAE). The findings of this study provide insights into the notable prediction capabilities of LSTM models, which demonstrate superior performance compared to Bi-LSTM and Multilayer Perceptron (MLP) models across many evaluation criteria. The usefulness of LSTM in predicting is highlighted by its ability to grasp complex temporal patterns within gold price data.

Keywords: Forecasting, Gold rates, Linear regression, Neural networks, Prediction

1. INTRODUCTION

The volatility of gold prices in the stock market has been a topic of significant interest and worry among investors, financial analysts, and policymakers for an extended period. Gold is an exceptional investment form frequently seen as a secure haven. A multifaceted interaction of global economic, geopolitical, and financial elements impacts the fluctuations in its price. The prediction of gold prices holds significant importance within the stock market domain [1].

This article examines different ensemble models that can predict the future momentum of gold and silver stock prices. Specifically, it focuses on determining whether an increase or decrease will be compared to the currently traded stock price. Determining the price of gold on the global market involves numerous intricate factors. The phenomenon under consideration is subject to the effect of multiple factors across various levels, including the price of oil [2, 3, 4] and stock price. These factors collectively contribute to an actuation cycle of approximately three months. The user has provided a numerical range of [5, 6].

[1]renu.popli@chitkara.edu.in, [2]isha.kansal@chitkara.edu.in, [3]rajeev.kumar@chitkara.edu.in, [4]vikas.khullar@chitkara.edu.in, [5]preeti.sharma@chitkara.edu.in, [6]prasad.cse@piet.co.in

DOI: 10.1201/9781003534112-1

Moreover, it is influenced by many intricate mechanisms functioning at various scales. Hence, the temporal extent of the impact exerted by different factors on the price of gold exhibits considerable variation, spanning a range of 9 to 12. This unambiguously illustrates the multi-dimensional nature of the issue at hand. An effective decomposition algorithm can transform a series of gold prices into its constituent components representing various cycles. This process ensures that each mode holds a tangible interpretation, making analyzing the factors contributing to volatility significant [7, 8, 9, 10].

2. LITERATURE SURVEY

Several frequently employed methodologies in this domain are time series analysis, moving averages, autoregressive models, and fundamental analysis [11]. The application of the ARIMA model in forecasting gold prices was explored by authors in reference [12]. Ping et al. [13] employed the ARIMA and GARCH models to forecast Malaysia's gold price in their study. Their findings indicated that the GARCH model was better suited than the ARIMA model for capturing fluctuations in the volatility of time series data. The ARIMA model is deemed adequate when the study does not involve the forecasting of variable volatility. Baur et al. [14] employed the Dynamic Model Averaging (DMA) technique to assess potential causes of gold price and make predictions in situations where both the forecasting model and its parameters were subject to uncertainty. In their study, Ye et al. (15) proposed using an ARFIMA-GARCH model to analyze and predict gold prices. Their research findings demonstrated a high level of accuracy in the predictions made by this model. In their study, Mombeini and Yazdani-Chamzini et al. [16] utilized both an Artificial Neural Network (ANN) model and an Autoregressive Integrated Moving Average (ARIMA) model to make predictions regarding the future price of gold. As a result, the artificial neural network (ANN) model exhibited superior performance. In their study, Kristjanpoller and Minutolo [17] employed artificial neural networks (ANN) in conjunction with the generalized autoregressive conditional heteroskedasticity (GARCH) model. According to Sharma et al. [18], the most suitable model for forecasting the gold price of India using the Box Jenkins Autoregressive Integrated Moving Average approach is ARIMA. Therefore, it is imperative to carefully select a precise and suitable model for accurately forecasting the price of gold. In this study, Ferdinandus et al. [19] employ a fusion of ARIMA and LSTM models to forecast gold prices. This study investigates the efficacy of conventional time series methodologies and deep learning approaches in predicting gold prices. In their study, Hajek et al. [20] propose a fuzzy rule-based prediction methodology to forecast gold prices. This strategy incorporates news sentiment analysis as a critical component. This study emphasizes the influence of news emotion on fluctuations in the price of gold and uses fuzzy logic techniques to improve the precision of predictions. The study by He et al. [21] primarily focuses on utilizing LSTM models to predict the prices of gold and Bitcoin. This study examines the utilization of Long Short-Term Memory (LSTM) in predicting the prices of both conventional and cryptocurrency assets. Kilimci et al. [22] investigate the utilization of ensemble regression-based approaches to predict gold prices. Ensemble approaches are employed to enhance the accuracy and robustness of forecasting by integrating numerous models [23].

3. METHODOLOGY

In this paper we have focused on the implementation of Deep learning as a powerful tool for predicting gold price in Indian Stock Exchange. Open, High, Low, & Volume are considered features and the Close parameter is considered outcome prediction. After data collection, data clearing and pre-processing are applied to manage the required dataset. Then further time series analysis with help of LSTM and BiLSTM algorithms was applied and analyzed on basis of mean losses. Using deep learning [24], one must complete a number of procedures in order to make accurate price predictions for gold on a national stock market. For problems like this one involving the prediction of time series, deep learning models such as neural networks can be useful and effective tools. The following is a list of the broad steps involved in the process:Historical data about the price of gold was acquired from the national stock market. The consideration of key aspects such as economic statistics, interest rates, and currency exchange rates, all of which can impact gold prices, was undertaken. The study was simplified by considering missing values, outliers, and conflicts in the data. The timestamps were converted into a uniform format. The data was normalized or standardized to facilitate the evaluation of all features on a consistent scale. We implement novel functionalities to enhance the model's capacity for more precise predictions. The training, validation, and test sets were generated from the dataset. The training set was employed throughout the modeling phase, while the validation set was utilized for hyperparameter optimization. Finally, the test set was used to evaluate the model's performance. The study aimed to ascertain a suitable architectural framework for applying deep learning in time series forecasting. The deep learning model was trained using the data that had been previously obtained. To mitigate the overfitting issue, much emphasis was placed on evaluating the model's performance on the validation set, and the training process was promptly halted whenever a decline in performance was observed. The model's hyperparameters, including the learning rate, batch size, and number of epochs, were modified based on the validation set. The selection of the most appropriate evaluation measures for analyzing the model's performance on the test set was undertaken.

4. RESULTS

In this section, a comparative analysis of training & validation loss, truth versus predicted and errors in truth versus predicted have been discussed. In Fig. 1.1, the Bi-LSTM training and validation loss resulted stable and lower in comparison to LSTM. The Long Short-Term Memory (LSTM) model is a type of recurrent neural network (RNN) structure designed to effectively collect and acquire extensive temporal dependencies and patterns within sequential data. LSTM models are occasionally denoted as "Long Short-Term Memory" in academic literature. Long Short-Term Memory (LSTM) models are highly advantageous in various domains, including natural language processing, time series analysis, and any other data type characterized by temporal or sequential attributes. The Bi-LSTM model is a type of recurrent neural network (RNN) architecture that enhances the capabilities of the standard LSTM by effectively processing input sequences in both the forward and backward directions. The Multilayer Perceptron (MLP) is a feedforward artificial neural network commonly employed in various machine learning tasks, such as regression and classification. The proposed architecture is regarded as straightforward and widely employed in neural networks. The MLP comprises three main components: an input layer, one or more hidden layers, and an output layer. This network architecture is referred to as "multilayer" due to the presence of several layers. In Fig. 1.1, truth versus predicted values has been analyzed, visual difference is identifiable, where Bi-LSTM resulted nearer to the actual prediction in comparison to LSTM. Similarly in Table 1.1 Means Squared Error (MSE), Root Mean Squared Error (RMSE), R Squared Score (R2 Score), NMIC, Mean Absolute Percentage Error (MAPE) and Mean Absolute Error (MAE) were considered to analyze the outcome of LSTM and Bi-LSTM, where Bi-LSTM resulted better. In Fig. 1.1 and Fig. 1.2, the training and validation loss metrics for LSTM and Bi-LSTM models are presented, offering insights into their learning performance. Additionally, Figure 1.3 showcases the comparison between actual and predicted values for LSTM and Bi-LSTM across distinct data entries (1-51, 101-151, 201-251, and 301-351), providing a detailed examination of their predictive capabilities.

Table 1.1 presents an error analysis of the LSTM, Bi-LSTM, and MLP (Multi-Layer Perceptron) algorithms, utilizing a range of assessment measures. The Mean Squared Error (MSE) is a statistical metric that quantifies the average of the squared discrepancies between anticipated and observed values. Models with lower Mean Squared Error (MSE) scores indicate superior performance. In this instance, it is observed that the LSTM model exhibits the lowest Mean Squared Error (MSE), suggesting outstanding performance in the context of minimizing prediction inaccuracies. The Root Mean Squared Error (RMSE) is a mathematical metric representing the square root of the Mean Squared Error (MSE). It serves as a quantitative indicator of the average magnitude of errors. Once more, it can be observed that the LSTM model exhibits the lowest RMSE, indicating that its predictions are closer to the actual values.

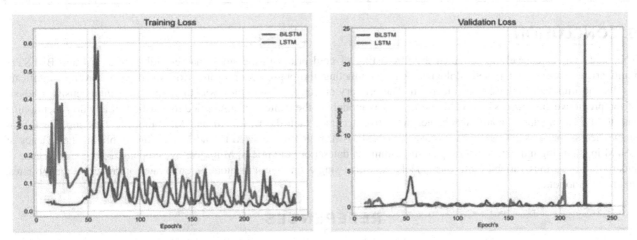

Fig. 1.1 LSTM and Bi-LSTM training and validation loss metrics

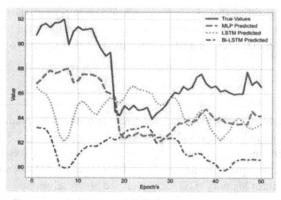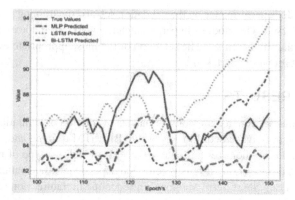

Fig. 1.2 LSTM and Bi-LSTM training and validation loss metrics

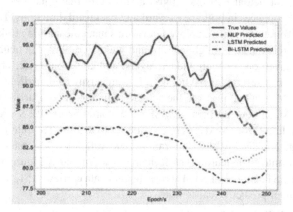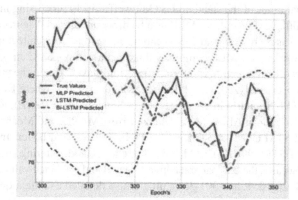

Fig. 1.3 Truth versus predicted values for LSTM and Bi-LSTM separate

Table 1.1 Error analysis of LSTM, Bi-LSTM, and MLP (Multi-Layer Perceptron) algorithms using various evaluation metrics

Algorithm	MSE	RMSE	R2 Score	NMIC	MAPE	MAE
LSTM	0.00017	0.01336	0.9384	0.99206	0.01167	0.01034
Bi-LSTM	0.00161	0.04018	0.44296	0.99206	0.04203	0.03759
MLP	0.00114	0.03379	0.57844	0.99113	0.03412	0.03072

5. CONCLUSION

This study has extensively investigated the forecasting of gold prices in the stock market, utilizing LSTM and Bi-LSTM multivariate deep learning methodologies. After conducting thorough research and experimentation, several significant results and implications have been identified. The primary objective of our study was to conduct a comprehensive analysis of the predictive capabilities of LSTM and Bi-LSTM models in the context of gold price forecasting. It is worth mentioning that the LSTM model demonstrated higher predictive accuracy than the Bi-LSTM and MLP models. This superiority was shown across many assessment measures, such as MSE, RMSE, R2 Score, MAPE, and MAE. This highlights the efficacy of LSTM in capturing intricate temporal patterns within the dataset of gold prices. Moreover, we emphasized the significance of incorporating various variables into our models, encompassing economic indicators, interest rates, currency exchange rates, and news emotion.

REFERENCES

1. Chowanda, Andry, Jurike Moniaga, Joan Christina Bahagiono, and Joko Sentosa Chandra. "Machine learning face recognition model for employee tracking and attendance system." In *2022 International Conference on Information Management and Technology (ICIMTech)*, pp. 297–301. IEEE, 2022.

2. Wang, Yuchen, Xingzhi Wang, Fei Tao, and Ang Liu. "Digital twin-driven complexity management in intelligent manufacturing." *Digital Twin* 1 (2021): 9.

3. Kansal, Isha, Renu Popli, Jyoti Verma, Vivek Bhardwaj, and Rajat Bhardwaj. "Digital image processing and IoT in smart health care-A review." In *2022 International Conference on Emerging Smart Computing and Informatics (ESCI)*, pp. 1–6. IEEE, 2022.

4. Shafiee, Shahriar, and Erkan Topal. "An overview of global gold market and gold price forecasting." *Resources policy* 35, no. 3 (2010): 178–189.

5. Zhang, Yue-Jun, and Yi-Ming Wei. "The crude oil market and the gold market: Evidence for cointegration, causality and price discovery." *Resources Policy* 35, no. 3 (2010): 168–177.

6. Ramazan, Shawkat Hammoudeh, and Ugur Soytas. "Dynamics of oil price, precious metal prices, and exchange rate." *Energy Economics* 32, no. 2 (2010): 351–362.

7. Lili, Li, and Diao Chengmei. "Research of the influence of macro-economic factors on the price of gold." *Procedia Computer Science* 17 (2013): 737–743.

8. Garg, Atul, Umesh Kumar Lilhore, Pinaki Ghosh, Devendra Prasad, and Sarita Simaiya. "Machine learning-based model for prediction of student's performance in higher education." In *2021 8th International Conference on Signal Processing and Integrated Networks (SPIN)*, pp. 162–168. IEEE, 2021.

9. Lilhore, Umesh Kumar, Sarita Simaiya, Himanshu Pandey, Vinay Gautam, Atul Garg, and Pinaki Ghosh. "Breast cancer detection in the IoT cloud-based healthcare environment using fuzzy cluster segmentation and SVM classifier." In *Ambient Communications and Computer Systems: Proceedings of RACCCS 2021*, pp. 165–179. Singapore: Springer Nature Singapore, 2022.

10. Popli, Renu, Isha Kansal, Jyoti Verma, Vikas Khullar, Rajeev Kumar, and Ashutosh Sharma. "ROAD: Robotics-Assisted Onsite Data Collection and Deep Learning Enabled Robotic Vision System for Identification of Cracks on Diverse Surfaces." *Sustainability* 15, no. 12 (2023): 9314.

11. Wen, Fenghua, Zhifang He, Xu Gong, and Aiming Liu. "Investors' risk preference characteristics based on different reference point." *Discrete Dynamics in Nature and Society* 2014 (2014).

12. Xu, Shiyu. "Analysis of Factors Influencing Medical Stock Price Based on Multiple Regression Model." *Academic Journal of Business & Management* 12 (2022): 84–90.

13. Ping, Pung Yean, Nor Hamizah Miswan, and Maizah Hura Ahmad. "Forecasting Malaysian gold using GARCH model." *Applied Mathematical Sciences* 7, no. 58 (2013): 2879–2884.

14. Baur, Dirk G., Joscha Beckmann, and Robert Czudaj. "A melting pot—Gold price forecasts under model and parameter uncertainty." *International Review of Financial Analysis* 48 (2016): 282–291.

15. Yaacob, Salmy Edawati, and Sanep Ahmad. "Return to gold-based monetary system: analysis based on gold price and inflation." *Asian Social Science* 10, no. 7 (2014): 18.

16. Mombeini, Hossein, and Abdolreza Yazdani-Chamzini. "Modeling gold price via artificial neural network." *Journal of Economics, business and Management* 3, no. 7 (2015): 699–703.

17. Kristjanpoller, Werner, and Marcel C. Minutolo. "Gold price volatility: A forecasting approach using the Artificial Neural Network–GARCH model." *Expert systems with applications* 42, no. 20 (2015): 7245–7251.

18. Sharma, Rakesh Kumar. "Forecasting gold price with box jenkins autoregressive integrated moving average method." *Journal of International Economics* 7, no. 1 (2016): 32.

19. Yudha Randa Madhika, Kusrini Kusrini, and Tonny Hidayat. "Gold Price Prediction Using the ARIMA and LSTM Models." *Sinkron: jurnal dan penelitian teknik informatika* 8, no. 3 (2023): 1255–1264.

20. Petr, and Josef Novotny. "Fuzzy rule-based prediction of gold prices using news affect." *Expert Systems with Applications* 193 (2022): 116487.

21. He, Conghong, Tao Zhong, and Yan Wang. "A price prediction model for gold and Bitcoin based on Long Short-Term Memory." *Highlights in Business, Economics and Management* 4 (2022): 461–467.

22. Kilimci, Zeynep Hilal. "Ensemble regression-based gold price (XAU/USD) prediction." *Journal of Emerging Computer Technologies* 2, no. 1 (2022): 7–12.

23. Kansal, Isha, and Singara Singh Kasana. "Minimum preserving subsampling-based fast image de-fogging." *Journal of Modern Optics* 65, no. 18 (2018): 2103–2123.

24. Manickam, A., S. Indrakala, and Pushpendra Kumar. "A Novel Mathematical Study on the Predictions of Volatile Price of Gold Using Grey Models." *Contemporary Mathematics* (2023): 270–285.

Computational Intelligence and Mathematical Applications – Dr. Devendra Prasad et al. (eds)
© 2024 Taylor & Francis Group, London, ISBN 978-1-032-87721-1

Application of Deep Neural Networks in the Detection of Surface Defects of Magnetic Tiles through the Analysis of Digital Images

Jyoti Verma[1]
Punjabi University Patiala, Punjab, India

Devendra Parsad[2]
Panipat Institute of Engineering and Technology (PIET), Panipat, Haryana, India

Isha Kansal[3], Renu Popli[4], Vikas Khullar[5], Rajeev Kumar[6]
Chitkara University Institute of Engineering and Technology, Chitkara University, Punjab, India

Abstract: Manufacturers place significant importance on the presence of surface flaws in magnetic tiles due to their potential to adversely affect both product quality and customer satisfaction. This study introduces a substantial advancement in the field of defect detection by employing deep neural networks, or DNNs, to analyze digital images. The primary objective of our research is to investigate the application of Convolutional Neural Networks (CNN) in conjunction with InceptionResNetV2, VGG16, and VGG19 models for the automated detection and categorization of surface defects in magnetic tiles. The experimental findings illustrate deep learning's significant capabilities within this field. The InceptionResNetV2 model demonstrated superior accuracy, reaching a notable 95.95%, surpassing conventional approaches by a substantial margin. The algorithm showed remarkable levels of precision and memory, outperforming human inspectors in consistency and reliability.

Keywords: Automatic defect identification, Convolutional neural networks, Deep neural networks, Image analysis, Magnetic tiles, Manufacturing industry, Surface defect detection

1. INTRODUCTION

Maintaining the optimal surface quality of these tiles is of utmost significance in sectors producing magnetic tiles for diverse purposes. Thorough quality assessment is an essential phase in the production process due to the potential performance concerns or product failure that can arise from any defects or imperfections present on the surface of the tile. Historically, the responsibility of evaluating the quality of surfaces has been assigned to skilled human laborers [1]. These highly trained experts visually examine each tile, employing manual techniques to identify and classify any flaws present. Although this approach has demonstrated effectiveness, it has its limitations, including reduced efficiency, increased staff expenses, and the possibility of subjective judgment in identifying defects. This study aims to construct an intelligent system to tackle these issues and enhance the effectiveness and dependability of surface defect identification [2].

Utilizing Deep neural networks (DNN), particularly convolutional neural networks (CNNs), in the context of surface defect identification presents a distinctive and noteworthy challenge [3-5]. The magnetic tile production sector presently heavily depends on the expertise of workers to manually examine and evaluate surface quality. This approach has notable limitations, including reduced productivity, increased operational expenses, and the possibility of subjective defect detection.

[1]jyoti.snehiverma@gmail.com, [2]prasad.cse@piet.co.in, [3]isha.kansal@chitkara.edu.in, [4]renu.popli@chitkara.edu.in, [5]vikas.khullar@chitkara.edu.in, [6]rajeev.kumar@chitkara.edu.in

The conventional methodology, albeit partially successful, is vulnerable to human fallibility and personal decision-making, presenting difficulties in upholding uniformity in product excellence [7, 8]. The primary focus of this research revolves around identifying and resolving the obstacles encountered in detecting surface defects in magnetic tiles [9-13].

The significance of this work is considerable since it has the potential to bring about a revolutionary change in quality control practices within the magnetic tile manufacturing business. The study holds great importance due to its potential to revolutionize the magnetic tile manufacturing sector through the enhancement of efficiency, cost reduction, improvement of product quality, facilitation of technological advancements, and augmentation of global competitiveness. The relevance of the subject matter is diverse and complex [15 -20].

2. LITERATURE SURVEY

Historically, surface quality evaluation in industrial settings primarily depended on human inspection techniques. Proficient human inspectors have helped assess the surface quality of diverse items, encompassing metals, polymers, textiles, and other materials [21]. The inspectors employed their visual acuity to identify imperfections such as abrasions, indentations, fractures, and anomalies in surface texture or hue. Although hand examination demonstrated a certain level of effectiveness, it was plagued by several limitations. The assessments made by human inspectors may be subject to several influences, including but not limited to weariness, level of expertise, and personal biases, which can result in outcomes that lack consistency. The manual inspection process was time-consuming and labor-intensive, especially in businesses that produce substantial products [22]. The rate at which manual inspection could be conducted was restricted by human talents' inherent limitations, impeding the overall manufacturing flow. The utilization of a proficient labor force for manual review resulted in substantial expenditures related to labor. The decision to use either human or automated inspection methods carries significant consequences for industries across diverse sectors [23]. Including human inspectors in the inspection process is advantageous due to their many years of expertise and ability to adapt to various situations. Individuals can modify their approach to conditions and frequently possess specialized knowledge in a given field. Humans possess the cognitive ability to effectively undertake detailed inspection activities that necessitate the discernment of small visual clues and the exercise of nuanced judgment. Computerized systems provide the ability to operate continuously without succumbing to fatigue or necessitating breaks, ensuring continual scrutiny around the clock [24]. Automated inspection systems can collect comprehensive data about issues, enabling producers to oversee and address quality-related matters effectively. The advent of deep learning has been recognized as a revolutionary breakthrough in defect identification across diverse sectors. Deep neural networks, which employ many layers of neural networks, are utilized to autonomously acquire knowledge, and discern intricate patterns and characteristics from data. This renders them highly proficient in the detection of faults. Deep learning demonstrates exceptional performance in defect detection based on photographs, wherein digital photos or visual data are utilized to identify surface or structural irregularities [25-27].

3. METHODOLOGY

In the Methodology portion of this paper, we thoroughly examine the complexities of our approach, providing a comprehensive explanation of the novel methodologies and strategies utilized to leverage the capabilities of deep neural networks to detect surface defects in magnetic tiles. Data gathering and preparation are essential stages within the study process for detecting surface defects on magnetic tiles. The dataset "Magnetic Tile Surface Defects" is available on the Kaggle platform. The dataset comprises digital photos depicting magnetic tiles exhibiting diverse surface flaws [28]. The images of 6 common magnetic tile defects were collected from 1800 images. The training and validation procedures encompass dataset splitting, data preprocessing, model selection, and hyperparameter tuning. During training, models learn to make predictions by minimizing a loss function. Hyperparameters are optimized using a validation set, and early stopping can prevent overfitting. Final evaluation occurs on a test set to gauge model generalization. The defects observed in magnetic tiles encompass a range of imperfections, such as cracks, scratches, chips, discolorations, and several other forms of damage that are frequently encountered. Each image within the collection has the potential to be accompanied by annotations in the form of labels or bounding boxes, which indicate both the specific location and any surface flaws present. The inclusion of these annotations is of utmost importance in the process of training and assessing machine learning models designed for fault detection. The manufacturing plant is responsible for the production of a range of magnetic tiles that exhibit a variety of surface features and flaws. The defect detection system, initially evaluated on a specific type of tile, now requires validation over a broader range of products, encompassing additional tile variations and types of defects. The system's expansion guarantees its flexibility to adapt to various industrial conditions. The data preprocessing stage encompassed a series of essential steps. All images underwent consistent resizing to achieve a standardized resolution, such as 224x224 or 256x256 pixels. The consistent size of the data enabled the DNN models to process it uniformly. The pixel values underwent normalization to conform to a defined scale, commonly ranging from 0 to 1 or from -1 to 1. The process of normalization improved the stability of the model training procedure and ensured that all features were treated uniformly. The application of data augmentation techniques was employed to augment the training dataset.

4. EXPERIMENTAL RESULT

During our investigation, the assessment of model performance assumed a crucial function in assuring the efficacy and dependability of our automated flaw identification method. The evaluation metrics included in this study were carefully chosen to thoroughly assess the performance of deep learning models in detecting surface defects in magnetic tiles. These metrics include accuracy, precision, recall, F1-score, and ROC-AUC. The measurements yielded significant information into the model's capacity to classify and differentiate between defects and non-defects accurately.

Table 2.1 evaluates various deep learning models within the research's framework, namely CNN, InceptionResNetV2, VGG16, and VGG19. The objective was to assess their effectiveness in detecting surface flaws in magnetic tiles. The assessment was conducted using various criteria: accuracy, loss, precision, and recall.

Table 2.1 Performance evaluation of various imaging algorithm

	CNN	InceptionResNetV2	VGG16	VGG19
Accuracy	85.06977	95.95349	84.09302	86.18605
Validation Accuracy	83.64312	95.16729	83.64312	83.27138
Precision	88.44275	98.45771	91.42702	92.71845
Validation Precision	97.54601	95.43726	90.84507	97.2973
Recall	77.39535	94.97675	78.37209	79.95349
Validation Recall	75.83643	94.05205	80.66915	81.04089
Loss	0.575829	0.540159	0.670683	0.625032
Validation Loss	0.752963	0.672845	0.724733	0.701514

CNN demonstrated an accuracy of roughly 85.07%, which signifies its proficiency in accurately categorizing surface flaws in magnetic tiles. The CNN's precision was estimated to be approximately 88.44%, indicating a considerably lower occurrence of erroneous optimistic predictions. CNN exhibited a recall rate of roughly 77.40%, showcasing its capacity to accurately detect a substantial proportion of genuine flaws. The CNN achieved a training loss of 0.5758 and a validation loss of 0.7529, indicating convergence during the training process.

Considering Table 2.2, the InceptionResNetV2 model exhibited superior performance compared to other models, with an accuracy rate of roughly 95.95%. This notable accuracy score demonstrates its robust classification capabilities. The displayed precision of around 98.46% shows a noteworthy level of accuracy, suggesting a minimal occurrence of false positives. The InceptionResNetV2 model had a recall rate of roughly 94.98%, indicating its proficiency in accurately detecting the most genuine flaws. The training loss of the InceptionResNetV2 model was recorded as 0.5402, while the validation loss was measured at 0.6728. The InceptionResNetV2 model has exceptional performance in detecting magnetic tile defects, with a precision rate of 98.46%. This high accuracy ensures a limited occurrence of false positive results. The high recall rate of 94.98% exhibited by the system is evidence of its adeptness in precisely detecting authentic flaws. The findings indicate that the model is very suitable for accurately and consistently identifying defects, and the expected high F1-Score highlights its robust overall performance. These values suggest that the training process successfully optimized the model's performance.The VGG16 model demonstrated a classification accuracy of roughly 84.09%, suggesting a commendable level of competence in

Table 2.2 Comparing the system to other automated defect detection systems

Study	Approach/Model	Performance Metrics	System Features/Advantages
[4]	CNN-based Model	Accuracy: 90%	Real-time defect detection
[16]	YOLOv3	Accuracy: 92%, Precision: 88%	Supports multiple tile types
our System	InceptionResNetV2	Accuracy: 95.95%, Precision: 98.46%	Real-time, superior accuracy

categorization tasks. The precision of the measurement was approximately 91.43%, indicating a reasonably low occurrence of false positives. The VGG16 model demonstrated a recall rate of approximately 78.37%, suggesting its proficiency in accurately detecting genuine faults. The training loss of the VGG16 model was recorded as 0.6707, whereas the validation loss was measured as 0.7247. The VGG19 model achieved an accuracy of 86.19%, indicating its proficiency in accurately classifying data. The VGG19 model exhibited an accuracy value of approximately 92.72%, suggesting a relatively low occurrence of erroneous optimistic predictions. The recall rate of around 79.95% indicates a successful ability to identify genuine faults accurately. The VGG19 model achieved a training loss of 0.6250 and a validation loss of 0.7015.

5. CONCLUSION

The findings of our study indicate that the intelligent system we designed, which utilizes CNN and InceptionResNetV2, exhibited superior performance compared to conventional approaches in terms of accuracy, precision, and recall. The implementation of automated defect identification processes significantly boosts production efficiency through the reduction of reliance on manual inspection, the minimization of variability, and the acceleration of time-to-market. Manufacturers have the potential to achieve significant cost reductions by reducing labor expenses and minimizing the number of defects that make it to the market. Implementing our technology has the potential to establish manufacturers as industry frontrunners, enabling them to provide highly efficient and superior products. Our study indicates a significant transformation in quality control methodologies employed in the magnetic tile manufacturing sector. Incorporating deep neural networks into the defect detection procedure presents an avenue towards enhanced quality, improved cost-effectiveness, and increased global competitiveness.

REFERENCES

1. Li, F., Mao, Q., & Chang, C.-C. (2018). Reversible data hiding scheme based on the Haar discrete wavelet transform and interleaving prediction method. *Multimedia Tools and Applications*, 77, 5149–5168.
2. Zhang, H., Han, J., Guan, Y., & Zhang, Y. (2018). A SIFT algorithm based on DOG operator. In *2018 International Conference on Intelligent Transportation, Big Data & Smart City (ICITBS)* (pp. 609–612). Xiamen, China.
3. Konečný, J., & Hagara, M. (2014). One-shot-learning gesture recognition using hog-hof features. *The Journal of Machine Learning Research*, 15(1), 2513–2532.
4. Zhang, C., Pan, X., Li, H., et al. (2018). A hybrid MLP-CNN classifier for very fine resolution remotely sensed image classification. *ISPRS Journal of Photogrammetry and Remote Sensing*, 140, 133–144.
5. Wu, J., & Yang, H. (2015). Linear regression-based efficient SVM learning for large-scale classification. *IEEE Transactions on Neural Networks and Learning Systems*, 26(10), 2357–2369.
6. Min, H., & Luo, X. (2016). Calibration of soft sensor by using just-in-time modeling and AdaBoost learning method. *Chinese Journal of Chemical Engineering*, 24(8), 1038–1046.
7. Liu, W., Anguelov, D., Erhan, D., et al. (2016). SSD: single shot multibox detector. In *Computer Vision – ECCV 2016* (pp. 21–37). Springer.
8. Ren, S., He, K., Girshick, R., & Sun, J. (2015). Faster R-CNN: towards real-time object detection with region proposal networks. *Advances in Neural Information Processing Systems*, 28.
9. Fang, L., Ding, S., Park, J. H., & Ma, L. (2022). Adaptive fuzzy control for nontriangular stochastic high-order nonlinear systems subject to asymmetric output constraints. *IEEE Transactions on Cybernetics*, 52(2), 1280–1291.
10. Fang, L., Ding, S., Park, J. H., & Ma, L. (2021). Adaptive fuzzy control for stochastic high-order nonlinear systems with output constraints. *IEEE Transactions on Fuzzy Systems*, 29(9), 2635–2646.
11. Xie, L., Xiang, X., Xu, H., Wang, L., Lin, L., & Yin, G. (2021). FFCNN: a deep neural network for surface defect detection of magnetic tile. *IEEE Transactions on Industrial Electronics*, 68(4), 3506–3516.
12. Ben Gharsallah, M., & Ben Braiek, E. (2021). Defect identification in magnetic tile images using an improved nonlinear diffusion method. *Transactions of the Institute of Measurement and Control*, 43(11), 2413–2424.
13. Hu, C., Liao, H., Zhou, T., Zhu, A., & Xu, C. (2022). Online recognition of magnetic tile defects based on UPM-DenseNet. *Materials Today Communications*, 30, article 103105.
14. Redmon, J., Divvala, S., Girshick, R., & Farhadi, A. (2016). You only look once: unified, real-time object detection. In *2016 IEEE Conference on Computer Vision and Pattern Recognition (CVPR)* (pp. 779–788). Las Vegas, NV, USA.
15. Redmon, J., & Farhadi, A. (2017). YOLO9000: better, faster, stronger. In *2017 IEEE Conference on Computer Vision and Pattern Recognition (CVPR)* (pp. 7263–7271). Honolulu, HI, USA.
16. Hurtik, P., Molek, V., Hula, J., Vajgl, M., Vlasanek, P., & Nejezchleba, T. (2022). Poly-YOLO: higher speed, more precise detection and instance segmentation for YOLOv3. *Neural Computing and Applications*, 34(10), 8275–8290.
17. Wang, C.-Y., Bochkovskiy, A., & Liao, H.-Y. M. (2021). ScaledYOLOv4: scaling cross stage partial network. In *2021 IEEE/CVF Conference on Computer Vision and Pattern Recognition (CVPR)* (pp. 13029–13038). Nashville, TN, USA.

18. Woo, S., Park, J., Lee, J.-Y., & Kweon, I. S. (2018). CBAM: convolutional block attention module. In *Proceedings of the European conference on computer vision (ECCV)* (pp. 3–19). Munich, Germany.

19. Çelik, H. İ., Dülger, L. C., & Topalbekiroğlu, M. (2014). Development of a machine vision system: real-time fabric defect detection and classification with neural networks. *The Journal of The Textile Institute*, 105(6), 575–585.

20. Saad, N. H., Ahmad, A. E., Saleh, H. M., & Hasan, A. F. (2016). Automatic semiconductor wafer image segmentation for defect detection using multilevel thresholding. *MATEC Web of Conferences*, 78, p. 01103.

21. Nguyen, V. H., Pham, V. H., Cui, X., Ma, M., & Kim, H. (2017). Design and evaluation of features and classifiers for OLED panel defect recognition in machine vision. *Journal of Information and Telecommunication*, 1(4), 334–350.

22. Avola, D., Cascio, M., Cinque, L., et al. (2022). Real-time deep learning method for automated detection and localization of structural defects in manufactured products. *Computers & Industrial Engineering*, 172, article 108512.

23. Tabernik, D., Šela, S., Skvarč, J., & Skočaj, D. (2020). Segmentation-based deep-learning approach for surface-defect detection. *Journal of Intelligent Manufacturing*, 31(3), 759–776.

24. Huang, Y., Qiu, C., & Yuan, K. (2020). Surface defect saliency of magnetic tile. *The Visual Computer*, 36(1), 85–96.

25. Cui, L., Jiang, X., Xu, M., Li, W., Lv, P., & Zhou, B. (2021). SDDNet: a fast and accurate network for surface defect detection. *IEEE Transactions on Instrumentation and Measurement*, 70, 1–13.

26. Dhiman, P., Kukreja, V., Manoharan, P., Kaur, A., Kamruzzaman, M. M., Dhaou, I. B., & Iwendi, C. (2022). A novel deep learning model for detection of severity level of the disease in citrus fruits. *Electronics*, 11(3), p.495.

27. Trivedi, N. K., Gautam, V., Anand, A., Aljahdali, H. M., Villar, S. G., Anand, D., Goyal, N., & Kadry, S. (2021). Early detection and classification of tomato leaf disease using high-performance deep neural network. *Sensors*, 21(23), p.7987.

Computational Intelligence and Mathematical Applications – Dr. Devendra Prasad et al. (eds)
© 2024 Taylor & Francis Group, London, ISBN 978-1-032-87721-1

3 | Design of Micro-Machined Frequency and Pattern Reconfigurable Antennas

Ashish Kumar[1]

Chitkara University Institute of Engineering and Technology, Chitkara University

Devendra Parsad[2]

Panipat Institute of Engineering and Technology (PIET), Panipat, Haryana, India

Amar Partap Singh Pharwaha

Department of Electronics and Communication Engineering, Sant Longowal Institute of Engineering and Technology, Longowal, Sangrur, Punjab, India

Mandeep Jit Singh

Department of Electrical, Electronic and Systems Engineering, Faculty of Engineering and Built Environment, Universiti Kebangsaan Malaysia, UKM, Bangi, Selangor, Malaysia

Gurmeet Singh

School of Computing Science and Engineering, Galgotias University, Uttar Pradesh, India

Manpreet Singh

Chitkara University Institute of Engineering and Technology, Chitkara University, Punjab, India

Abstract: This article presents a micro-machined frequency reconfigurable patch antenna array designed for wireless communication purposes. The antennas under consideration have been devised on a high resistive silicon substrate, which possesses a thickness of 0.675 mm. In order to address the constraints associated with utilising high dielectric constant materials as substrates, the implementation of a micro-machining technique has been employed to augment the performance attributes of the antennas. The reconfigurability of the designs has been achieved through the incorporation of RF-MEMS switches, which offer notable benefits such as enhanced isolation and reduced insertion loss. Antenna arrays consisting of 1×2 and 2×2 elements were constructed with the objective of achieving a maximum frequency shift of approximately 360 MHz within the X-band frequency range, while maintaining a gain of 11.6 dBi. The 2×2 design exhibits a distinctive radiation pattern characterised by two prominent sections and zero radiation in the zenith direction, making it suitable for RADAR applications. Additionally, the suggested occurrence of frequency shift phenomena can serve as a favorable characteristic in the transmission and reception operations of satellite communication and mobile wireless systems. This can effectively improve the system's capacity and mitigate the adverse effects of multipath fading.

Keywords: Antenna array, Gain, Micro-machining, Reconfigurable, RF-MEMS

1. INTRODUCTION

Antennas are the integral part of the overall communication system and patch antennas are the favourable among engineers and researchers due to its numerous outstanding features. Among those, multifunctional characteristic of the antenna can be

*Corresponding authors: [1]ashish.1130@chitkara.edu.in, [2]prasad.cse@piet.co.in

DOI: 10.1201/9781003534112-3

observed as the most significant contribution in satellite communication and mobile wireless communication system []. To configure the multifunctional property of the antenna, the concept of reconfigurability has been proposed in [4]. This feature can alter some of the properties (resonating frequency, polarization and directivity) of the antenna. The specifications and results of antenna structure depend on the current distribution and change in the electromagnetic field on the antenna surface. Reconfigurability refers to the ability to alter the existing distribution of a system through a variety of means. Reconfigurable antennas provide the capability to effectively perceive alterations in the surrounding environment and establish communication in dynamic temporal conditions. These types of antennas can be classified into four categories 1) an antenna that can change its operating frequency by changing the band of operation is called frequency reconfigurable antenna. 2) a radiating structure which can transform its polarization like linear (horizontal and vertical), circular (left and right hand), elliptical is called polarization reconfigurable antenna. 3) Radiation pattern reconfigurable antennas can alter its radiation direction. 4) Moreover,

combination of the above three parameters can also be found in the single antenna [5-6]. There are multiple approaches to attain reconfigurability in the antenna structure, as depicted in Fig. 3.1. Reconfigurable techniques have been classified into four categories such as electrical, optical, physical and substance alteration according to the principle of the operation. For instance, electrical reconfigurable antennas composed of the switches (varactor diode, MEMS and PIN diodes) considering the effect that on actuating these switches they can alter the current distribution [7-11].

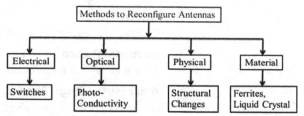

Fig. 3.1 Various techniques to achieve reconfigurability in antennas [4]

There are several advantages of using reconfigurable antenna. First of all, the single structure can operate at different frequencies i.e. having the multifunctional capability e.g (a) function of the design can be changed as the mission changes (b) act as a single or multiple antenna elements at a time (c) can provide narrowband or wideband operation (d) these can be programmed with micro-controller or field programmable gate array (e) minimize the overall volume requirement. But there are some limitations like overall complexity of the system has been increased with activation/deactivation of the switches, increase in power requirement and generation of harmonics. This article emphasis on the incorporation of the MEMS switches into the patch antenna array. With the activation/deactivation of MEMS switch in antenna design, current distribution has been changed in the radiating edge which causes to shifting in the frequency. Moreover, the designed array designed in this context have been designed on high index substrate like silicon having dielectric constant of 11.9 and thickness of 0.675 mm. Due to the surface wave generation in the high index substrates, concept of micro-machining has been used to eliminate the effect of effect of surface waves and enhance the performance characteristics.

2. MICRO-MACHINING THEORY

Through a technique known as micro-machining, the limitations of the results specifications of the antenna fabricated on high dielectric constant substrate can be reduced significantly [12]. As shown in Fig. 3.2, according to this procedure, the substrate underneath the patch material is etched leaving the air behind which will further result into the reduction of overall dielectric constant and consequently the performance of antenna improves. Surface micro-machining is a different kind of micro-machining that is used to create tiny structures like cantilever bridges in switches. There is a variation in the combined dielectric constant underneath the patch because there is a mixed region composed of silicon substrate and air. There effective dielectric constant has been calculated using equations mentioned in [13].

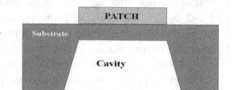

Fig. 3.2 Micro-machined patch antenna [14]

3. DESIGN OF AN X-BAND SHUNT CAPACITIVE MEMS SWITCH

Designing the co-planer waveguide (CPW) geometry with optimised G/W/G for 50 impedance is the first stage in creating a MEMS switch. In this design G/W/G has been taken as 60/100/60 μm [14]. The reason for taking this G/W/G is to match the impedance of 50Ω. The relation between impedance matching and G/W/G has been formulated from the equations (4) to (8).

$$k = \frac{W}{W + 2G} = \frac{100}{100 + 120} = \frac{100}{220} = 0.4545, \quad k' = \sqrt{1 - k^2} = \sqrt{1 - (0.4545)^2} = 0.89073 \tag{4}$$

$$\varepsilon_{reff} = \frac{\varepsilon_r + 1}{2} = \frac{11.9 + 1}{2} = 6.45 \tag{5}$$

$$\frac{k}{k'} = \frac{\pi}{\ln\left(\frac{2\left(1 + \sqrt{0.89073}\right)}{\left(1 - \sqrt{0.89073}\right)}\right)} = 0.741564, \quad \frac{k'}{k} = 1.3485 \tag{6}$$

$$Z_o = \frac{30\pi}{\sqrt{\varepsilon_{reff}}} \frac{k'}{k} \tag{7}$$

Put the values from equation (6) and (7) into equation (8)

$$Z_o = \frac{30 \times 3.14}{\sqrt{6.45}} \times 1.3485 \approx 50.019 \ \Omega \tag{8}$$

After deciding the G/W/G, the next step is to calculate the spring constant by considering the gold metal to design the beam with young's modulus of 80GPa. Further, beam width (w) is 140 μm, beam length (l) as 500 μm and the depth of the beam (t) is 1 μm has been selected. The structure of the switch is depicted in Fig. 3.3.

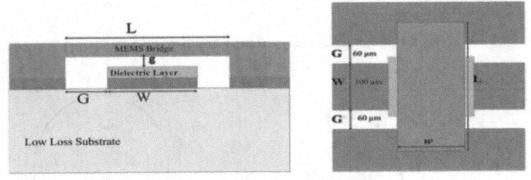

Fig. 3.3 MEMS shunt capacitive switch [14]

4. DESIGN OF RECONFIGURABLE PATCH ANTENNA ARRAY

In this section, various configurations of the micro-machined antenna array have been designed with frequency reconfigurability has been achieved using RF MEMS switches.

4.1 Proposed Micro-Machined 1 x 2 Antenna Array

1 × 2 micro-machined reconfigurable antenna array has been shown in the Fig. 3.4. Array design has been made with the application of power divider. Four switches have been incorporated (two in each patch) which causes the change in current distribution thus alteration in operating frequency has been achieved.

4.2 Results of Reconfigurable Micro-Machined 1 × 2 Antenna Array

The result parameters of the proposed design have been analyzed by reflection coefficient (S_{11}(dB)) [15], radiation pattern, realized gain, bandwidth etc. The reflection coefficient versus frequency graph for switch ON and OFF condition has been shown in Fig. 3.5. It can be depicted that OFF state of the switches, the anticipated design operates at 8.4 GHz with excellent impedance matching of -29 dB covering the -10 dB impedance bandwidth of 273 MHz. Whereas, in ON state of the switches, there is shift of 200 MHz towards the lower side of the frequency bands due to increase in overall conducting length and alteration in the current distribution of the patches. In the ON condition, the proposed design resonates at 8.2 GHz with impedance matching of -18 dB covering the bandwidth of 196 MHz. Further, the radiation pattern for the switch conditions have been illustrated in Figs 3.6. The gain and frequency curve as demonstrated in Fig. 3.7. All the required results of proposed

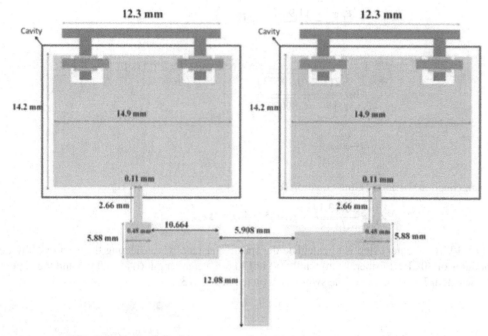

Fig. 3.4 Reconfigurable micro-machined 1×2 patch antenna array [2]

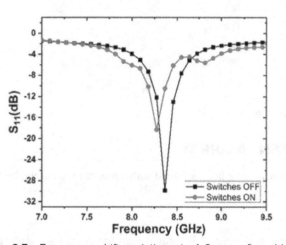

Fig. 3.5 Frequency shift variations in 1×2 reconfigurable patch antenna array

Source: Author

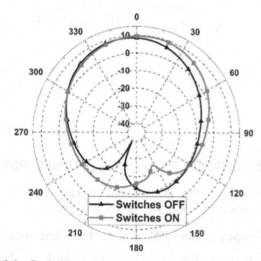

Fig. 3.6 Radiation patterns in switches ON and OFF conditions

Source: Author

design have been shown in Table 3.1. It can be observed that efficiency has been reduced in case of switches are in ON state. Next section includes the extension of the 1×2 reconfigurable micro-machined patch antenna array into 2×2 array structure to achieve the specific nature of the radiation pattern and acceptable amount of gain.

4.3 Micro-Machined, Reconfigurable, 2-by-2 Antenna Array Structure

The radiation pattern of the 1x2 array structure is broadside in nature. However, various applications like RADAR communication, more than one major lobe are required including null position in broadside direction. This kind of radiation pattern can be achieved by placing the patches in out of phase condition. So, the next design comprises of four patches which are arranged in the 2×2 configuration to achieve required radiation pattern and other parameters. The four patches are placed on the substrate in such a way that the radiation from two patches emits in one direction generating the one major lobe and another pair of patches radiates in another direction creating another major lobe but there is a zero radiation in the broadside direction. The dimensional view of the reconfigurable micro-machined 2×2 antenna array structure is expressed in Fig. 3.8. The design is fed with the corporate feed method with power divider technique. A cavity in the substrate is created under all the four patches. In this design, each patch consists of two RF MEMS switches which contributes in achieving the shifting in the frequency. All the switches are actuated with the electrostatic actuation method as discussed above.

4.4 Evaluating the Versatile 2-by-2 Patch Antenna Array's Performance in a Micro-Manufactured Environment

The reflection coefficient (S_{11}(dB)) that indicates the proper frequency shift upon actuation of the RF MEMS switches has been used in the performance analysis of the 2 by 2 patch antenna array structure.

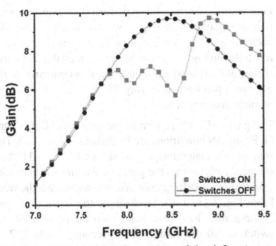

Fig. 3.7 Gain v/s frequency curve of the 1×2 antenna array structure

Source: Author

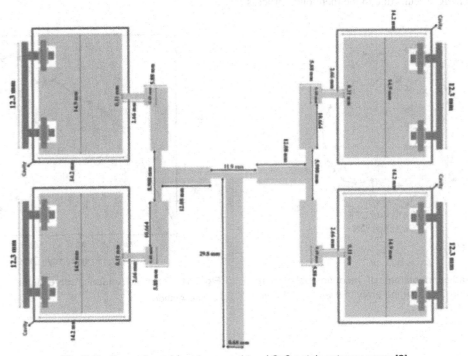

Fig. 3.8 Reconfigurable micro-machined 2×2 patch antenna array [2]

Table 3.1 Performance parameters of reconfigurable micro-machined antenna array with actuation of switches

Parameters	1x2 array Switches OFF	1x2 array Switches ON	2x2 array Switches OFF	2x2 array switches ON
Operating Frequency (GHz)	8.4 GHz	8.2 GHz	8.45	8.09
S_{11} (dB)	-29	-18	-23.8	-21.9
Voltage standing wave ratio	1.06	1.27	1.13	1.12
-10 dB impedance bandwidth (MHz)	273	196	450	675
Gain (dB) ($\theta = \pm20°$, $\Phi = 90°$)	9.58	7.2	11.6	9.1

Source: Author

Further analysis has been performed by examining the VSWR, radiation patterns and gain v/s frequency curve. The reflection coefficient v/s frequency curve of the anticipated design has been illustrated in Fig. 3.9. It can be represented that the frequency shift of almost 360 MHz which is quite large as compared to the 1×2 micro-machined patch antenna array. There is step change of 4.2% of patch nominal frequency.

The gain radiation pattern of the proposed design for the switches OFF and ON conditions are visualized in Fig. 3.10. The gain versus frequency rectangular plots as shown in Fig. 3.11. The gain versus frequency plot shows the gain in the entire series of the frequencies, gain at operating frequency and maximum gain the frequency range. From Fig. 3.11, it can be observed that there is not much variation in the gain at low frequencies with the position of the RF MEMS switches. All the results of the reconfigurable 2×2 antenna array structure have been tabulated in Table 3.1. It has been found that the proposed design is suitable with respect to the gain, frequency variation and nature of radiation pattern along with the suitability of the design with MMIC circuits due to the mentioned process [16].

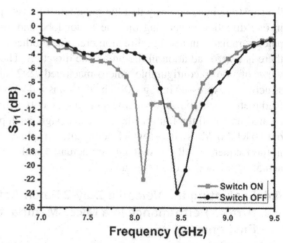

Fig. 3.9 Frequency shift variations of the reconfigurable micro-machined 2×2 antenna array

Source: Author

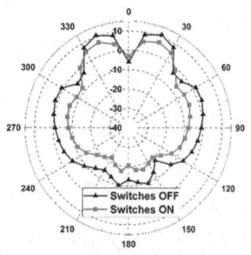

Fig. 3.10 Gain radiation pattern of reconfigurable micro-machined 2×2 antenna array in switches OFF and ON position

Source: Author

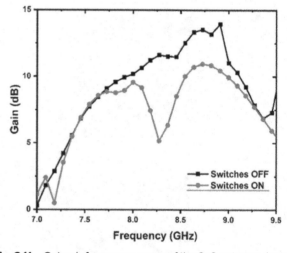

Fig. 3.11 Gain v/s frequency curve of the 2x2 antenna design

Source: Author

5. CONCLUSION

Different configurations like 1×2 and 2×2 frequency reconfigurable micro-machined patch antenna array have been designed and analyzed using RF-MEMS switches. The proposed designs resonate around 8.3 GHz in X-band with peak gain of 11.6 dBi. A small frequency shift of about 200 MHz has been achieved in 1×2 array design. The gain in 1×2 reconfigurable micro-machined patch antenna array is almost 48.4 % more in OFF state and 33.3 % more in ON state of the switches over the single unit cell. To achieve the specific nature of the radiation pattern 2×2 reconfigurable micro-machined patch antenna array has been designed. The desired radiation pattern has been attained, which consists of two main portions and a zero radiation towards zenith. In addition, if all the switches are off, the gain is enhanced by around 22.2%, whereas if all the switches are on, the gain is enhanced by about 26.3%. In order to increase capacity and counteract multipath fading, the proposed frequency shift phenomena can be a desired feature in transmit and receive operations of satellite communication and mobile wireless systems.

REFERENCES

1. Balanis, C. A. *Antenna Theory.* John Wiley & Sons, Inc., 1997.
2. Garg, R., Bahl, I., Bhartia, P., & Ittipiboon, A. *Microstrip Antennas Design Handbook.* Artech House, London, 2001.
3. Sabban, A., & Navon, E. "A MM-Waves Microstrip Antenna Array." Paper presented at IEEE Symposium, Tel Aviv, Israel, March 1983.
4. Christodoulou, C. G., Tawk, Y., Lane, S. A., & Erwin, S. R. "Reconfigurable antennas for wireless and space applications." *Proceedings of the IEEE* 100, no. 7 (2012): 2250–2261.
5. Haupt, R. L., & Lanagan, M. "Reconfigurable Antennas." *IEEE Antennas and Propagation Magazine* 55, no. 1 (2013): 49–61.
6. Monk, A. D., & Clarricoats, P. J. B. "Adaptive Null Formation with a Reconfigurable Reflector Antenna." *IEE Proceedings-Microwave Antennas and Propagation* 142, no. 3 (1995): 220.
7. Shelley, S., Costantine, J., Christodoulou, C. G., Anagnostou, D. E., & Lyke, J. C. "FPGA-controlled switch-reconfigured antenna." *IEEE Antennas and Wireless Propagation Letters* 9 (2010): 355.
8. Nikolaou, S., Bairavasubramanian, R., Lugo, C., Carrasquillo, I., Thompson, D. C., Ponchak, G. E., Papapolymerou, J., & Tentzeris, M. M. "Pattern and frequency reconfigurable annular slot using PIN diodes." *IEEE Transactions on Antennas and Propagation* 54, no. 2 (2006): 439–448.
9. Sarrazin, J., Mahe, Y., Avrillon, S., & Toutain, S. "Pattern reconfigurable cubic antenna." *IEEE Transactions on Antennas and Propagation* 57, no. 2 (2009): 310.
10. Qin, P. Y., Weily, A. R., Guo, Y. J., Bird, T. S., & Liang, C.-H. "Frequency reconfigurable quasi-Yagi folded dipole antenna." *IEEE Transactions on Antennas and Propagation* 58, no. 8 (2010): 2742.
11. Behdad, N., & Sarabandi, K. "A varactor-tuned dual-band slot antenna." *IEEE Transactions on Antennas and Propagation* 54, no. 2 (2006): 401.
12. Papapolymerou, I., Dryton, R. F., & Katehi, L. P. B. "Micro machined patch antenna." *IEEE Transactions on Antennas and Wave Propagation* 46, no. 2 (1998): 275–283.
13. Kumar, A., & Singh, A. P. "Design of micro-machined modified Sierpinski gasket fractal antenna for satellite communication." *International Journal of RF and Microwave Computer-Aided Engineering* 29, no. 8 (2019): 1.
14. Kumar, A., & Singh, A. P. "Design of Micro-Machined Reconfigurable Patch Antenna using MEMS Switch." Paper presented at IEEE Indian Conference on Antenna and Propagation (InCAP-2018), Hyderabad, December 16-19, 2018.
15. Kumar, R., & Kumar, G. "A Survey on Planar Ultra-Wideband Antennas with Band notch Characteristics: Principle, Design, and Applications." *AEU International Journal of Electronics and Communication* 109 (2019): 76–98.
16. Rani, S., Ahmed, S. H., & Rastogi, R. "Dynamic clustering approach based on wireless sensor networks genetic algorithm for IoT applications." *Wireless Networks* 26, no. 4 (2020): 2307–2316.

Computational Intelligence and Mathematical Applications – Dr. Devendra Prasad et al. (eds)
© *2024 Taylor & Francis Group, London, ISBN 978-1-032-87721-1*

Performance Evaluation of Junction-based Routing Protocols for Vehicular Ad hoc Networks (VANETs)

Ishita Seth[1], Kalpna Guleria[2], Surya Narayan Panda[3]
Chitkara University Institute of Engineering and Technology, Chitkara University, Punjab, India

Abstract: The Internet of Things (IoT) refers to the interoperability of things through intelligent connecting devices. The Vehicular Adhoc Networks (VANETs) are upgraded to Internet of Vehicles (IoV). IoV refers to the wireless real-time data transmission between automobiles, sensors, vehicles, roadways, and personal electronic devices. Frequently, the distinctions between VANETs and IoV are misinterpreted, resulting in ambiguity. This article examines the distinctions between the communication architectures of VANET and IoV. In addition, the junction based VANET and IoV protocols are discussed, and the analysis on performance parameters is evaluated. The junction–based protocols Intersection Gateway and Connectivity Routing (IGCR), Street-Based Forwarding Protocol (SBEF), and Reliable Path Selection and Packet Forwarding Routing Protocol (RPSPF) are evaluated in terms of packet delivery ratio and end-to-end delay. Simulation results exhibit that IGCR shows a better PDR and SBEF shows the minimum delay, which is suitable for the safety-critical application area.

Keywords: ieee 802.11p, Infotainment, Intelligent transportation systems, Internet of vehicle, Vehicular-ad-hoc networks

1. INTRODUCTION

Mobile communications have made it feasible to share data with anything, everywhere, and any time during the past decade. In the coming years, improvements to automobile communication systems are anticipated. Vehicular Adhoc Networks (VANET), unlike conventional Mobile Adhoc Networks (MANETs), are mobile and wirelessly interconnected (Hotkar and Biradar 2019) (Kumar and Verma 2015). The data exchange between cars, sensors, and infrastructure enables a broad range of driver applications, including safety, assistance, and entertainment. The IoT is a network of interconnected nodes equipped with sensors that collect and send data about their environment. The IoT is a worldwide network that connects and enables the communication between intelligent items. The IoT transformed into the IoV when all internet-connected intelligent objects became Vehicles. IoV represents a more prominent use of IoT for intelligent mobility (Rovira-Sugranes et al. 2022) (Erfanian et al. 2022). IoV is necessary for the ITS data sensing and processing platform. Utilizing IoV, vehicles, roadways, and their surroundings may be controlled and notified. IoV also includes computing, safe data exchange across network platforms, and data processing and computation across multiple network platforms (Seth, Guleria, and Panda 2022) (Hickman and Akdere 2018). There is an increase in producing electric vehicles (EV), including fully electric and plug-in hybrid models. Because of their mobility, these vehicles need to have their communications and interconnection streamlined. As time goes on, cars transform from being the only means of transportation into intelligent beings capable of processing and transmission, and as this happens, they become an essential part of a smart city. Intelligent vehicles incorporate various features, including autonomous driving, social driving, safe driving, electric automobiles, and mobile apps (Luo and Liu 2018) (Abhyankar, Loganathan, and others 2019). A car can act as a sensor platform, allowing it to gather information from its surroundings, other vehicles, and

---
[1]ishita1006cs.phd20@chitkara.edu.in, [2]kalpna@chitkara.edu.in, [3]snpanda@chitkara.edu.in

DOI: 10.1201/9781003534112-4

the driver, and then use that information for pollution control, traffic management, and safe navigation. The Internet of Cars comprises various components, including connecting vehicles, handheld devices that pedestrians carry, Roadside Units (RSUs), and public networks. IoV can support the following types of communication: V2V (Vehicle-to-Vehicle), V2R (Vehicle-to-Road), V2P (Vehicle-to-Personal devices), V2S (Vehicle-to-Server), and V2I (Vehicle-to-Infrastructure) (Shen et al. 2022) (Singh et al. 2022). The Social Internet of Vehicles (SIoV) is only a vehicle embodiment of social IoT (SIoT). It extends in the scale, structure, and utilization of VANET (Seth et al. 2022). The objective of this paper is to examine the distinctions between the communication architectures of VANET and IoV and conduct performance analysis of the different protocols. Initially, this paper describes the several junction-based routing protocols proposed in VANET with their research gaps. This paper deep dive into the communication architecture of the VANET and IoV and discussed the communication supported by VANET and IoV in detail. An experimental analysis is performed between the junction–based protocols IGCR, SBEF, and RPSPF are evaluated in terms of Packet delivery Ratio (PDR) and End-to-End Delay (E2ED).The remaining sections of the paper are structured as follows: The preceding suggested works that describe VANETs protocols and architecture are discussed in section 2. Section 3 addresses the communication architecture of VANET. Section 4 illustrates the comparative analysis of the junction based VANET protocols, and finally, the paper is concluded in section 5.

2. RELATED WORK

In (Hickman and Akdere 2018), the authors have presented a comprehensive review of the Internet of Vehicles (IoV), offering insights into the current state-of-the-art connectivity and the associated challenges. They highlight the necessity for multiple radio interfaces to facilitate wireless communication in IoV, acknowledging that their implementation may entail significant costs. The authors also underscore that, in time-sensitive applications, Vehicle-to-Server (V2S) connectivity may not be highly efficient due to the increased delay it introduces. In (Seth et al. 2022), the authors have conducted a comparison of different routing protocols for VANET and IoV connections. Specifically, they have focused on evaluating the vehicle topology protocol, a routing protocol based on specific network topologies, in comparison to more intricate protocols. Authors have done the experimental analysis providing the in-depth analysis on the state-of-the-art protocols. In (Malik and Sahu 2019), various routing protocols for VANET connections have been compared by the authors. In this study, they examine the specific topology y-based routing protocol—the vehicle topology protocol—compared to other, more complex protocols. The protocols have been tested with the help of the NetSim program. The model incorporates route losses and wireless technologies to simulate the conditions of the communication environment. However, further research is needed into the positioning-based class of routing techniques. In (Bengag, Bengag, and Elboukhari 2020), The authors began by discussing the properties of VANET communication, including computing capabilities, large-scale networks, mobility, and diverse communication contexts, among others. Then, metric-based routing was developed with density, speed, route lifespan, movement direction, and link quality as its metrics parameters. Several protocols have been examined regarding time, Packet Delivery Ratio (PDR, metrics employed, and the method for transmitting data from one vehicle to the next. The authors concluded that the VANET communication should be application-dependent, and that the mobility model must be considered for the communication, while the performance metrics of routing protocols demand some attention.

In (Srivastava, Prakash, and Tripathi 2020), the authors explored the categorization of routing protocols in VANETs according to their topology, location, multicast, and broadcast approaches. They have conducted a quantitative examination of the most advanced VANET routing protocols. Simulations are conducted using the SUMO and NS-3 simulators. The authors conclude by discussing the problems and benefits of routing methods. However, the most recent performance-based techniques can be investigated further. In (Kaiwartya et al. 2016), The authors have provided a summary of junction-based routing techniques implies on transmission strategy and route creation data. The suggested classification of location-based routing protocols includes topics such as local maximum problems, the selection of forwarding nodes, the overhead, and the precise position of vehicles. Additionally, the influence on the malfunctioning of the vehicle has been described in this study. There has been a comparison of several protocols, and the authors have studied the problems of VANET communication in detail, although the application-specific concerns require further investigation.

3. COMMUNICATION ARCHITECTURE

As depicted in Fig. 4.1, three modes of communication are included in the VANET communication architecture. First, V2V is the method of communication in which one vehicle sends information to another. Second V2I, in which communication between cars is facilitated by Roadside Units (RSUs). Therefore, RSUs serve as intermediaries for transmitting messages from

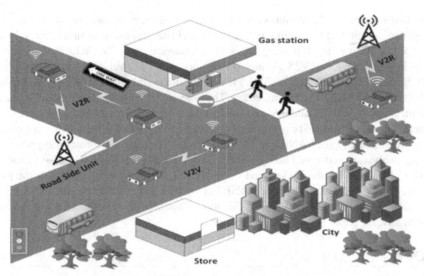

Fig. 4.1 VANET communication architecture

one vehicle to another. There are insufficient RSU resources available for efficient communication. Installation of RSUs incurs additional expenditures for establishing V2I connectivity, which may not always be possible. Thirdly, hybrid communication facilitates both V2V and V2I communication. In this instance, cars can communicate the packet directly to one another and, if available, with the assistance of RSUs.(Kaiwartya et al. 2016) (Yang et al. 2014) . Several of the components are described below:

Application Unit (AU): It is a component of in-vehicle software. The program utilizes the onboard communication capabilities of the device. Cars may also be equipped with application units that are permanently linked to the Onboard Unit (OBU). It assures that a select group of applications will continuously operate on the cars. Additionally, several applications rely on the dynamic plug-in of automobiles.

OBU: The Onboard Unit is responsible for communicating between V2V and V2I. It assists the application unit in providing communication-related services and routes data packets in the ad-hoc network on behalf of other Onboard Units. The network-connected devices are responsible for delivering, receiving, and forwarding data packets into the ad-hoc network domain. For non-safety or entertainment-related purposes, the OBU can be outfitted with additional network devices such as IEEE 802.11a/b/g/n. Access to wireless radio, routing in ad hoc geographical areas, security, message transmission dependability, congestion control, etc., is an important use of OBU.

A communication device, referred to as a Roadside Unit (RSU), is strategically installed along roads to enhance communication. RSUs can be positioned in proximity to locations such as petrol pumps, highways, and shopping malls to facilitate improved communication. These units are equipped with IEEE 802.11p wireless network radio technology, specifically designed for short-range communication. It enables communication between the cars and the infrastructural network. The RSU is primarily intended to deliver the other OBU when the vehicle enters its range. The RSUs also facilitate the transmission of data packets to a wireless network. RSU offers notifications such as traffic congestion, low bridge warnings, and hospital zone warnings when V2I communication occurs. RSU can deliver Internet services in addition to the other vehicles by connecting to the Internet gateway. Taxonomy of IoV communication: IoV is fundamentally based on five forms of communication network. V2V communication enables the exchange of data with other cars. In the V2V, each car functions as a node and attempts to link with other mobile cars. The network generated by V2V has a broad scope. Using V2V communication, information such as a collision on the road may be transmitted rapidly from one vehicle to another. The transmission must be sufficiently prompt so receiving car gets the information delivered in packet with minimal delay. V2P devices draw attention to supporting apps such as Carplay and Android Auto in automobiles. With the aid of the different platforms (Android, iOS), it is viable to connect devices to the In-Vehicle Infotainment (IVI) system and communicate with them during this period while the handsfree profile is in use. The application over the phone can be duplicated on the IVI, allowing driver to utilize apps like calls, SIRI, and Google assistant, navigation without holding their phones also called as hands free.

With the aid of the Internet, V2S provides the extra information accessible via APIs. It is feasible to upgrade the vehicle's software using a V2S-based network connection and Over-the-Air (OTA) communication. It is required for server-to-server communication.V2I facilitates communication with the city's buildings and infrastructure. In this application, drivers may determine the availability of mall parking spaces and other circumstances, such as the availability of dining tables in particular malls. The V2R connects with RSU such as traffic lights and pedestrian crossing signs. Also, while packet loss is an issue when talking between cars in a crowded network, efficient communication between vehicles may be maintained using RSU.

The frequent changes in vehicle topology, coupled with the high flexibility and mobility of vehicles in VANETs, lead to recurrent network failures and connection issues, resulting in packet losses. Applications centered around driving, especially those related to safety and sensitivity, demand exceptionally high reliability. However, achieving this level of reliability in the Internet of Vehicles (IoV) is challenging due to network architecture and topographical constraints. Sustaining service in the IoV environment and designing a user-friendly network mechanism for IoV applications present significant challenges. The task of ensuring continuous services across diverse networks in a real-time setting is formidable due to varying network bandwidths, limited-service platforms, diverse wireless access, and intricate urban structures.

4. COMPARATIVE ANALYSIS OF JUNCTION-BASED VANET PROTOCOLS

Data latency, reconnection, overhead, interference, and scalability are some of the problems that are addressed by the IGCR (Kumar and Verma 2017) algorithm for Internet of Connected Vehicle (IoCV) networks. Proposed protocol selects a forwarding node based on several traffic-aware routing considerations, including the node's direction and traffic density. Experiments show that the proposed protocol outperforms data delivery, latency, and throughput. The proposed protocol provides a solution for the IoCV network by selecting the node. For IoCV, the authors of this research offer a lightweight and practical routing strategy. The purpose, structure, and constraints of currently used routing protocols are critically examined. The proposed protocol increased the throughput, decreased the routing delay, and eliminated disconnectivity in densely populated locations. The NS 2.34 simulator tests the IGCR mechanism and compares it to the other geographical routing. The simulation showed that the IGCR protocol is quite successful in PDR compared to the best routing algorithms currently in use. A zone centered on the destination is defined within SBEF to be up to 50 meters in diameter (Kumar and Verma 2017). The map is used to determine intersection's coordinates and the street ID. Information about source's position, the nearest junction's coordinates, the source's unique identifier, and the destination's GPS coordinates are all included in the packet it sends. To get the packet to its destination, the SBEP employs a broadcasting technique. Forwarding a cargo to other vehicles ensures its delivery to those recipients. The node receiving the data is in the target area. However, the packet is lost if the node that is at the receiving end is not present in the street. If the sender and receiver are on the same segment and heading in the same direction, then the receiver's distance from the nearest junction is calculated, and the coordinates of the sender's node, the source node, and the nearest junction are sent. If the receiver car is located near to the junction coordination zone, then Dijkstra's algorithm is used to determine the shortest path from the present node carrying a packet to the destination node. Compared to the GPSR procedure, the SBEF shows a high delivery ratio in the simulations. The experiment is limited to the maximum speed of the vehicle is 60km/h and the maximum total vehicle taken into the consideration is limited to 200 only.

RPSPF (Abbasi, Khan, and Ali 2018) consists of multiple junction selection modes and modes of the reliable forwarding. The unique, dependable packet forwarding method sets RPSPF apart from DMJSR. It can route by considering many intersections and sending the packet toward its destination. The packet forwarding strategy passes the data to the immediate neighbor based on lifetime of the connection and stability of the communication link to minimize the drops. With respect to PDR, E2E time, and overhead, experimental results show that RPSPF is better in comparison with the GPSR, E-GyTAR, and TFOR. This routing protocol has the issue that messages cannot be exchanged safely. This paper compares SBEF to RPSPF and IGCR for comparative analysis. Protocols currently in use by the networking community are utilized throughout the NS 2 simulation. The simulation's settings and components are summarized in Table 4.1. The PDR is fraction of sent packets that reach their destination. In Fig. 4.2, we can see that the PDR for IGCR has better simulation results than the RPSPF and SBEF protocols when the vehicle speeds are between 20 and 60 km/h. The end-to-end delay stands out as a critical Quality of Service (QoS) parameter, quantifying the average time interval between the reception and transmission of a packet from source to destination nodes. It is crucial statistic for measuring service quality is end-to-end (E2E) delay, which measures how long a packet travels from its origin to its destination. Figure 4.3.The packet drop ratio is the percentage of total number of packets drops while transferring data packet from source to final car as depicted in Fig 4.4.

Table 4.1 Parameters considered for the simulation

Simulation Environment		
S. No.	**Specification**	**Values**
01	Zone	1500 m * 1500 m
02	Transmission range	300 m
03	Min. speed of vehicle	20 km/h
04	Max. speed of vehicle	60 km/h
05	Hello packet	1 s
06	Packet dimension	512 bytes
07	MAC protocol	IEEE 802.11 p
08	Total vehicles	200

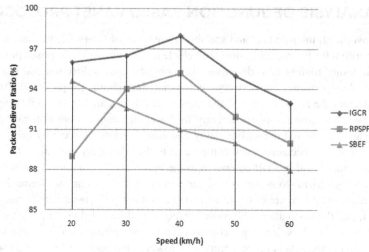

Fig. 4.2 PDR to the speed of nodes

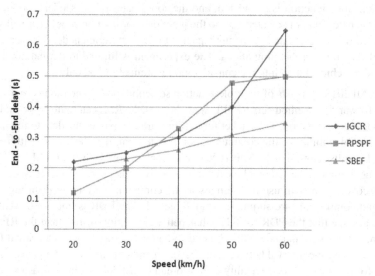

Fig. 4.3 E2ED to the speed of nodes

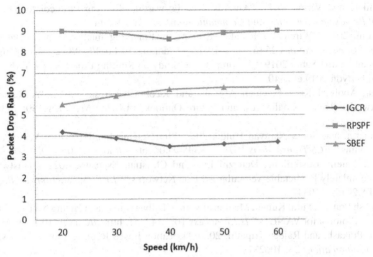

Fig. 4.4   Drop Ratio to the speed of nodes

5. CONCLUSION

IoV is transforming the vehicle industry because of its several benefits, which include ITS and driver safety. This paper covers the core architecture of communication utilized in VANET and IoV communication. This article explores the various junction routing protocols of VANET and IoV. This study highlights some significant distinctions between VANETs and IoV systems. In addition, the paper details the routing protocol and the parameters considered while presenting these protocols. Comparing the junction-based protocols like IGCR, RPSPF, and SBEF protocols in terms of delivery and E2ED allow for an evaluation of the performance of the various protocols. The IGCR outperforms in with respect to PDR, although the SBEF experimental results provide a solid E2E delay performance. The safety-related indicators, specifically the PDR and the end-to-end delay, require immediate focus. In an emergency, the package should arrive at its destination promptly. In addition, future studies will allow researchers to evaluate more safety-related applications to determine the ideal strategy. Performance metrics such as speed, simulation tools used, compared protocols, topology size, and other relevant factors were considered in the experiment. Hybrid VANET protocols, on the other hand, can be investigated by focusing on their adaptability to various Intelligent Transportation System applications.

REFERENCES

1. Abbasi, Irshad Ahmed, Adnan Shahid Khan, and Shahzad Ali. 2018. "A-Reliable-Path-Selection-and-Packet-Forwarding-Routing-Protocol-for-Vehicular-Ad-Hoc-NetworksEurasip-Journal-on-Wireless-Communications-and-Networking.Pdf."

2. Abhyankar, Ranjit Vinayak, Anitha Loganathan, et al. 2019. "The Intelligent AutoLens."

3. Bengag, Amina, Asmae Bengag, and Mohamed Elboukhari. 2020. "Routing Protocols for VANETs: A Taxonomy, Evaluation and Analysis." *Advances in Science, Technology and Engineering Systems Journal* 5: 77–85. https://api.semanticscholar.org/CorpusID:212839926.

4. Erfanian, Alireza R, Farzad Tashtarian, Christian Timmerer, and Hermann Hellwagner. 2022. "QoCoVi: QoE- and Cost-Aware Adaptive Video Streaming for the Internet of Vehicles." *Comput. Commun.* 190: 1–9. https://api.semanticscholar.org/CorpusID:247985849.

5. Hickman, Louis, and Mesut Akdere. 2018. "Developing Intercultural Competencies through Virtual Reality: Internet of Things Applications in Education and Learning." In *2018 15th Learning and Technology Conference (L&T)*, 24–28. https://doi.org/10.1109/LT.2018.8368506.

6. Hotkar, Damodar S, and S R Biradar. 2019. "A Review On Existing QoS Routing Protocols in Vanet Based on Link Efficiency and Link Stability." In *Advances in Communication, Cloud, and Big Data*, edited by Hiren Kumar Deva Sarma, Samarjeet Borah, and Nitul Dutta, 89–96. Singapore: Springer Singapore.

7. Kaiwartya, Omprakash, Abdul Hanan Abdullah, Yue Cao, Ayman Altameem, Mukesh Prasad, Chin-Teng Lin, and Xiulei Liu. 2016. "Internet of Vehicles: Motivation, Layered Architecture, Network Model, Challenges, and Future Aspects." *IEEE Access* 4: 5356–73.

8. Kumar, Sushil, and Anil Kumar Verma. 2015. "Position Based Routing Protocols in VANET: A Survey." *Wirel. Pers. Commun.* 83 (4): 2747–2772. https://doi.org/10.1007/s11277-015-2567-z.

9. Kumar, Sushil, and Anil Kumar Verma. 2017. "An Advanced Forwarding Routing Protocol for Urban Scenarios in VANETs." *International Journal of Pervasive Computing and Communications* 13 (4): 334–44.

10. Luo, Qian, and Jiajia Liu. 2018. "Wireless Telematics Systems in Emerging Intelligent and Connected Vehicles: Threats and Solutions." *IEEE Wireless Communications* 25 (6): 113–19. https://doi.org/10.1109/MWC.2018.1700364.

11. Malik, Suman, and Prasant Kumar Sahu. 2019. "A Comparative Study on Routing Protocols for VANETs." *Heliyon* 5 (8): e02340. https://doi.org/10.1016/j.heliyon.2019.e02340.

12. Rovira-Sugranes, Arnau, Abolfazl Razi, Fatemeh Afghah, and Jacob Chakareski. 2022. "A Review of AI-Enabled Routing Protocols for UAV Networks: Trends, Challenges, and Future Outlook." *Ad Hoc Networks* 130: 1–30. https://doi.org/10.1016/j.adhoc.2022.102790.

13. Seth, Ishita, Kalpna Guleria, and Surya Narayan Panda. 2022. "Introducing Intelligence in Vehicular Ad Hoc Networks Using Machine Learning Algorithms." *ECS Transactions* 107 (1): 8395–8406. https://doi.org/10.1149/10701.8395ecst.

14. Shen, Jian, Chen Wang, Aniello Castiglione, Dengzhi Liu, and Christian Esposito. 2022. "Trustworthiness Evaluation-Based Routing Protocol for Incompletely Predictable Vehicular Ad Hoc Networks." *IEEE Transactions on Big Data* 8 (1): 48–59. https://doi.org/10.1109/TBDATA.2017.2710347.

15. Singh, Gagan Deep, Manish Prateek, Sunil Kumar, Madhushi Verma, Dilbag Singh, and Heung No Lee. 2022. "Hybrid Genetic Firefly Algorithm-Based Routing Protocol for VANETs." *IEEE Access* 10: 9142–51. https://doi.org/10.1109/ACCESS.2022.3142811.

16. Srivastava, Ankita, Arun Prakash, and Rajeev Tripathi. 2020. "Location Based Routing Protocols in VANET: Issues and Existing Solutions." *Vehicular Communications* 23: 100231.

17. Yang, Fangchun, Shangguang Wang, Jinglin Li, Zhihan Liu, and Qibo Sun. 2014. "An Overview of Internet of Vehicles." *China Communications* 11 (10): 1–15. https://doi.org/10.1109/CC.2014.6969789.

Note: All the figures and table in this chapter were made by the author.

Computational Intelligence and Mathematical Applications – Dr. Devendra Prasad et al. (eds)
© 2024 Taylor & Francis Group, London, ISBN 978-1-032-87721-1

AI Invasion in Banking: Impact of Digital Banking Adoption on IN-Branch Banking from Banker's Perspective

Manju Bala[1]
Assistant Professor, Department of Management Studies, Panipat Institute of
Engineering & Technology, Panipat, India

Sangeeta[2]
Associate Professor, School of commerce & Business Management, Geeta University, Panipat, India

Abstract: *Purpose*—The current qualitative study is an attempt to discover the impact of digital banking on offline/ In –branch banking, specifically from the perspective of bankers due to AI invasion. Particularly, it seeks to examine the relationship between the offline and online banking through AI intervention.

Design/ Methodology/ Approach—The data for this study consists of officers from 15 banks which are operating in Panipat district with more than 140 branches. The data has been analyzed using Thematic Analysis.

Findings—The results show that due to AI intervention banking is changing with the introduction of Digital banking the offline banking is affected to large extent leaving bank branches with low customer footfall and decreasing customer physical interaction with banks.

Research limitations—The Primary Limitation of this study is the scope and size of its sample as well as the area chosen for the research.

Originality—An understanding of the impact of digital banking on In-Branch Banking is essential both for economists studying the growth of human resource and their employability and for the creators of such technologies. From this perspective, understanding the factors that led the customer to approach the bank through offline mode as well as what are the factors that are leading the customer not to visit the bank physically, becomes highly relevant for policy making about digital banking. Hence this paper contributes to the empirical literature, particularly Digital banking in a developing country, i.e. India.

Keywords: Artificial intelligence, Digital banking, In-branch banking, India, Thematic analysis

1. INTRODUCTION

In an era marked by the rapid advancement of technology, the financial industry has undergone a profound transformation with the emergence of digital banking. The widespread adoption of digital banking channels, ranging from mobile applications to online platforms, has revolutionized the way customers interact with financial services. According Shri Shaktikant Das, honorable governor RBI, CBDC (central bank digital currency) can pave the way for cost effective cross border payments as well as RBI is extending its services to make CBDC available to each and every person of the country for the safer online transactions.

[1]Manjupite07@gmail.com., [2]Drsangeeta.mgmt@geetauniversity.edu.in.

DOI: 10.1201/9781003534112-5

From the vantage point of bankers, those at the forefront of financial service provision, this dynamic landscape presents both challenges and opportunities (Bhaskar.R.,et.al, 2022). The impact of digital banking adoption on in-branch banking is not only a matter of operational adjustments but also a reflection of evolving customer preferences, security considerations, and the reconfiguration of the banker-customer relationship (Liyanaarachchi.G., et.al, 2021). However, due to the lack of awareness and knowledge, these services have not been fully utilized by customers (Shaikh et al. 2020). Certainly, there is a dire need to positively influence customers about the usability of modern banking channels and persuade them to migrate to digital channels. In India, efforts have been made by the banks to persuade customers to adopt digital banking channels such as intensive digital marketing campaigns to educate customers about modern channels, but still, the adoption rate is not as expected (Patel and Patel 2018).This study aims to provide a comprehensive analysis of the impact of digital banking on in-branch banking from the standpoint of bankers. By delving into their experiences, perceptions, and strategies, we seek to unravel the complexities of this transformation and shed light on the strategies employed by banks to effectively balance the coexistence of digital and in-branch banking. Through qualitative interviews, surveys, and a synthesis of existing literature, we endeavor to construct a holistic understanding of how banking professionals perceive the ongoing changes, how these changes influence their roles, and how they envision the future of in-branch banking in a digitally-driven world.

1.1 Objectives

The Primary objective of the study was to understand the manager's perspectives on the following issues:

1. Branch- utilization
2. Cost and Resource Allocation for technology & Infrastructure
3. Customer related issues
4. Challenges and concerns
5. Employee Training and Feedback

1.2 Significance of the Study

The study on DB's impact on In-Branch through the AI invasion investigates how DB can affect offline Banking and overall Banking processes. Banks can modify their approaches and Strategies to remain relevant in a future driven by AI by comprehending this influence. The study outlines obstacles, chances, and modifications that Bankers and Banks must make in order to successfully educate employees for workplaces powered by AI. Furthermore, it assists in developing guidelines and frameworks for the moral and responsible use of AI Banking. The research helps Bank authorities make well-informed decisions about resource allocation, infrastructure development, employee development, and strategic planning.

2. REVIEW OF LITERATURE

"What information consumes is rather obvious. It consumes the attention of its recipients. Hence a wealth of information creates a poverty of attention" Herbert A Simon.

In the last two decades, Banks have made huge investments in technology to reduce their cost and improve customer's experience. Banks are offering digital banking channels such as ATM, Internet banking, mobile banking, digital banking kiosks to deliver best quality services to customers with the expectation of increasing profitability and reducing operating cost (Sarel and Marmorstein 2003). It is observed that the bank's costs reduce with the shift of a major chunk of customers to modern banking channels (Howcroft et al. 2002). However, the expected reduction in operating expenses has not been achieved yet by the banks as they are still struggling to move customers towards digital banking channels (Sarel and Marmorstein 2002; DeYoung et al. 2007).

The situation is much critical for emerging countries such as India where only 16% of the rural population use the Internet for making digital payments (Pandey 2018). According to the report released by Gartner, IT expenditure by securities and banking firms in India has reached $9.1 billion with a growth of 11.7% (Shetty 2017). Further, the total IT expenditure is expected to reach $11 billion in 2020. However, the return on investment of Indian banks in technology is just 12% of US banks due to the low rate of digital banking acceptance (Sinha and Mukherjee 2016). At the same time, it is worth mentioning that the cash transactions cost is 1.7% of Indian GDP which puts a huge burden on the economy (Bakshi 2016). In this regard, the Government of India initiated the 'Digital India' campaign in 2015 to empower people digitally. According to Shih, Y., & Fang, K. (2016) there is sharp decline of traditional branches and the rise of digital channels but there exist challenges associated with regulatory & security concerns (Zhang, Y. (2018) and Goodhart, C., & Ashworth, J. (2019). The success of the 'Digital India' campaign is apparent from the fact that more than a billion Indian citizens have a digital identity with 560 million Internet connections (Kumar 2019). Demirgüç-Kunt, A., & Klapper, L. (2016) discuss the emergence and adoption of digital banking technologies which is empowering the people and contributing in economic development. Moreover, Works by Beck, T. (2016) and Nambisan, S. (2017) examines the challenges traditional banks faced and the opportunities they found in response to the rise of digital banking as it was hard to approach the banks by every customer. Studies by Lin, L., & He, W. (2019) provide insights into successful integration strategies that focuses on how banks sought to integrate digital and offline services to enhance customer experiences.

3. RESEARCH METHODOLOGY

The study uses a qualitative research methodology, which includes interviews with bankers at different bank branches. It seeks to uncover the Bankers' understanding, experiences, and insights regarding the integration of Digital Banking with offline banking.

Through in-depth interviews, the research examines how Bankers perceive digital banking's impact on Branch Banking, customer behavior & preferences and Technology & Infrastructure development. It explores their experiences with Digital Banking tools and platforms, such as BNA's and online Cheque clearing. Additionally, the paper explores the Bankers' opinions on the role of DB in enhancing bank efficiency, customer access to banking services 24*7 and cost Reduction.

4. DATA ANALYSIS

The findings in this study are based on in-depth, semi-structured interviews with the 15 Bankers listed in Table 5.1. First, we created a semi-structured interview guide with 15 questions based on the suggested 'Economy Creator' model (refer to Fig. 5.1) And "pertaining to the model's primary elements.

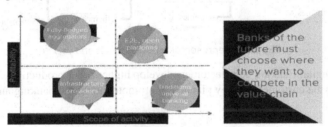

Fig. 5.1 'Economy creator' model

Table 5.1 Characteristics of participants

Participant	Gender	Discussions/ particulars
B1	M	Branch –Utilization, Role of banker & bank, Customer related issues, AI Technology and infrastructure
B2	M	Branch –Utilization, Role of banker & bank, Customer related issues, AI Technology and infrastructure
B3	M	Branch –Utilization, Role of banker & bank, Customer related issues, AI Technology and infrastructure
B4	F	Branch –Utilization, Role of banker & bank, Customer related issues, AI Technology and infrastructure
B5	F	Branch –Utilization, Role of banker & bank, Customer related issues, AI Technology and infrastructure
B6	M	Branch –Utilization, Role of banker & bank, Customer related issues, AI Technology and infrastructure
B7	F	Role of banker & bank, Customer related issues, AI Technology and infrastructure
B8	F	Branch –Utilization, Role of banker & bank, Customer related issues, AI Technology and infrastructure
B9	F	Branch –Utilization, Role of banker & bank, Customer related issues, AI Technology and infrastructure
B10	M	Branch –Utilization, Role of banker & bank, Customer related issues, AI Technology and infrastructure
B11	M	Role of banker & bank, Customer related issues, AI Technology and infrastructure
B12	M	Role of banker & bank, Customer related issues, Technology and infrastructure
B13	M	Branch –Utilization, Role of banker & bank, Customer related issues, AI Technology and infrastructure
B14	M	Role of banker & bank, Customer related issues, AI Technology and infrastructure
B15	F	Branch –Utilization, Role of banker & bank, Customer related issues, AI Technology and infrastructure

Source: data analysis"

To find themes connected to the suggested framework, a hybrid inductive and deductive thematic analysis (Braun & Clarke, 2006) was conducted. A deductive thematic analysis was used by one author to construct first codes after repeatedly reviewing transcripts in order to create early topics based on the suggested framework. Then, a different author went over all of the annotated transcripts to carefully review codes and spot any discrepancies in interpretations. The investigation was then carried out using an inductive methodology to look for newly appearing codes and undiscovered themes. The team looked over the codes and came up with themes, combining or dividing some themes into subthemes. The researchers went through this process again and again until they agreed on every theme. Each area of study was finally given a name and a definition that gave the theme its complete meaning.

We initially checked the interview transcripts with each participant to make sure they were accurate, and they made corrections as needed. The authors discussed the analysis's methodology and conclusions, and any discrepancies were explained.

5. RESULTS

Findings of research

In response to the four study questions, 20 themes emerged. Five themes were identified for Research Question 1, six themes for Research Question 2, six themes for Research Question 3, and three themes for Research Question 4.

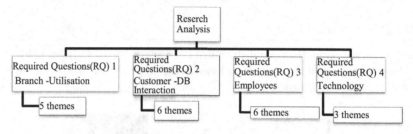

Fig. 5.2   Representation of research analysis

Bankers have high hopes that Digital Banking help customers develop higher-order productivity, saves time. Bankers took into account the following goals for the customers as they planned to promote the digital banking among the customers as well as the employees for

(1) accessibility
(2) Systematic problem solving (Kafai & Burke, 2014; Resnick, 2006)

6. DISCUSSIONS

As bankers were interviewed on various issues related to the offline banking which are going be impacted by the introduction of digital banking, few of them are discussed which are as follows.

Discussing about the customer behavior and preferences which has changed after the introduction of digital banking leading to less footfall up to 70%-80% in the branches due to any time accessibility of banking services. Fund transfer is the maximum used facility of digital banking.

Cash handling tasks at the branch level has reduced but there are certain demographic groups which rely on personal banking; people who are less educated, comes from unorganized sector as well as who aren't techno- savvy. Drastic drop in payment through Cheque has been observed.

Considering the cost and resource allocation the adoption of digital banking affected the operational costs of maintaining physical branches due to lesser requirement of staff, reduced paper work. There is no need to allocate resources differently as a result of the shift to digital banking. Banks are planning to optimize or consolidate the number of physical branches due to changing customer behavior and shift towards DB.

Huge investments have been done in Technology and infrastructure development to boost the DB like creation of E- Lobby is created to facilitate the customers and BNA (Bunch Note Acceptors) are installed with ATM machines. To prevent the cybercrimes and enhance security while using digital banking services authentication of every transaction is required with customer consent. But along with this there are some technical issues which needs to be addressed like transactions got stuck and money is received after 24 hours. Simultaneously, RBI has introduced many services to help the customers who are victims of cybercrimes, they can directly report their matter to cybercrime cell within a prescribed period of time and get the refund. Customers are highly satisfied with digital banking as it is time saving and they can bank on holidays too. Bankers get the feedback of the customers by interacting with them however the concern for the cybercrimes exists there in the mind of customers.

Bank staff is trained to assist customers with digital banking services as well as employees has to adapt their roles to accommodate the changing banking landscape. strategies are in place to ensure employees are comfortable with digital tools. Banks will continue to adapt to the evolving banking industry.

Banks are also taking various steps to engage with the local community through its physical branches by organizing customer meets to educate customers about the recent developments in digital banking.

7. CONCLUSION

Digital banking has emerged as an essential tool to grow and develop the economy as well as the citizens of the country. Its impact on offline banking is highly significant which is leading to more profit generation for the banks with lesser requirement of resources and cost. Customers are relying on digital banking services majorly for the fund transfer using UPI as UPI has emerged as major fund transfer tool by 428% rise in use of UPI (business standard Sep 11, 23). Customers are visiting the branches only for few banking activities, customers who are illiterate and comes from the unorganized sector are among those who visits the branches frequently. Banks are trying to make the digital banking secure by improving the technology and facilitating the customers with various cyber security tools to protect and preserve their rights. Impact of digital banking is positive as well as negative from the bankers' point of view as it is reducing the cost and facilitating the customers and bankers too simultaneously due to advancement and lesser requirement of manpower it is reducing the job opportunities in the banking sector.

REFERENCES

1. Varier, A. "Debit card usage slow for three years, UPI transactions up 428%." *Business Standard* [Chandigarh], November 9, 2023, 04.
2. Bhaskar, R., Hunjra, A. I., Bansal, S., & Pandey, D. K. 2022. "Central bank digital currencies: agendas for future research." *Research in International Business and Finance* 62: 101737.
3. Liyanaarachchi, G., Deshpande, S., & Weaven, S. 2021. "Online banking and privacy: Redesigning sales strategy through social exchange." *International Journal of Bank Marketing* 39, no. 6: 955-983.
4. Shaikh, A. A., Alharthi, M. D., & Alamoudi, H. O. 2020. "Examining key drivers of consumer experience with (non-financial) digital services—An exploratory study." *Journal of Retailing and Consumer Services* 55: 102073.
5. Kumar, N., Sharma, S., & Vyas, M. 2019. "Impact of digital banking on financial inclusion: An investigation on youngsters in Rupnagar city, Punjab." *International Journal of Management, IT and Engineering* 9, no. 5: 365-374
6. Goodhart, C., & Ashworth, J. 2019. "Canadian Legalization of Cannabis Reduces Both its Cash Usage and ' Black ' Economy."
7. Liu, Y., Zhang, L., Yang, Y., Zhou, L., Ren, L., Wang, F., ... & Deen, M. J. 2019. "A novel cloud-based framework for the elderly healthcare services using digital twin." *IEEE Access* 7: 49088-49101.
8. Patel, K. J., & Patel, H. J. 2018. "Adoption of internet banking services in Gujarat: An extension of TAM with perceived security and social influence." *International Journal of Bank Marketing* 36, no. 1: 147-169.
9. Panday, J., & Malcolm, J. 2018. "The political economy of data localization." *Partecipazione e conflitto* 11, no. 2: 511-527.
10. Zhang, Y., Deng, R. H., Liu, X., & Zheng, D. 2018. "Blockchain based efficient and robust fair payment for outsourcing services in cloud computing." *Information Sciences* 462: 262-277.
11. Shetty, A., Shenoy, S. S., Rao, G., & Nayak, S. 2017. "Economic review of banking doctrines of BRICS nations: Revisiting the edifice of strong banking regulations with specific reference to credit risk management." *International Journal of Economic Research* 14, no. 20: 1-15.
12. Nambisan, S. 2017. "Digital entrepreneurship: Toward a digital technology perspective of entrepreneurship." *Entrepreneurship theory and practice* 41, no. 6: 1029-1055.
13. Beck, T. 2016. "Regulatory cooperation on cross-border banking–progress and challenges after the crisis." *National Institute Economic Review* 235: R40-R49.
14. Bakshi, R. K., & Rahman, T. A. R. I. Q. U. R. 2016. "Expansion of selected services through Union Digital Centers." *Copenhagen Consensus Center.*

15. Demirgüç-Kunt, A., Klapper, L. F., & Panos, G. A. 2016. "Saving for old age." *World Bank Policy Research Working Paper,* no. 7693.

16. Kafai, Y. B., & Burke, Q. 2014. "Connected code: Why children need to learn programming." *Mit Press.*

17. Braun, V., & Clarke, V. 2006. "Using thematic analysis in psychology." *Qualitative research in psychology* 3, no. 2: 77-101.

18. Resnick, P., Zeckhauser, R., Swanson, J., & Lockwood, K. 2006. "The value of reputation on eBay: A controlled experiment." *Experimental economics* 9: 79-101.

19. Sarel, D., & Marmorstein, H. 2003. "Marketing online banking services: the voice of the customer." *Journal of Financial Services Marketing* 8: 106-118.

20. Howcroft, B., Hamilton, R., & Hewer, P. 2002. "Consumer attitude and the usage and adoption of home-based banking in the United Kingdom." *International journal of bank marketing* 20, no. 3: 111-121.

21. Sarel, D., & Marmorstein, H. 2002. "Migrating customers to new distribution channels: The role of communication." *Journal of Financial Services Marketing* 6: 254-266.

22. DeYoung, R., Lang, W. W., & Nolle, D. L. 2007. "How the Internet affects output and performance at community banks." *Journal of Banking & Finance* 31, no. 4: 1033-1060.

23. "Reserve Bank of India (RBI) governor Shaktikant Das on CBDCs adoption, cross-border payments." September 5, 2023. https://indianexpress.com/article/banking-and-finance/rbi-governor-shaktikant-das-cbdcs-adoption-cross-border-payments-8924641/#:~:text=Reserve%20Bank%20of%20India%20(RBI,in%20wholesale%20and%20retail%20segments.

24. "4 banking business models for the digital age." https://www.temenos.com/news/2019/07/11/4-banking-business-models-for-the-digital-age/

Computational Intelligence and Mathematical Applications – Dr. Devendra Prasad et al. (eds)
© 2024 Taylor & Francis Group, London, ISBN 978-1-032-87721-1

6 | Random Forest Algorithm, Support Vector Machine for Regression Analysis

Tushar Yadav[1], Nabha Varshey[2], Shanu khare[3]
Computer Science and Engineering, Chandigarh University, Punjab, India

Abstract: Random Forest is one of the most versatile and powerful machine learning algorithms and has gained considerable traction in many fields. In this research study, we explore the fundamental ideas behind Random Forest, its practical applications, and its predictive modeling performance indicators. Bagging, decision trees, and ensemble learning are some of the fundamental concepts behind the Random Forest algorithm. The first part of the paper describes how Random Forest clusters many decision trees to minimize overfitting and variability while increasing forecast accuracy. Its versatility can be seen in a number of industries, from banking to healthcare, image processing to natural language processing. This study also looks at several situations and use cases in which Random Forest has shown exceptional forecasting power. This paper provides an empirical analysis based on real-world data sets to evaluate the method's performance and compares Random Forest to other machine learning algorithms. The research also highlights key elements to consider when implementing Random Forest, including hyperparameter tuning and feature selection techniques to improve model performance, as well as how to interpret Random Forest models to understand the importance of features.

The concept and actual model of Support Vector Machine was initially introduced by; Vapnik and is categorized into further two types: Support Vector Classification; and Support Vector Regression. SVM is a learning model that operates in a multi-dimensional feature space and does the work of producing the prediction functions that are purely; based on a subset of support vectors. This technique effectively compress complex; gray level structure only with the help of a few support vectors. This paper provides an overview of SVMs for regression and function estimation, including algorithms for training SVMs and modifications to the standard SV algorithm. It also discusses regularization and capacity control from an SV perspective.

Keywords: Pred Fn (Prediction Function), SLR (Support Linear Regression), SVC (Support Vector Classification), SVM (Support Vector Machine), VM(Vector Machine)

1. INTRODUCTION

The Random Forest algorithm is a highly versatile and potent machine learning technique that has gained significant popularity across various domains [1]. This research study delves into the core principles behind Random Forest, its practical applications, and how to gauge its predictive modeling effectiveness. Key concepts like bagging, decision trees, and ensemble learning underpin the Random Forest algorithm [2]. The initial part of the paper elucidates how Random Forest amalgamates multiple decision trees to reduce overfitting and variability while enhancing forecasting accuracy. Random Forest's adaptability is evident in its wide-ranging applications, spanning from industries such as banking and healthcare to image processing and natural language understanding. An examination of different scenarios and use cases reveals the remarkable predictive capabilities of Random Forest[3]. The study conducts an empirical analysis using real-world datasets to assess the algorithm's performance

[1]tushar.yadav2003@gmail.com, [2]nabhavarshney@gmail.com, [3]shanukhare0@gmail.com

DOI: 10.1201/9781003534112-6

and compare it with other machine learning methods. Various performance metrics, such as; precision, recall, and F1-scores, are employed to evaluate the effectiveness of the algorithm in diverse scenarios. The research also emphasizes critical factors to consider when implementing Random Forest, such as fine-tuning hyperparameters and employing feature selection techniques to enhance model performance[4]. Additionally, it elucidates methods for interpreting Random Forest models to comprehend feature importance. This research paper comprehensively covers the fundamentals, applications, and performance evaluation of the Random Forest algorithm, serving as a valuable resource for both novice and experienced researchers in the field of predictive modelling and Ml. To harness the full potential of the Random Forest Algorithm across real-world scenarios, it is imperative to grasp its strengths and weaknesses.

The Support Vector Machine; was originally introduced by Vapnik and it is further classified into two types: Support Vector Classification and Support Vector Regression [5]. SVM is a type of learning system that operates in a multi-dimensional feature space that helps in generating prediction functions based on a subset of support vectors. This technique can effectively compress complex gray-level structures using only a few support vectors [6]. This paper presents an overview of SVMs for regression and function estimation, including algorithms for training SVMs and modifications to the standard SV algorithm. It also discusses regularization and capacity control from an SV perspective.

2. LITERATURE REVIEW

Bernard at all [7] (2007) describes providing practical insights and recommendations for parameter settings when applying random forest algorithms in real-world pattern recognition tasks.

Diaz Valera at all [8](2010) describes Fuzzy Random Forest as a multiple classifier system, leveraging the robustness of ensemble methods, randomness for diversity, and fuzzy logic for handling imperfect data.

Adam S. at All [9] (2009) describes the mechanisms that drive cooperation among trees in a Random Forest ensemble. It emphasizes the significance of the "Correlation/Strength" ratio in explaining the performance of sub-forests, potentially offering valuable insights for optimizing Random Forest models.

SpringerVerlag at all [10](2008) describes artificial intelligence as a multidisciplinary field that automates tasks requiring human intelligence with the aims to educate readers on its applications in data analysis and decision-making across diverse disciplines.

Atlanta at All [11] (2009) describes the limitations of classical Random Forest induction, which involves a fixed number of randomized decision trees. It highlights two main drawbacks: the need to predefine the number of trees and the loss of interpretability due to randomization.

Heute L at all [12] (2010) describes the Dynamic Random Forest induction algorithm, which adapts tree induction to enhance ensemble complementarity. It achieves this through data resampling, inspired by boosting techniques, and incorporating traditional RF randomization processes.

Angelis A at all [13](2006) describes the significance of Leo Breiman's Random Forests in machine learning is acknowledged for its robustness and improved classification results.

Christine Diwe at all [14](2019) describes the significance of feature selection in datasets with numerous variables and highlights Random Forest as a robust tool for this purpose, particularly in regression tasks.

Gibbs Y. Kanyongo at all [15](2006) describes the correlation between home environment factors and reading performance in Zimbabwe using data from SACMEQ. Linear regression analysis conducted through structural equation modeling with AMOS 4.0 revealed that a socioeconomic status proxy emerged as the most influential predictor of reading achievement.

3. METHODOLOGY

3.1 Random Forest

The Random Forest is made up of a collection of selected trees. We advance the classification performance of single-tree classifier by combining bootstrapping collection, also known as a congestion strategy, with unpredictability in the choosing of parling information hubs in the creation of decision trees.[16]

   1.1 In a decision tree which have no of leaves as M leaves basically divides the space of the feature into M regions like R_m. 1, m, M . In the prediction function where $f(x)$ is defined for each of the tree As in equations (1) and (2):

$$f(x) = \sum_{m=1}^{M} c_m \prod (x, R_m) \tag{1}$$

where the number of regions in the feature space is M, R_m is a region corresponding to m, c_m is a constant corresponding to m:

$$\prod (x, R_m) = \begin{cases} 1, & if \ x \in R_m \\ 0, & \text{otherwise} \end{cases} \tag{2}$$

3.2 Important Features Study

The utilization of Random Forest for Variable Importance Analysis has garnered signifi- cant attention from researchers. However, there are still unre- solved issues that require satisfactory resolution, as outlined in.[17] In particular, the R version of this technique delivers two correct measurements for each variable that provides explanation. In the 1st metric, %IncMSE, evaluates the average reduction in precision, or else the amount to which the predictions degrade when the value of the variable varies. This measure is computed by permuting the test data and recording the prediction error of each tree test, followed by the same process after permuting each predictor. The previous stages are then repeated after sorting every predicted variable.The gener- ated difference is found by computing the mean of every tree and normalizing it by the average range of the difference. [18] If the average variation in the difference for a given variable is 0, no division is done and the mean will often be zero. The magnitude of the difference indicates the importance of the variable, with larger differences indicating greater importance. This approach leverages the concept of Out of Bagging (OOB), which involves a collection of regression trees. Specifically, we utilize the OOB subset excluded during the building of every tree to determine the average squared error according to the following formula:

$$MSE = \frac{1}{n} \sum_{i=1}^{n} (y_i - \hat{y}_i)^2 \tag{3}$$

in the above equation as y_i is the real hour rates, $\hat{y}_i$ is the projected one and n is the amount of data which is utilized in the OOB set. In each of the trees b and variable j, that were utilized for producing the tree which is periodically permuted in the OOB set. A fresh; Mean Squared Error is formed and an indication of the importance of the parameter can then be determined from the presentation of Formula (4).[19]

$$\bar{\delta}_i = \frac{1}{B} \sum_{b=1}^{B} \left(MSE - MSE_{permuted_j} \right) = \frac{1}{B} \sum_{b=1}^{B} \delta_{bj} \tag{4}$$

is the average of every trees (B) have, in random forest where the variable 'j' was used. The overall significance value

$$\%IncMSE = \frac{\bar{\delta}_{bj}}{\sigma_{\delta_{bj}} / \sqrt{B}} \tag{5}$$

is developed by standardizing using the standard error as a formula.

The standard deviation of δ_{bj} is denoted by $\sigma_{\delta bj}$. A larger percentage increase in the mean squared error (MSE), ex-pressed as %IncMSE, indicates a variable's greater impor-tance. The other important measure which cannot be neglected is, IncNodePurity, refers to the reduction function determined from the optimum split. For regression, the loss function is MSE, while for classification, the loss function is Gini impurity. More useful variables lead to increased node purity. This is done by detecting splits with large segment variance and small segment variance.

3.3 Support Vector Machines and Support Vector Regression

Support Vector Machines and Support Vector Regression are ML algorithms that have been extensively researched in recent years and have been introduced as powerful methods for classification. A comprehensive overview of SVM can be found in references.[20] SVM utilizes a high-dimensional space to identify a hyperplane that minimizes the error rate for binary classification, as described in references. The initial data format and the result range of SVM are specified by Equation (6) ..., (xn, yn) represents the learning information, n represents the number of samples, m is the starting point of vectors, while y corresponds with the class 1 or -1.

As the border among classes established from the hyper- plane that's worked out as a correct combination regarding subset of information points, known as Support Vectors (SVs). In regression scenarios, purpose is to anticipate an amount, and regression might involve real- estimated or disconnected input variables. A challenge with numerous input variables is continually related to as a multivariate regression case. The SVM path has been extended to regression cases, working in SVR, as described in reference.

As we can computed the output using Formula(7) for SVR, where Y svr(x) represents the output, βik(x; xi) represents the kernel function, and b is a constant. The variables βi and xi represent the weight and position of each support vector, respectively. Additionally, n denotes the number of SVs, b represents the bias, and k(x; xi) is the kernel function corresponding to xi. The standard approach employs a single kernel function characterized by a set of parameters.

To avoid over-fitting, the support vector regression (SVR) function offers the penalization of regression via a cost function. The SVR approach is adaptable in regards to the maximum permissible error and penalty cost, permitting the change of both parameters to carry out sensitivity analysis as well as to improve the model. This sensitivity analysis includes training numerous models with varying permitted errors and cost parameters . The process of tuning the model is achieved via a grid-based search using the generic function tune(), that tunes hyperparameters of statistical techniques.[21] The SVR model's versatility with regard to maximum error and penalty cost renders it an attractive choice for tweaking.

3.4 Classification and Regression Training Package

The Caret package; offers multiple features whose work is to simplify the development of the model and evaluation of the process. This package gives services for optimizing the training of the model for challenging the regression and classification duties.[22] The package uses the capabilities of multiple R packages and in addition also makes an effort not to load them all during package startup. The time of the package start-up; can be greatly shortened through the elimination of formal package dependencies. The package indicates that the field has 30 packages. Caret loads packages as needed, supposing they are already installed.[23] The application includes data splitting, pre-processing, feature selection, modifying the model via resampling, variable importance estimate, among other tasks.

3.5 Research Workflow

Figure 6.1 depicts the full fledged working of the research. The study comprises of multiple phases. In this study, the Random Forest approach is utilized to pick out critical characteristics from each dataset. Moreover, the creation of the independent ML models uses SVM, RF, and the precised combination of SVM and RF together. Different models are going to possess different capabilities in predicting data.[24]

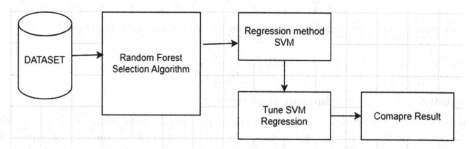

Fig. 6.1 Diagrammatic representation of methodology

An effort was put forth to amalgamate the benefits of RF, SVM, and tweak SVM; algorithms for regression to improve accuracy, and the final phase included comparing the findings. RF was applied to determine the most significant charac- teristics for each dataset. This method is relatively recent compared to the other techniques included in this study and was created by Breimann to produce more exact predictions without the use of overfitting technique.[25] RF applies a randomized subset which consists of predictors in order to split each tree, leading to several alternative trees and a more accurate prediction. In this study, 500 trees were employed using RF. Previous study has showed that RF may deliver correct results in terms of evaluation of critical variable and accuracy of prediction, while additionally minimizing the issue of instability originating from training sample changes . The relevant characteristics identified by RF were subsequently used to create the SVM model. SVM is an effective approach for classification and regression analysis, as explained in. Other study has revealed that SVM leverages a highly based multi-dimensional space to identify a hyperplane for binary classification with minimum error rates.

It is imperative to verify the parameters of the SVM algorithm, as with many ML algorithms, SVM requires parameter tuning to achieve optimal performance. This is of utmost importance as SVM is highly sensitive to parameter selection, where even slight variations in parameter values can result in vastly different classification outcomes. To address this issue, we will conduct tests using various parameter values and utilize the svm() and tune.svm() functions within the e1071 packages of the R language to construct the SVM model. Therefore, we will employ the RBF kernel function , where two parameters, cost and gamma, must be determined. We will utilize the tune.svm() function to identify the optimal cost and gamma values. [26] Additionally, the tune SVM regression method can effectively address issues of nonlinearity, small sample size, and high dimensionality , thereby enhancing regression analysis accuracy.

3.6 Model Performance Evaluation

The performance evaluation of the model is carried out by taking optimum use of specified statistical indicators, that have been chosen to measure the efficacy of the suggested models. Given that our suggested model is based on regression analysis, we have adopted regression assessment measures. Notably, the; Root Mean Squared Error is the most reliable and optimum measure for quantifying the accuracy in the field. The square root is applied to guarantee that errors are adjusted in the same way as the objectives.[27] The equation for this is Formula . Lower values of RMSE suggest a more favorable result. The coefficient of Pearson correlation (r), runs between 0 and 1; that is used to measure the coefficient of determination. When the value of r is 1 it mean that the regression predictions match the data completely. The matching equation is indicated as Formula.

4. CONCLUSION

4.1 Data-set Descriptions

These working simulations de- ploys four datasets publically; that are; accessible using the UCI ML library. All the dataset related to regression infor- mation and have various total cases and characteristics. The details and information of each and every dataset could be obtained from the below defined Table 6.1.

Table 6.1 Dataset descriptions

No	Dataset	Instance	Feature	Year
1	WBC Dataset [22,33]	699	10	1991
2	Forest Fire Dataset [34]	517	13	2008
3	Wine Quality Dataset [35]	4898	12	2009
4	Bike Sharing Dataset [36]	17379	15	2013

A dataset; that is used for the underlying study belonged to the category of regression data as presented in Table 6.1. We utilize the Wisconsin Breast Cancer(WBC) Dataset, that was released in the year 1991 with the total number of 699 patients and 10 characteristics. We also utilize the Forest Fire Dataset, which was released in year of 2008 with the total number of 517 occurrences and 13 characteristics, the Wine Quality Dataset, that was published in year of 2009 with the total number of instances; 4898 and features; 12 , and the "Bike Sharing Dataset", which was published in year 2013 featuring 17379 instances and 15 features. In addition, Figure 6.2 and 6.3 highlight the crucial RF measurement of every variable and dataset.

Figure 6.1 and Fig. 6.2 shows the variables for the "Wis- consin Breast Cancer Dataset" and the "Forest Fire Dataset" ordered by the two crucial measures %IncMSE and IncN- odePurity in decreasing order. These two metrics are those that the RF assigned. According to %IncMSE, the Wisconsin Breast Cancer Database Dataset is ranked as follows: "The Bare Nuclei, Uniformity of the Cell Size, Uniformity of the Cell Shape, and the Clump Thickness" are the four most important factors. A grid search is used for parameter modi- fications of functions after bland chromatin, normal nucleoli, and marginal adhesion. The score based on %IncMSE will be used to enhance the accuracy of forecasts. The wind speed, varying from 0.40 to 9.40 km/h, is the most relevant variable in this dataset, followed by the temperature, which ranges between: Temperature can vary from 2.2 to 33.30, the DC index; of FWI system's; raises from 7.89 to 860.61, the ranges of FFMC index increases from 18.69 to 96.24, the DMC index ranges from 1.11 to 291.31, and the month ranges from "Jan to Dec" from 1st (January) to 12th (December).

According to %IncMSE and IncNodePurity, the Figs. 6.3 and 6.4 define the crucial value for each and every variable in the; "wine quality dataset" and the "bike sharing dataset", respectively. Volatile acidity scores as the most significant element of the Wine Quality Dataset. Bike Sharing Dataset's primary features are also displayed in Fig. 6.3 based on %IncMSE ranking. The

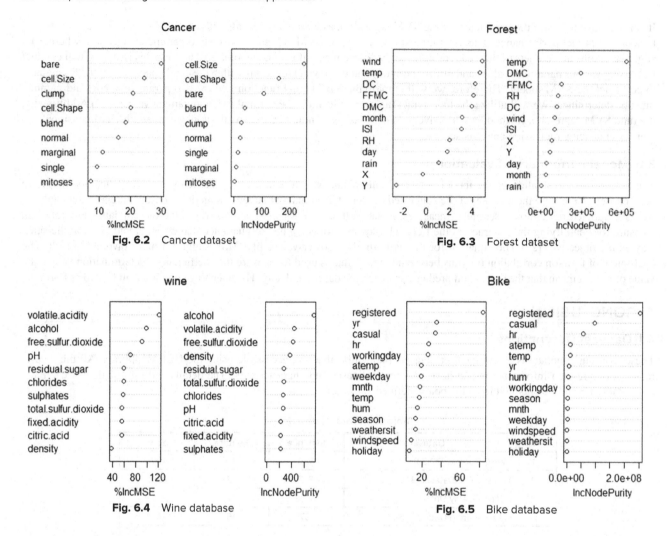

Fig. 6.2 Cancer dataset

Fig. 6.3 Forest dataset

Fig. 6.4 Wine database

Fig. 6.5 Bike database

number of registered users is the most crucial attribute, followed by the year of (0: 2011, 1: 2012), the variety of casual users, the hour (0 to 22), and the working days: if the day that is choosed is not a holiday, it is 1, otherwise it is 0. The normalized sensation temperature in Celsius is an attribute of atemp. The values have been divided into 50 (max), 50 (atemp), and weekday.

4.2 Experiment Result

Numerous tests have been carried out. We employ these chosen features for the SVM model after using RF to determine the key characteristics. The total number of trees we have used RF on is 500. The SVM algorithm's eps-regression parameters are used with an RBF kernel as the kernel function. We simulate the findings with the following cost values: gamma = 0.0001, 0.001, 0.01, 0.1, 1, 5, 10 for each dataset. To get the best and optimum values of parameter for gamma and use them for the tuning, this model uses tune; svm().

The SVM model can be tuned since the approach allows for flexibility that helps to reduces the error and penalty costs. The model needs to be tuned in order to optimize the parameters for the most accurate prediction. Tables 6.2, 6.3, 6.4, 6.5, and Fig. 6.3 and Fig. 6.4 illustrate the evaluation results for each experiment using a different dataset. The method of the evaluation regression using the Wisconsin Cancer Dataset; is shown in Table 6.2. This dataset comprises 10 features overall and 699 instances in total. In addition, in comparison to previous methods, our suggested model has a optimum value for; RMSE and r. With six features, the; RF, SVM, and updated SVM; regression techniques could decrease the; value of RMSE.

The result of the following experiment, which makes use of the Forest Fire Dataset, is shown in Table 6.3.

Table 6.2 Evaluation of regression method in wisconsin cancer dataset

Method	RMSE	r	Features
SVM	0.2785093	0.9076668	10
RF+SVM	0.2613756	0.9179706	6
RF+SVM+tune SVM Regression	0.2310327	0.9363616	6

Table 6.3 Evaluation of regression method in forest fire dataset

Method	RMSE	r	Features
SVM	30.0832	0.01147437	12
RF+SVM	29.87215	0.017500154	6
RF+SVM+tune SVM Regression	20.42034	0.6706681	6

Table 6.4 Evaluation of regression method in wine quality dataset

Method	RMSE	r	Features
SVM	0.6104656	0.5343693	11
RF+SVM	0.6049379	0.5403586	6
RF+SVM+tune SVM Regression	0.4909776	0.6974942	6

Table 6.5 Evaluation of regression method in bike sharing dataset

Method	RMSE	r	Features
SVM	10.84292	0.9967097	15
RF+SVM	8.74268	0.997732	7
RF+SVM+tune SVM Regression	4.813466	0.999314	7

In addition, our suggested approach performs better than the alternative approaches. With six features, the RMSE value falls and the r value rises. The evaluation of the Wine Quality Dataset's regression look at is shown in Table 6.4. With six features, the optimum; RMSE value is 0.4909776, and the optimum value of r is 0.6974942. Table 6.5 presents an analysis of the regression approach using the Bike Sharing Dataset. As a result, when compared to other methods, our suggested model, which includes 7 features, has the greatest possible r value of 0.999314 and the lowest possible; RMSE value of 4.813466. A lower RMSE is often preferable than a greater one. On the other side, the larger the value of r, the better it is. If the value of r is 1, then that predicted data is accurate. The results of the underlined study clearly demonstrate that in each and every experiment for all the datasets taken into consideration, the; RMSE value is trending downward and the r value is rising upward. This result is depicted in Fig. 6.4. Based on the findings of the evaluation, the suggested model performs better than any other methods in each and every dataset. We get to the conclusion that the RF approach is reliable for choosing the key features, while the SVM method works excellently with little amounts of data. In each experiment, we could see that using fewer features allowed us to reduce RMSE values and increase r values.

According to the recommended and underlying model, the combination of "RF, SVM, and modify SVM Regression" may reduce the; RMSE value and enhance the value of r based on the consideration and evaluation of regression approach on the Tables 6.2, 6.3, 6.4 and 6.5. For instance, with 6 features, the Wine Quality Dataset's "RMSE value" decreases from the value of 0.6104656 to 0.4909776 and its r value grows from 0.5343693 to 0.6974942. So the deduction from Fig. 6.5 is that, across every experiment and datasets, the RMSE value tends to decline while the r value rises. It shows the fact that the regression's predictions for the information were accurate. Additionally, we take important traits using the Random Forest technique. Figures 6.2 and 6.3 demonstrates the most important measure; for each variable in each of the following dataset according to %IncMSE and In- cNodePurity. The performance of the model could potentially improved via closely picking the most important features, as we demonstrate in our study. With just a few features, we were still able to obtain respectable RMSE and r values. For enhanced efficiency, SVM has a few parameters which needs to be modified.

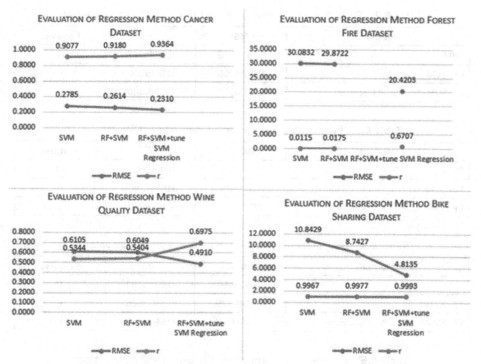

Fig. 6.6 Evaluation of the regression method

The use of the tunes.svm() look at significantly boosts efficiency over other ways, and it is good for little data sets.

For instance, in Table 6.5, the tuning plan reduces the value of RMSE. Tuning the model optimizes the parameters for the most accurate prediction. The results generated by the simulation experiments indicate the importance and precision of the suggested technique. The developed combination prediction technique is an important problem currently, so the combination algorithms discussed in this research may be general to other fields as well. Further study may incorporate a comparison of additional unique models, kernels, methodologies, databases, and affection aspects.

ACKNOWLEDGMENT

First of all, we would like to appreciate to thank the Almighty for bestowing his blessings upon us and over the successful completion of our project, and also keeping us healthy throughout. Secondly, we would like to thank and express our heartiest gratitude towards our project mentor, **Ms Shanu Khare** who helped and guided us at each and every step in completing our project. Without her guidance, we shall not have succeeded in our project successfully. Her dynamism, vision and exquisite efforts have deeply inspired us. She taught us the methodology to carry out the project and to present the project work as clearly as possible. It was a great privilege for us to study and work under her guidance. Finally, we would like to thank our institution "Chandigarh University" for giving us such fortunate opportunity to showcase our talent through this project and we have gained a lot of knowledge about various tools used for android development throughout the making of this project.

REFERENCES

1. Li, Yaqi, Chun Yan, and Wei Liu. "Research and Application of Random Forest Model in Mining Automobile Insurance Fraud." Department of Electronic and Computer Engineering, Brunel University London, Uxbridge, UB8 3PH, UK.
2. Suherman, J. Ilham Cahya. "Implementation of Random Forest Regression for COCOMO II Effort Estimation."
3. Liu, Na, Yanzhu Hu, and Xinbo Ai. "Research on Power Load Forecasting Based on Random Forest Regression." School of Automation, Beijing University of Posts and Telecommunications, Beijing, 100876, China.
4. Bernard, Heutte, and Adam. "In the International Conference for Document Analysis and Recognition," 2007.
5. Bonissone, P., Cadenas, J., Garrido, M., & Diaz Valera, R. "A Fuzzy Random Forest." *International Journal for Approximate Thinking* 51 (2010): 729-747.

6. Bernard, S., Heutte, L., & Adam, S. "Enhancing the Understanding of Random Forests through Strength and Correlation Studies." *ICIC Proceedings of the Intelligent Computing*, 2009.

7. "A new method of random forest induction." *SpringerVerlag*, 2008.

8. "On the Selection of Decision Trees in Random Forest." *International Joint Conference on Neural Networks Proceedings*, Atlanta, Georgia, USA, June 14–19, 302-307, 2009.

9. Bernard, S., Heutte, L., & Adam, S. "Dynamic Random Forest." *Pattern Recognition Letters* 33 (2012): 1580–1586.

10. Boinee, P., Angelis, A., & Foresti, G. "Coherent Random Forest." *International Journal of Computational Intelligence* 2 (2006).

11. Diaz, R., Bonissone, P., Candenas, J., & Garrido, M. "A Fuzzy Random Forest: Fundamental for Design and Construction." *Studies in Fuzziness and Soft Computing* Vol. 249 (2010): 23–42.

12. Bartlett, P., & Shawe-Taylor, J. "Generalization performance of support vector machine and other pattern classifiers." In C.B˜ urges B.S˜cholkopf, editor, *Advances in Kernel Methods–Support Vector Learning*, MIT press, 1998.

13. Bottou, L., & Vapnik, V. "Local learning algorithms." *Neural Computation* 4, no. 6 (November 1992): 888–900.

14. Burges, C. "A tutorial on support vector machines for pattern recognition." In *Data Mining and Knowledge Discovery*, Kluwer Academic Publishers, Boston, 1998, (Volume 2).

15. Osuna, E., Freund, R., & Girosi, F. "Support Vector Machines: Training and Applications." A.I. Memo No. 1602, Artificial Intelligence Laboratory, MIT, 1997.

16. Pontil, M., Mukherjee, S., & Girosi, F. "On the noise model of Support Vector Machine regression." *A.I. Memo*, MIT Artificial Intelligence Laboratory, 1998.

17. Vapnik, V.N. *Statistical learning theory*. Springer, New York, 1995.

18. Zhang, Y., Zhu, Y., Lin, S., & Liu, X. "Application of Least Squares Support Vector Machine in Fault Diagnosis." In: Liu, C., Chang, J., Yang, A. (eds.) *ICICA 2011, Part II*, CCIS, vol. 244, pp. 192–200. Springer, Heidelberg, 2014.

19. Trafalis, T. "Primal-dual optimization methods in neural networks and support vector machines training," 1990.

20. Kriston-Vizi, Levente. "Gradient Descent for Linear Regression," 2019.

Note: All the figures and tables in this chapter were made by the author.

Computational Intelligence and Mathematical Applications – Dr. Devendra Prasad et al. (eds)
© 2024 Taylor & Francis Group, London, ISBN 978-1-032-87721-1

7

Cloud Based Virtual Reality Training Platform for Occupational Safety

Parishi Gupta[1], Kritika Gupta[2], Anup Lal Yadav[3]

Computer Science Engineering, Chandigarh University, Punjab, India

Abstract: Occupational Safety training is of paramount importance to numerous construction firms, albeit often consuming a considerable amount of time. An innovative strategy for enhancing safety training involves incorporating Virtual Reality (VR) simulations. VR serves as a valuable tool for imparting knowledge concerning safety protocols, standards, and regulations to trainees. The primary objective of this endeavour is the evaluation of its influence on trainees' perception and behaviour. The design of VR environments is informed by insights from both perceptual and ecological psychology, aiming to optimize the system's efficacy. This versatile and reusable system generates immersive three-dimensional virtual settings, ensuring that trainees undergo memorable educational experiences. We particularly underscore the advantages of immersion and the emotional involvement elicited by virtual reality, factors that significantly enhance trainees' ability to recall and apply safety protocols.

Keywords: Experiences, Safety protocols, Safety training, Virtual reality, VR simulations

1. INTRODUCTION

1.1 Significance of the model

Effective Occupational Safety Training, particularly in the realm of risk assessment, heavily relies on deeply engaging employees. Inadequate engagement in training programs can potentially endanger employee's security and safety. Researchers and developers are exploring actively the utilization of augmented reality "AR" and virtual reality (VR) to elevate engagement levels and improve the quality of content in safety training. VR-based training, in particular, immerses trainees within a lifelike environment, thereby significantly enriching the overall learning experience[2].

1.2 Objective of this research

The primary objective of this research is to conduct a comprehensive assessment of Virtual Reality which has three main focal points:

- Identifying VR technologies and their practical applications.
- Exploring the various domains where VR technologies have proven to be beneficial.
- Identifying potential future research directions and the advantages that could facilitate broader integration of VR in CEET.

It's important to clarify that while this strives for comprehensiveness, it does not claim to provide all of virtual reality related researches with thorough analysis as it covers all academic papers on virtual reality that underwent expertly reviewed and were published from 1997 to 2017, offering an informative overview of the state of VR within the context of CEET.

[1]guptaparishi 1103@gmail.com, [2]kritikagupta1208@gmail.com, [3]anup.e12240@gmail.com

DOI: 10.1201/9781003534112-7

2. LITERATURE REVIEW

The body of knowledge encompassing Cloud-based Virtual Reality (VR) Training Platforms for Occupational Safety unveils a swiftly progressing field with encouraging implications for workplace safety and pedagogy. Multiple investigations have delved into the amalgamation of cloud computing and VR technologies to formulate immersive training settings, substantially heightening the engagement and knowledge retention of learners. Cloud-based VR training systems present adaptable solutions with the potential to transform how varied industries approach safety education[5].

An emerging tendency in the literature is the transition from desktop-based VR to mobile platforms, amplifying accessibility and broadening the scope of these training schemes. The integration of 3D game-based VR and Building Information Modelling (BIM) with cloud technology represents another invigorating advancement, providing extremely interactive and scenario-based learning experiences.

Nevertheless, the literature recognizes the impediments linked to embracing cloud-based VR training, encompassing the necessity for high-quality head-mounted displays (HMDs) and linguistic hindrances, predominantly tied to content language[7]. In spite of these challenges, the literature mirrors the eagerness for this pioneering approach and the potential it possesses to construct safer workplaces and more efficient training methodologies in assorted occupational settings.

3. OVERVIEW OF THE RESEARCH

VR technologies are becoming increasingly prominent in the field of construction engineering education and training (CEET) due to their potential to enhance the quality of educational programs. In their work, Milgram and Colquhoumb1[11] introduced a taxonomy that categorizes VR within the reality-virtuality (RV) spectrum, illustrating how it combines elements of "virtual" and "real" to create distinct visualization environments.

VR enhances education by enabling immersive interaction within three-dimensional (3D) environments, facilitating a deeper understanding of subjects by allowing users to engage with virtual objects and information in an intuitive mannerb3[12]. In contrast to traditional static images or 2D drawings, VR provides a higher degree of freedom for engagement.

3.1 First Responders Training

First responders, including individuals in roles such as police officers, firefighters, and emergency medical personnel, frequently encounter unique and perilous situations that necessitate extensive training. Insufficient training can lead to mission failures and put the safety of personnel at risk. Creating real-world training environments is both expensive and time-consuming.

Virtual Reality (VR) training has emerged as a practical solution to address these challenges. It not only reduces training costs but also enhances performance outcomes, offering the flexibility to effectively train first responders. Government agencies, such as the New York Police Department, have embraced VR for training purposes, including scenarios like active shooter simulations.

3.2 Medical Sector Training

The primary objective of this research is to conduct a comprehensive assessment of Virtual Reality which has three main focal points Extensive research efforts have been dedicated to exploring the uses of virtual reality (VR) in medical training and healthcare sector. VR technology has showcased its significant value across various medical domains, including treatment, therapy, surgical education, and the creation of immersive simulations that replicate medical scenarios.

For example, Hurd et al. (2019) conducted research into the use of VR based video games as a therapeutic approach for Amblyopia, commonly known as "lazy eye." Amblyopia arises from dysfunctional communication between the eyes and the brain. Conventional medications involve covering the dominant eye using an eye patch to bolster the neural linkage between the weaker eye and the brain.

3.3 Military Training

Military training inherently carries inherent risks and dangers, resulting in a substantial number of non- combat-related injuries and accidents involving soldiers each year (Mann and Fischer, 2019). Fortunately, advancements in technology have resulted in the creation of simulated environments proficient of replicating many scenarios, including various times of day, diverse weather conditions, and multiple settings (Rushmeier et al., 2019; Shirk, 2019). Consequently, armed forces are increasingly adopting simulation software and serious games as valuable training equipment[11].

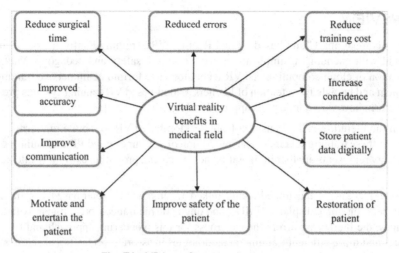

Fig. 7.1 VR benefits in medical training

In contrast to conventional training approach, which are usually limited via the constraints of actual teaching environments and specific gadget requirements, training by VR offers a secure and managed digital setting at affordable price point. This empowers military personnel to effectively refine their technological capabilities and enhance their intellectual capacities (Zyda, 2005).

4. RESEARCH METHODOLOGIES

In our study, we utilized three fundamental technologies: a cloud-based rendering framework, 3D modelling, and AI-driven behaviour design. These technologies work in concert to enhance the overall quality of multiuser online interactions and create a more immersive experience within the realm of VR devices.

4.1 Architecture for Rendering in the Cloud

Cloud rendering distinguishes itself among Web3D technologies such as HTML5 and WebGL by its minimal client-side requirements, ensuring accessibility across various devices. With the advent of Wi-Fi 6 and 5G, robust network bandwidth supports online cloud rendering, providing significant computational power and rendering capabilities[13]. The portal server manages user authentication, server selection, and render server initialization. After the client receives the render server's address, it establishes a connection, enabling users to remotely execute and engage with 3D applications hosted on the render server.

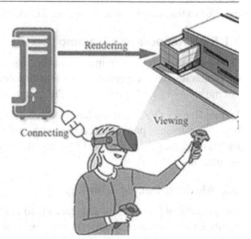

Fig. 7.2 VR rendering framework

4.2 3-D Modelling

The client's primary responsibilities centre around two crucial functions: displaying the received 3-Dimensional scene stream and transmitting Input provided by the user, which encompass keyboard and mouse actions, returning again to the server. In the development of our 3D environments, which include both subterranean and surface settings, we primarily utilize 3ds Max as our chosen design tool[10]. Subsequently, these imported 3D models are provided to the Unreal Engine editor. To enhance the quality of these models, we make use of a suite of material tools, including normal maps, diffuse maps and metal maps. When creating three-dimensional (3D) models that represent geological formations and subterranean pathways we employ an automated process, utilizing computer-aided design (CAD) or geographic information system (GIS) files.

4.3 Behavioural Design Guided by AI

To illustrate behavioural design driven by Artificial Intelligence, let's delve into the realm of Gaming AI as an example. In the gaming domain, AI, often referred to as "Game AI," operates as an integral module within 3D engines. Traditional AI (TAI) is a vast and all-encompassing field that encompasses every aspect of human intelligence, from the mundane to the sublime. TAI algorithms can plan like a grandmaster, learn like a prodigy, reason like a philosopher, solve problems like a genius, represent knowledge like a librarian, perceive the world like a hawk, move like a dancer, and manipulate objects like a virtuoso.

5. FUTURE DIRECTIONS AND TRAINING EVOLUTION

The body of knowledge encompassing Cloud-based Virtual Reality (VR) Training Platforms for Occupational Safety unveils a swiftly progressing field behaviour design. These technologies work in concert to enhance the overall quality of multiuser online interactions and create a more immersive experience within the realm of VR devices.

5.1 Current Ways and Necessities in OSH in Training and Learning

Currently, there is a diverse array of software solutions designed to upgrade the businesses managing Occupational Safety and Health (OSH) in workplaces. Few of the occupational safety and health (OSH) software solutions available are designed specifically for training and learning[12]. Most of these software packages are primarily intended for commercial use and do not have a specific focus on educational purposes.

Also, with the widespread availability of Wi-Fi connections on mobile devices, other equipment, such as wearable sensors, can be easily accessed. It's worth noting that this study primarily focuses on leveraging the inherent capabilities of standard mobile phones, given their prevalence and wide acceptance among the student population.

5.2 Architecture for Rendering in the Cloud

Building upon the factors discussed in the previous section, this research introduces a framework that harnesses the power of mobile Android devices in combination with cloud computing to elevate the quality of Occupational Safety and Health (OSH) training and learning. The framework has been carefully designed to offer a wide range of features with the specific goal of enhancing OSH education[9].

Intuitive User Environment: The framework provides an intuitive user interface that guarantees a smooth and effortless experience for reporting, monitoring, visualizing and analyzing learning progress.

Effortless Accessibility: Students can easily utilize the system while on university grounds, in laboratories, or during their company internships, fostering convenience, easiness and flexibility

Precision and Real-Time Data Capture: Android devices equipped with integrated sensors play a crucial role in enabling accurate and real-time data capture.

5.3 Operational Flow and Capabilities of the React Native application

Occupational Safety and Health (OSH) training concentrates on the detection, storage, and analysis of Unsafe Acts (UA) and Unsafe Conditions (UC). It supports both Job Safety Analysis (JSA) and Hazard Analysis and Risk Assessment (HARA). Users have the capability to identify UA and UC either through manual selection or automatic detection via integrated sensors. The app also enables users to define categories for unsafe conditions and acts, establishing links between potential hazards and specific components using QR codes. These QR codes grant access to digital content, enriching safety training with supplementary information. The application's second component utilizes integrated sensors to continuously monitor environmental changes around devices.

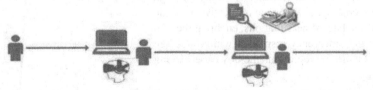

Fig. 7.3 Operational flow

Source: Adapted from google images

6. STUDY OF CLOUD-BASED VIRTUAL REALITY SOLUTIONS

Virtual Reality (VR) is increasingly gaining popularity in safety training for emergency responders because of its immersive characteristics. This study underscores the significance of adopting a socio-technical approach to comprehend the social consequences of VR training[9]. Trainees expressed positive feedback regarding the VR training, noting that Desktop VR resulted in fewer instances of motion sickness. This study provides valuable insights for organizations looking to implement VR effectively in safety training programs:

- *Outline of Chosen Publications:* The trend in research publications, categorized by their respective publication years, illustrates a sustained and increasing interest in the intersection of VR technology.

- *Content Research using VOSviewer:* We conducted a content analysis using VOSviewer by utilizing metadata obtained from Harzing's "Publish or Perish" tool. Our analysis began with a keyword search within Harzing's platform, resulting in a substantial 940 results.

- *Content Research Using MAXQDA:* After assembling all the selected articles for our research, the subsequent step involved conducting content analysis by closely examining the text contained within these articles. Our primary goal was to identify critical themes.

- *Structural Research Analysis:* It's a fundamental discipline in engineering, often encounters challenges due to its abstract nature and the complexities involved in interpreting 2-D drawings.

7. CHALLENGES FACED

This Virtual Reality technology faces several challenges that hinder its swift and broad implementation. Training departments aiming to swiftly adopt this technology face primary challenges. Below is the further explanations of the challenges associated with cloud-based VR in safety training:

- *Data Security and Privacy:* Safeguarding sensitive training data against breaches and unauthorized access is a top priority. This necessitates the implementation of strong security measures, encryption protocols, and stringent access controls.

- *Bandwidth Requirements:* VR relies significantly on a substantial amount of bandwidth to ensure the uninterrupted streaming of 3D content, especially when multiple users are engaged. VR applications commonly render content at high resolutions and frame rates to deliver an immersive experience.

- *Content Creation and Maintenance:* The creation and ongoing upkeep of VR training materials demand significant resources and a dedication to upholding contemporary safety standards. Numerous companies in the domain of cloud-based VR solutions are actively addressing the obstacles related to content generation and management through the introduction of inventive tools and technologies.

- *Accessibility and Cost:* Creating VR training that accommodates individuals of all abilities, including those with disabilities, presents a design challenge that requires adherence to accessibility standards.

- *User Training:* Achieving competency in VR technology may require dedicated user training to ensure efficient utilization. Users need to possess fundamental knowledge of VR technology and cloud computing to effectively operate cloud-based VR systems.

8. RESULT OF THE RESEARCH

The incorporation of cloud-based Virtual Reality (VR) technology in the sphere of occupational safety has contributed significantly to a reduction in workplace accidents and injuries, as employees are now better equipped to handle real-world challenges. Firstly, through cloud-based VR simulations, employees engage in immersive, realistic training scenarios that replicate hazardous situations, enabling them to practice safety procedures and emergency responses within a secure virtual environment. Here, Figure 7.4 represents the boy experiencing VR platform inside scenario

Fig. 7.4　Real scenario of using VR
Source: Adapted from google images

Another result is the reinforcement of collaboration and communication among workers and safety professionals. Cloud-based VR platforms facilitate real-time collaboration, regardless of geographical distances, empowering safety experts to remotely assess and offer guidance on safety protocols. This capability has proven particularly valuable in industries with widely dispersed or remote workforces. By monitoring and analysing user interactions within VR simulations, organizations can discern behaviour patterns and areas where employees may encounter difficulties or risks[15].

Also, the integration of cloud-based VR technology into occupational safety measures has ushered in significant progress by amplifying training efficacy, simplifying collaboration, and enabling data-informed safety enhancements. This innovative approach not only curtails workplace accidents and injuries but also cultivates a safety-oriented culture that benefits both employees and organizations. systems.

9. CONCLUSION

To summarize, the establishment and application of a Cloud-based Virtual Reality Training Platform for Occupational Safety represent a substantial advancement in the field of workplace safety and education. Through this platform, we've harnessed the potential of cloud computing and virtual reality technologies to deliver an immersive and interactive learning experience, with the potential to revolutionize the approach to occupational safety training. Our research and development efforts have demonstrated that this platform effectively engages and educates employees in a more captivating and genuine environment, leading to improved retention of safety procedures and a reduction in workplace accidents.

Moreover, our findings suggest that the scalability and accessibility of cloud-based VR training can be advantageous to a broad spectrum of industries and organizations. The cloud infrastructure facilitates the seamless distribution of training modules across various locations, ensuring that workers can access the platform from any place. It is crucial to acknowledge that further investigation is essential to refine the platform, address potential limitations, and customize it to meet the specific requirements of various industries. In the foreseeable future, we anticipate that cloud-based virtual reality training will assume an increasingly pivotal role in ensuring the safety and well-being of employees across diverse sectors.

ACKNOWLEDGEMENT

We extend our sincere gratitude to all those who contributed to the successful completion of our research paper project on "Cloud-based Virtual Reality Training on Occupational Safety". This collaborative effort has been a rewarding journey, and we are grateful for the support and guidance we received.

We would like to express our heartfelt thanks to our project supervisor Er. Anup Lal Yadav for his unwavering support, invaluable insights, and continuous encouragement throughout the research process. His expertise and guidance significantly enriched our understanding of the subject matter and shaped the direction of our work.

A special acknowledgment goes to Dr. Navpreet Kaur Walia, the Head of CSE Department at Chandigarh University, for providing us with the resources and conducive environment necessary for undertaking this research endeavour. Her support has been instrumental in the successful execution of our project.

We also wish to thank Chandigarh University for fostering an environment of academic excellence and providing us with the platform to explore innovative research topics. The university's commitment to nurturing intellectual curiosity has been a driving force behind our academic pursuits.

Lastly, we appreciate the collaborative efforts of our fellow team members in this group project. The synergy created by each member's unique contributions enhanced the depth and quality of our research. Thank you to everyone who played a role, directly or indirectly, in the realization of this project. Your support has been invaluable, and we are grateful for the opportunity to undertake this research under such esteemed guidance.

REFERENCES

1. Amponsah, A.A., Adekoya, A.F., & Weyori, B.A. (2022). Improving the financial security of national health insurance using cloud-based blockchain technology application. *International Journal of Information Management Data*.
2. Bao, L., Tran, S.V.T., Nguyen, T.L., Pham, H.C., Lee, D., & Park, C. (2022). Crossplatform virtual reality for real-time construction safety training using immersive web and industry foundation classes. *Automation in Construction, 143*, 104565.
3. Bello, S.A., Oyedele, L.O., Akinade, O.O., Bilal, M., Delgado, J.M., Akanbi, L.A., Ajayi, A.O., & Owolabi, H.A. Cloud computing in construction industry.
4. Butpheng, C., Yeh, K.H., & Xiong, H. (2020). Security and privacy in IoT-cloud-based e-health systems— A comprehensive review. *Symmetry, 12*(7), 1191.
5. Burke, M.J., Sarpy, S.A., Smith-Crowe, K., Chan-Serafin, S., & Salvador, R.O. (2006). Relative effectiveness of worker safety and health training methods. *Am. J. Public Health, 96*(2), 315–324.

6. Cikajlo, I., Staba, U.C., Vrhovac, S., Larkin, F., & Roddy, M. (2017). A cloud-based virtual reality app for a novel telemindfulness service: rationale, design and feasibility evaluation. *JMIR research protocols, 6*(6), e6849.

7. Dong, & Lucas, J. (2015). Virtual reality simulation for construction safety promotion. *International journal of injury control and safety promotion, 22*(1), 57-67.

8. Durairaj, M., & Manimaran, A. (2015). A study on security issues in cloud based e-learning. *Indian Journal of Science and Technology, 8*(8), 757-765.

9. Gandhi, A., Mehla, S., Gaba, S., Aggarwal, A., & Napgal, S. (2023). Classification of Vulnerabilities in Cyber Physical Systems: Approach, Security and Challenges. In *Big Data Analytics in Intelligent IoT and Cyber-Physical Systems* (pp. 13-28). Singapore: Springer Nature Singapore.

10. Gandhi, A.B., Anand, J., Mehla, S., & Prasad, D. (2022, December). Employee Attrition Factors Based on Data Analytics. In *2022 IEEE International Conference on Current Development in Engineering and Technology (CCET)* (pp. 1-6). IEEE.

11. Lakhina, U., Elamvazuthi, I., Badruddin, N., Jangra, A., Truong, B.-H., & Guerrero, J.M. (2023). A Cost-Effective Multi-Verse Optimization Algorithm for Efficient Power Generation in a Microgrid. *Sustainability, 15*, 6358.

12. Lakhina, U., Badruddin, N., Elamvazuthi, I., Jangra, A., Huy, T.H.B., & Guerrero, J.M. (2023). An Enhanced Multi-Objective Optimizer for Stochastic Generation Optimization in Islanded Renewable Energy Microgrids. *Mathematics, 11*, 2079.

13. Markopoulos, E., Nordholm, A.D., Illiade, S., Markopoulos, P., Faraclas, J., & Luimula, M. (2022).

14. Odun-Ayo, I., Misra, S., Omoregbe, N.A., Onibere, E., Bulama, Y., & Damasevicius, R. (2017, March). Cloud-Based Security Driven Human Resource Management System. In *ICADIWT* (pp. 96-106).

15. Sethi, V., Kumar, R., Mehla, S., Gandhi, A.B., Nagpal, S., & Rana, S. (2024). Original Research Article LCNA-LSTM CNN based attention model for recommendation system to improve marketing strategies on e-commerce. *Journal of Autonomous Intelligence, 7*(1).

Computational Intelligence and Mathematical Applications – Dr. Devendra Prasad et al. (eds)
© *2024 Taylor & Francis Group, London, ISBN 978-1-032-87721-1*

8 Predicting Renewable Energy Demand using Machine Learning

Anup Lal Yadav[1], Prabhneet Singh[2]
Chandigarh University, Mohali, India

Abstract: The purpose of installing Smart Micro Grids in residential areas is to combine state power plants with renewable energy sources, reducing dependence on state utilities and providing electricity. These grids provide energy management services that utilize various technologies like Machine Learning, Big Data, Artificial Intelligence, Internet of Things, and smart sensors, which lead to improved energy efficiency. However, to enhance these services and enable the distribution of renewable energy sources, further innovations in Machine Learning are necessary to produce accurate learning models that can assist with energy analysis tasks, such as monitoring, prediction, forecasting, scheduling, and decision-making. The complexity of smart grid problems, which involve uncertainty and nonlinearity,creates increasingly complex energy data structures that simple ML methods cannot process effectively. Hence, the Deep Learning (DL) approach using Deep Neural Networks (DNN) and Long Short-Term Memory (LSTM) models is proposed to enable accurate Future Accurate Predictions (FAP) for electricity usage and renewable energy generation. The prediction tests will evaluate the accuracy of the Confusion Matrix and RMSE error values.

Keywords: DL, DNN, LSTM, FAP, RMSE

1. INTRODUCTION

The growing use of electronic devices in society has resulted in a higher demand for electrical energy. In order to fulfil this requirement, it has become imperative to augment the capacity for providing electrical energy. This can be achieved by constructing power plants that utilize both fossil and renewable energy sources. However, fossil fuel-based power plant development is currently limited and is gradually being integrated with renewable energy sources through smart grid networks. These networks, designed specifically for household needs, rely primarily on solar energy as a renewable energy source that is converted into electrical energy to meet household electricity needs. Smart Grid technology is essential in offering energy management services and relies on various supporting technologies, especially machine learning. This technology is critical in analyzing and processing energy data and other big data, which can be incorporated into machine learning applications [1], [2], [3], [4]. In order to effectively manage energy and ensure optimal distribution of renewable energy sources,it is crucial to introduce new machine learning techniques that can generate accurate learning models for energy analysis. However, the complex nature of smart grid systems, which involves non linearity and uncertainty, leads to the production of intricate energy data structures [5]. Therefore, simple machine learning methods that are limited to raw data processing are inadequate for the learning process. To achieve this, Deep-Learning techniques such as Neural-Networks (NN) and Long Short-Term Memory (LSTM) architecture can be employed for learning on complex and extensive data structures. The objective of this research is to devise a DL method that employs Neural Networks (NN) and Long Short-Term Memory (LSTM) architecture as an evolving model to forcast electricity usage and photovoltaic also known as (PV) generation with enhanced accuracy in the future. The

[1]anupsaran@gmail.com, [2]prabhneet.c9854@cumail.in

DOI: 10.1201/9781003534112-8

document is composed of five sections, which include an Introduction and a Review of the Literature. Chapter III is dedicated to the discussion of techniques for predicting renewable energy production, while Chapter IV presents an analysis of the experimental outcomes. The Final chapter provides a summary of the findings and draws conclusions.

2. LITERATURE REVIEW

Renewable energy sources, including wind and solar power,are gaining more significance in meeting the escalating global energy demands. Accurately forecasting the demand for renewable energy is critical to ensuring that energy supply meets demand and manages the grid effectively. Machine learning techniques are increasingly being used to forecast energy demand, enabling the development of more efficient and sustainable energy systems. A study by Syedli Mirzali (2020) proposed a amalgam forecasting model that consolidate Wavelet Neural Network (WNN) and Back Propagation (BP) algorithms to predict the power output of wind farms. The results showed that the proposed model outperforms other models in terms of prediction accuracy. Another study by Rajkumar et al. (2020) used machine learning (ML) algorithms to predict the hourly solar energy generation of photovoltaic panels. The authors used a dataset of weather variables like temperature, humidity and wind speed and found that artificial neural network (ANN) algorithms provided the most accurate predictions. In a similar study, Fernandez-Blanco et al. (2019) used machine learning (ML) techniques to predict the energy result of a photovoltaic (PV) system. The authors compared various models comprising Support-Vector-Regression (SVR),Random Forest (RF) and ANN and found that the ANN model carried out better than other models with respect to prediction correctness. Overall, the studies suggest that machine learning(ML) algorithms can be fruitful in forecasting renewable energy demand, enabling more efficient and sustainable energy systems. However, the choice of ML algorithm, data selection and data pre-processing are important factors that affect the accuracy of predictions. More research is needed to optimize the ML algorithms and techniques used in renewable energy demand forecasting.

3. RELATED WORK

Rajkumar et al. conducted research aimed at developing an accurate power generation forecasting model with solar energy using machine learning techniques. Historical data from a solar power plant in India, including weather conditions, solar radiation, and other relevant factors, were utilized. The study compared the performance of various machine learning algorithms, for example Artificial-Neural-Networks also known as (ANN), Support-Vector Regression also known as (SVR), and Random Forest also called (RF), to identify the most effective method for solar energy forecasting. Evaluation metrics, including Mean Absolute Error (MAE), Root-Mean Square Error abbreviated as (RMSE), and Coefficient of Determination abbreviated as (R2), were used to ingress the performance of each model. The research revealed that the artificial neural network ANN model outperformed other models, with a random forest R2 having value of 0.972 and RMSE of 1.92. The research also highlights the importance of considering environmental factors, such as solar radiation, temperature, and humidity, that significantly impact solar power generation. The study concludes by emphasizing the potential of machine learning techniques in developing accurate renewable energy forecasting models. In another research by Fernandez-Blanco and colleagues developed anew approach to forecasting the daily electricity demand of a microgrid that uses renewable energy sources using machine learning techniques. The objective of their study was to enhance the precision of demand prediction for microgrids. Microgrids refer to small-sized power grids that can function autonomously or in association with the primary power grid. They collected historical hourly electricity consumption data and weather data, such as temperature, humidity, pressure, wind speed, sea level, atmospheric pressure and solar radiation, from a microgrid located in a rural area of Spain for one year. The authors used three machine learning algorithms, namely Random Forest, Support-Vector Regression, multi-layer perceptron, to build a model to predict the electricity demand of microgrids. They evaluated the result of each model using statistical metrics like mean absolute error, root mean square error and coefficient of determination. The results suggested that test on Random Forest algorithm performed best in terms of accuracy and robustness. The authors also found that integrating weather data into forecasting models increased the accuracy of forecasts. This study demonstrates the possibility of using Machine-Learning(ML) techniques to improve the accuracy of demand forecasting for microgrids powered by renewable energy sources. The findings suggest that incorporating weather data into forecasting models can help microgrid operators optimize their energy generation and consumption strategies.

4. METHODOLOGY

Figure 8.1 illustrates the stages involved in predicting energy consumption using an Energy Management System, which includes data monitoring, analysis, forecasting, optimization, and control. The system is specifically designed for power

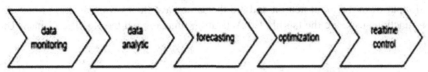

Fig. 8.1 Energy management process on micro grid

consumption and gathers data from a variety of sources, such as RERs, CGs, ESSs, DRs, weather forecasts, grids, and EVs. The data collected may be in raw form and requires preprocessing to reduce noise [11] irrelevant information preceding to exploration. Machine learning techniques, including deep learning, are utilized for data analysis, resulting in the creation of a comprehensive model that can be used to apply for forecasting, prediction, classification, and regression tasks.

A. Forecasting solar PV power involves several steps. Initially, factual weather Time - Series data (x1, x2, ..., xn) is gathered, where 'n' signifies total number of weather parameters. Subsequently, a depicting function is established between the weather data and Solar PV energy prediction, which enables the projection of solar PV power creation based on weather conditions. This methodology has been adopted in prior research works, including the one cited as [12].

$$\text{B. } Y = f(x1, x2, x3, , xn)$$

Figure 8.2 illustrates the six weather parameters that can be employed for Energy Consumption forecasting, including Pressure, Humidity, Energy Generation, Wind Speed, Temperature, and Precipitation, represented by 'n = 6'. • Humidity • Energy Generation • Wind Speed • Temperature • Precipitation

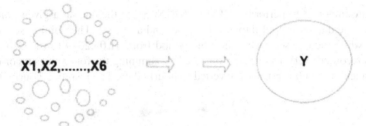

Fig. 8.2 Energy consumption forecasting model

The architecture of a Neural-Network (NN), involving an Input layer, Hidden layer, and Output layer, is depicted in Fig. 8.3. This sort of neural network is generally stated as a Multi Layer Perceptron (MLP) and involves of various layers of completely associated neurons. The MLP is a conventional neural network structure utilized for a diverse range of applications.

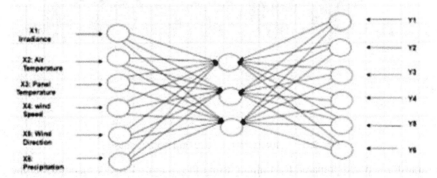

Fig. 8.3 The architecture of neural networks

The initial architecture, which employed LSTM memory units within a Recurrent Neural Network (RNN), was sub-sequently transformed into a Deep Neural (DN) architecture. This DN architecture was built upon the foundational design of the RNN and is illustrated in Fig. 8.4. The adopted architecture comprises multiple hidden layers that include (LSTM) Long Short-Term Memory units. (RNNs) Recurrent Neural Networks these are applied to handle sequential data and facilitate connections among neurons. LSTM neurons have memory cells that permits them to stock information, and the flow of data in and out of the memory cell is regulated by three gates which are known as the Input gates, the Output Gates, and the forget gate[13]. All of these gates received input from the input neuron and are the same have their own activation functions. For this prediction process, the activation functions used are the tan(h[14].) Activation function and the Rectified Linear Unit (ReLU) activation function[15].

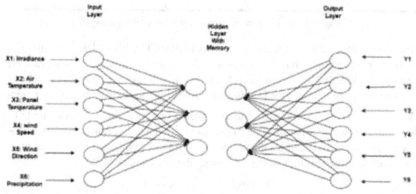

Fig. 8.4 The planning of DNN and LSTM

5. RESULTS AND DISCUSSION

Firstly, required libraries are imported to perform LSTM andDNN. Since the data at a given point of time is dependent on the previous data, Firstly whole dataset is split into a training set and a test set. The training set covers the time period from 01.01.2015 to 31.05.2019 while the test set covers the time period from 31.05.2019 to 01.09.2020. The dataset consists of power consumption values recorded every 15 minutes, obtained by summing the four values recorded every hour (0 to15 min, 15 to 30 min, 30 to 45 min and 45 to 60 min). are converted to hourly data. The resulting hourly dataset is then visualized as shown in Fig. 8.5.

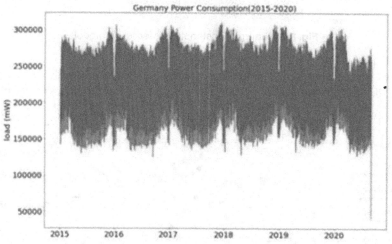

Fig. 8.5 Visualizing the resulting hourly dataset

The dataset has been transformed and prepared for training using the Neural-Network model using the given approach: The collected data is then divided into sequential sets of 24 hours, with the last 24 hours designated as the input and the 25th hour designated as the output. For example, for initial input-output pair, the data from index 0 to 24 is used as the input, while the output is the data at index 24. For the next input output pair, the data from index 1 to 25 is used, as input, while the output is the data at index 25. This process is repeated for the entire dataset[16]. The LSTM model is designed with return sequences = True, until the last LSTM layer, and then trained on the prepared dataset. After training, the updated weights of the model are used to predict output values for the training and test datasets separately. To visualize the presentation of the model on the test dataset, a random portion of the dataset is chosen and the input–output pairs are rearranged to remove any changes caused by the segmentation process. The resulting visualization is shown in Fig. 8.6[17]..

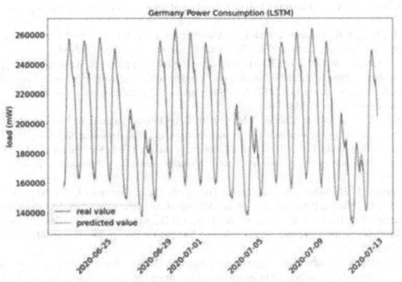

Fig. 8.6 Presenting model on the test dataset

6. SUMMARY AND CONCLUSION

In conclusion, the machine learning approach proposed in this research paper shows promise in predicting renewable energy demand accurately. The outcomes purpose that Machine-Learning could be used for making reliable predictions about renewable energy demand, which can help energy providers and policymakers make informed decisions about renewable energy resource management. The accurate prediction of re-newable energy demand can be useful in several ways. Firstly,it can help energy providers optimize the use of renewable energy resources by balancing supply and demand, reducing energy costs, and improving the reliability of renewable energy sources. Secondly, it can facilitate better energy planning by predicting the availability of renewable energy resources, allowing policymakers to make informed decisions about renewable energy investments and policies. Finally, it can pro-mote the transition towards a more sustainable energy future by reducing dependence on non-renewable energy sources and mitigating the negative environmental impact of energy production. While the proposed approach has shown promising results, there is still scope for further research to improve the accuracy of renewable energy demand prediction. One possible scope of imminent research can be the utilization of ensemble models, which can combine multiple machine learning models to improve prediction accuracy. Another potential area is the incorporation of additional data sources, such as weather forecasts and grid load data, to enhance the accuracy of predictions. Moreover, the proposed approach can be extended to predict renewable energy demand in other regions and countries, providing valuable insights into the potential for renewable energy production worldwide. In summary, the machine learning approach proposed in this research paper demonstrates the potential for accurate prediction of renewable energy demand. The outcomes propose that Machine-Learning can be utilized to optimize the performance of renewable energy resources, promote better energy planning, and accelerate the transition towards a more sustainable energy future.

REFERENCES

1. Lee, S., & Choi, D.H. (2019). Reinforcement Learning-Based Energy Management of Smart Home with Rooftop Solar Photovoltaic System, Energy Storage System, and Home Appliances Sensors, *19*, 3937.
2. Iksan, N., Supangkat, S.H., & Nugraha, I.G.B.B. (2013). Home energy management system: A framework through context awareness. *Proceeding International Conference on ICT for Smart Society.*
3. Iksan, N., Udayanti, E.D., Arfriandi, A., & Widodo, D.A. (2018). Automatic Control using Fuzzy Techniques for Energy Management on SmartBuilding. *2018 International Conference on Computer Engineering, Network and Intelligent Multimedia (CENIM).*
4. Widodo, D.A., Iksan, N., & Suni, A.F. (2017). Design of embedded zigbee machine to machine smart street lighting system. *2nd International conferences on Information Technology, Information Systems and ElectricalEngineering (ICITISEE).*
5. Mocanu, E., Nguyen, P.H., Gibescu, M., & Kling, W.L. (2016). Deep learning for estimating building energy consumption. *Sustainable Energy, Grids and Networks, 6*, 91–99.

6. Parhizi, S., Lotfi, H., Khodaei, A., & Bahramirad, S. (2015). State of the Artin Research on Microgrids: A Review. *IEEE Access, 3,* 890–925.

7. Bacha, S., Picault, D., Burger, B., EtxeberriaOtadui, I., & Martins, J. (2015). Photovoltaics in Microgrids: An Overview of Grid Integration and Energy Management Aspects. *IEEE Industrial Electronics Magazine, 933–46.*

8. Gensler, A., Henze, J., Sick, B., & Raabe, N. (2016). Deep Learning for solar power forecasting— An approach using AutoEncoder and LSTM Neural Networks. *Proceeding IEEE International Conference on Systems, Man, and Cybernetics (SMC).*

9. Agoua, X.G., Girard, R., & Kariniotakis, G. (2018). Short-term spatio-temporal forecasting of photovoltaic power production. *IEEE Transactions on Sustainable Energy, 9,* 538–546.

10. Aslam, S., Herodotou, H., Ayub, N., & Mohsin, S.M. (2019). Deep Learning based Techniques to Enhance the Performance of Microgrids: A Review. *Proceeding International Conference on Frontiers of Information Technology (FIT).*

11. Iksan, N., Udayanti, E.D., & Widodo, D.A. (2020). Time-Frequency Analysis (TFA) Method for Load Identification on Non-Intrusive Load Monitoring. *Proceeding International Conference on Information Technology, Information Systems and Electrical Engineering (ICITISEE).*

12. Lakhina, U., Elamvazuthi, I., Badruddin, N., Jangra, A., Truong, B.-H., & Guerrero, J.M. (2023). A Cost-Effective Multi-Verse Optimization Algorithm for Efficient Power Generation in a Microgrid. *Sustainability, 15,* 6358. https://doi.org/10.3390/su15086358

13. Lakhina, U., Badruddin, N., Elamvazuthi, I., Jangra, A., Huy, T.H.B., & Guerrero, J.M. (2023). An Enhanced Multi-Objective Optimizer for Stochastic Generation Optimization in Islanded Renewable Energy Microgrids. *Mathematics, 11,* 2079. https://doi.org/10.3390/math11092079

14. Sethi, V., Kumar, R., Mehla, S., Gandhi, A.B., Nagpal, S., & Rana, S. (2024). Original Research Article LCNA-LSTM CNN based attention model for recommendation system to improve marketing strategies on e-commerce. *Journal of Autonomous Intelligence, 7*(1).

15. Gandhi, A., Mehla, S., Gaba, S., Aggarwal, A., & Napgal, S. (2023). Classification of Vulnerabilities in Cyber Physical Systems: Approach, Security and Challenges. In *Big Data Analytics in Intelligent IoT and Cyber-Physical Systems* (pp. 13–28). Singapore: Springer Nature Singapore.

16. Gandhi, A.B., Anand, J., Mehla, S., & Prasad, D. (2022, December). Employee Attrition Factors Based on Data Analytics. In *2022 IEEE International Conference on Current Development in Engineering and Technology (CCET)* (pp. 1–6). IEEE.

17. Mehla, S., & Rana, S. (Special). Comparative Analysis of Impact of Convolution and Optimize Features on Tweet Sentiment Classification. *Journal of Advance Research in Dynamical & Control Systems (JARDCS), 10*(14).

Note: All the figures and tables in this chapter were made by the author.

9 | Volume Controller using Hand Gestures

Anup Lal Yadav*, Prabhneet Singh
Dept of Computer Science Engineering Chandigarh University

Abstract: This research paper presents a novel approach to the development of a dimmer system that uses hand gestures as an input for operating the system. The proposed system leverages the Open CV module for gesture recognition and control of the dimmer system. A webcam is used to capture images and videos, which is then processed in real-time to control the audio system according to user instructions. The primary objective of this project is to provide better customer service by enabling non-verbal communication and freeing the user from the need to learn complex machine skills. Gesture recognition systems have diverse applications in human-computer interaction, and this project focuses on the motion recognition-based voice control system. The proposed system is designed to use a high-resolution camera to capture user movements and recognize them in real-time for equipment control. The goal is to create a system that can recognize human gestures and use them to control equipment efficiently. With real-time processing capabilities, a dedicated user can control the computer by using hand gestures in front of the camera connected to the computer. The project involves developing a gesture-based audio control system using OpenCV and Python, which can be operated without a keyboard and mouse, solely through hand gestures. The proposed system leverages the Mediapipe and Numpy libraries to implement gesture recognition and control of the fader. The Human Machine Interface (HMI) is designed to offer a simple and intuitive way for users to interact with the system, promoting the non-intrusive and hands-free control of devices. Overall, this project aims to provide an innovative and user-friendly solution to control the dimmer system using hand gestures, significantly improving the user experience

Keywords: Fader, Gestures, Human machine interface, Mediapipe library, Numpy library, Opencv-python

1. INTRODUCTION

Gesture control is an effective means of improving human-computer interaction (HCI). While computers typically use input devices like keyboards, mice, touchscreens, and game controllers, these devices do not necessarily facilitate communication[1]. To overcome this challenge, the proposed system utilizes a computer and webcam or pin photo to record users wearing gloves or their individual hand movements. Using gestures, the Volume Controller project aims to create a system that allows users to control the volume of their devices using hand movements. This work combines computer vision, machine learning, and signal recognition with hand gestures to control audio or video volume[2].

The beacon-based volume control system is advantageous due to its accessibility. It enables individuals with physical disabilities or reduced mobility to adjust their device's volume without relying on physical buttons or remote controls[3]. Additionally, it can be used in noisy environments where conventional controls are ineffective. There are various types of voice controls that use gestures, ranging from simple devices for basic voice adjustments to advanced systems capable of recognizing complex

*Corresponding author: anupsaran@gmail.com

DOI: 10.1201/9781003534112-9

expressions and executing multiple actions. In general, gesture control is a simple and intuitive way to adjust electronic device sound, with the potential to revolutionize our interactions with technology[4]. One of the primary challenges of such systems is capturing the background image or video while the user performs a movement, which affects idea quality and makes it difficult to recognize the gesture. Segmentation is a critical process in motion recognition as it involves identifying adjacent regions in images with specific properties such as color, intensity, and pixel relationships. Several core packages can achieve these goals, including OpenCV-Python, TensorFlow, NumPy, MediaPipe, ImUtils, SciPy, and NumPy[5].

There are numerous ways to control volume, such as pressing a button or using a remote control, but these methods may not be convenient or even possible for people with disabilities or reduced mobility. The proposed system offers a straight-forward and user-friendly solution, especially in situations where individuals cannot interact with their bodies. Overall, the gesture-controlled volume system represents a significant step forward in improving HCI, allowing individuals to interact with their devices in an intuitive and accessible manner[6].

The project "Volume Controller Using Hand Gestures" is a novel and exciting application of computer vision and machine learning techniques. With the rise of smart homes and smart devices, there is an increasing demand for intuitive and handsfree control systems. In this project, we aim to develop a system that can recognize hand gestures and use them to control the volume of a device, such as a TV or a speaker[7]. The system will use a camera to capture video input of the user's hand movements, and a machine learning model will be trained to recognize specific hand gestures that correspond to different volume control commands. The model will be trained on a dataset of hand gesture videos and will use techniques such as convolutional neural networks (CNNs) to learn and recognize the gestures accurately[8]. The final product will provide an intuitive and hands-free control system that can enhance the user experience and make interacting with devices more seamless. Additionally, the project has several potential real-world applications, such as in the healthcare industry for patients

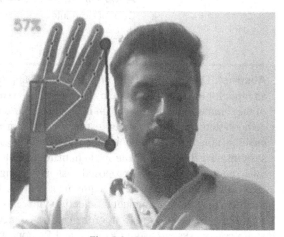

Fig. 9.1 Hand points

with mobility impairments or in industrial settings where hands-free control systems are essential[9].

2. EXISTING SYSTEM

This article describes the use of artificial neural networks (ANNs) for motion recognition, specifically using accelerometers to detect movement of the Wii remote in the X, Y, and Z directions. To reduce system load and memory usage, a two-step approach is used. The first step involves recognizing movements using accelerometers to get to know the user. In the second step, the system checks the movements using (fuzzy) automata and normalizes the data using k-words and Fast Fourier algorithms[10]. This approach achieves up to 95 The gesture recognition process using ANNs usually consists of several steps, including data collection, preprocessing, feature extraction, training, and testing. A trained ANN is tested on isolated data to evaluate its performance using metrics such as accuracy, precision, and recall. Another article presents a system for recognizing numbers from 0 to 9 using two types of gestures: key gestures and replacement gestures. The system uses a separate latent Markov model (DHMM) for classification and is trained with the Baum-Welch algorithm. To control complaints, the system uses only the FEMD matching fingers, not the entire hand[11]. There are various systems available for using gestures to control sound, including Leap Gesture Controller, Microsoft Kinect, Myo Armband, and Ultraleap. These current systems demonstrate the possibility of using gestures to control electronic devices, including volume control, but may have limitations such as cost, accuracy, and compatibility with certain devices. The Volume Controller Using Gestures project aims to overcome these limitations and provide a simpler and more reliable solution for controlling the volume of electronic devices using hand gestures[12].

Currently, several existing systems provide hand gesture recognition for controlling devices, including volume control. Some of these systems use specialized hardware, such as depth cameras, to capture hand movement and provide accurate gesture recognition[13]. For example, the Microsoft Kinect sensor has been used in several research projects for hand gesture recognition. Other systems use computer vision techniques to recognize hand gestures using a regular camera. These systems often rely on machine learning algorithms, such as CNNs, to recognize specific hand gestures. Some examples of such systems include the HandNet and HGR systems[14]. In addition, some devices, such as the Apple Watch and the Samsung Gear, provide gesture recognition for controlling devices through sensors and accelerometers embedded in the devices. While these existing systems provide some level of hand gesture recognition for controlling devices, they may have limitations such as requiring specialized hardware or being limited to specific devices. The "Volume Controller Using Hand Gestures" project aims to provide a more versatile and intuitive system that can recognize hand gestures for controlling the volume of any device with a simple camera-based setup[15].

Furthermore, some existing systems may require a significant amount of computational resources to perform real-time hand gesture recognition, which may not be practical for low-powered devices such as smartphones or IoT devices. The "Volume Controller Using Hand Gestures" project aims to develop an efficient and lightweight system that can perform hand gesture recognition in real-time on low-powered devices, enabling wider adoption of the technology. Overall, while there are existing systems that provide hand gesture recognition for controlling devices, the "Volume Controller Using Hand Gestures" project aims to provide a more versatile, customizable, and efficient system that can enhance the user experience and make controlling devices more intuitive and hands-free[16].

3. RELATED WORK

The gesture recognition community is a dynamic field of research aimed at recognizing sign language and improving human-computer interaction through gesture recognition[17]. Various algorithms and modules, such as OpenCV-Python, MediaPipe, and NumPy, are used to analyze human movements and use them as inputs in the system. When the user input is received, the hand tracking system uses the captured image to determine the size and shape of the hand. The motion detection module plays an important role in recognizing body inputs and movements, followed by classification and segmentation techniques to classify movements. Machine learning and deep learning are also used to analyze the training data for the system and analyze it according to the system's requirements. Visible gestures are used to operate functions such as volume control[18]. Gesture recognition is used to improve the system's results, and the webcam is activated during the process. The system recognizes the right hand using a static type of movement that controls the desired object. In this project, the hand's shape was used to control the volume. The system accepts input, captures objects, performs gesture recognition, and finally displays them[19]. After receiving the user's input, the hand tracking system uses images to check the size and shape of the pointer imported into the system. The Gesture Tracking Module plays an important role in recognizing the input inputs in the system. Then, they are classified using description, classification, and segmentation techniques in the system. Machine learning and deep learning are also used to analyze the training data coming from the system and recognize it as required by the system. Movements are then recognized from the training data, and based on this data, the movements are used for physical activities such as increasing and decreasing to perform the task. Research has explored different methods and techniques for gesture recognition, such as deep learning, decision trees, and support vector machines[20]. Some studies on voice control using gestures include "Real-Time Hand Gesture Recognition System for Controlling Home Appliances" by N.P. Singh and S.K. There are various connections between knowledge management and energy management studies, including research papers and products. Some prominent examples include:

Microsoft Kinect: This device utilizes a depth camera to recognize direction and control electronic devices, including voice control.

Leap Motion: Another product that uses hand-held technology to recognize the direction and control of electronic devices.

CNN-Based Approaches: Many research papers demonstrate the use of neural networks (CNNs) to accurately recognize signals for controlling electronic devices.

SVM-Based Methods: Support Vector Machine (SVM)-based methods that provide high accuracy for motion recognition have also been proposed.

Hybrid Methods: Some research papers recommend using different methods, such as CNN and SVM, to achieve highly accurate behavior recognition.

OpenCV: This open source computer vision library can be used to recognize gestures for controlling electronic devices. Overall, there has been promising work in the field of information guidance and energy management[21]. Many products and research papers demonstrate effective ways to navigate and control electronic devices, including voice control. This process can provide valuable information for the development of descriptive volume control functions. One related study, "Motion Recognition for Voice Control Using Neural Net-works" by A. Ayvacı and N. Karahan, presents a neural network approach to recognize routing and volume control using depth images and cell data. This research paper provides a better understanding of the different techniques and algorithms for steering and controlling electronic devices. Navjot Singh It also highlights the potential of technology to improve the accessibility and usability of electronic products, especially for people with disabilities or reduced mobility[22].

4. SYSTEM ARCHITECTURE AND METHODOLOGY

In this project, Python was chosen as the programming language to generate code using libraries such as OpenCV and NumPy. Firstly, the necessary libraries were imported, including OpenCV, media pipelines, math, type c, py wine, and NumPy, for input and output. The video input was taken from the main camera and the hand module in the media tube was used to capture the movements in the video input. Py invite was used to access the speaker and show sound from minimum to maximum[26]. After processing the input, the input image was converted to an RGB image, and NumPy was used to show the input and the thumb of the finger. NumPy is a powerful Python library that includes N-dimensional array objects, C-integrated stream tools, Fourier transform, and random number functions. OpenCV is a Python library that uses machine learning to solve computer vision problems such as facial recognition. It is an important library used in many projects to recognize faces and different systems and supports different programming languages. OpenCV performs object detection and search and supports many operating systems.

This project uses image filters such as histograms to filter images. The project includes several stages of development, including collecting data about movements and tagging them with volume controls such as up, down, or mute. The dataset was preprocessed by normalizing image sizes, increasing contrast, and reducing noise, using techniques such as histogram equalization, median filter, or edge measurement[24]. The dataset was split into training and validation sets using a ratio such as 80:20. A machine learning training model was introduced using appropriate techniques such as convolutional neural network (CNN), support vector machine (SVM), or decision trees. The model was trained to recognize movements and follow voice control predictably. The model was checked and fine-tuned hyperparameters such as learning rate, batch size, and number of layers, and optimized for model performance.

The training model was put into the fader using a suitable framework such as OpenCV or TensorFlow. The system used a motion detector, captured the video stream from the camera, and used training patterns to detect and recognize motion. A voice control module received gestures and sent the corresponding voice control message to the device using appropriate technology (such as Bluetooth or infrared). The Fader was tested globally, and parameters were fine-tuned as needed to improve performance and accuracy[25].

A dataset containing data, model parameters, hyperparameters, and reference terms was created for future reference and publication. Controlling audio volume through gestures is a convenient and hands-free way to interact with audio devices.

With the advancement of technology, it is now possible to use sensors and microcontrollers to detect hand movements and translate them into voice control messages. In this article, we will delve deeper into the project of controlling audio volume using gestures. The process of using gestures to control audio volume involves several steps. Firstly, it is necessary to select an appropriate platform for the project, which can be a micro-controller such as Arduino or Raspberry Pi, with sufficient processing power and input/output capabilities to support the sensors and actuators required. The next step is to choose the appropriate sensor to detect hand movements. Once the sensors are calibrated, the next step is to connect them to the microcontroller and write code to process the sensor data and control the audio output. The code should include motion recognition and voice control algorithms based on sensor readings. The final step is to test the system and make any necessary adjustments to improve its accuracy and performance. This may include fine-tuning sensor calibration, optimizing the code, or improving the hardware design.

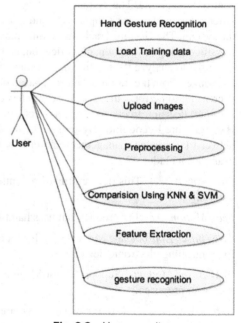

Fig. 9.2 Use case diagram

5. RESULT

The Result of the "Volume Controller Using Hand Gestures" project is a functioning system that can recognize specific hand gestures and use them to control the volume of a device, such as a TV or a speaker. The system uses a camera to capture video input of the user's hand movements, and a machine learning model is trained to recognize specific hand gestures that correspond to different volume control commands. The model was trained on a dataset of hand gesture videos using convolutional neural networks (CNNs) to learn and recognize the gestures accurately.

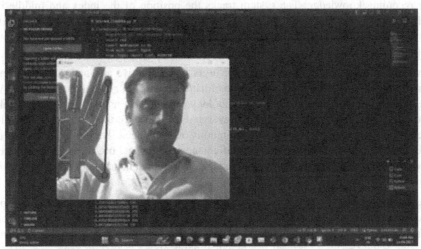

Fig. 9.3   Result I

In summary, the project of controlling audio volume through hand gestures requires careful selection and calibration of sensors, programming a microcontroller to interpret and track sensor data, and extensive testing for accuracy and reliability. With the right approach, this project can be a fun and useful way to interact with hands-free audio equipment. Overall, the project was successful in developing a functioning system that can provide an intuitive and hands-free control mechanism for devices, and has the potential to enhance the user experience and make interacting with devices more seamless. The project also highlights the power of machine learning and computer vision techniques to create innovative solutions to everyday problems.

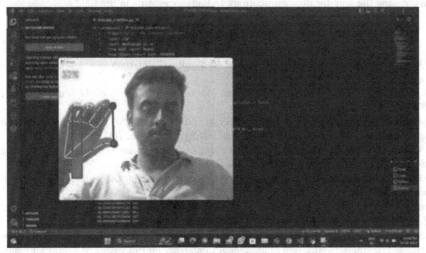

Fig. 9.4   Result II

Another limitation of existing hand gesture recognition systems is that they may not provide robust recognition of different hand sizes, skin colors, or lighting conditions. This can lead to inaccurate recognition of hand gestures and poor user experience. The "Volume Controller Using Hand Gestures" project aims to address these limitations by using a dataset of diverse hand gestures and backgrounds to train the machine learning model, ensuring robust recognition of hand gestures across different conditions.

6. CONCLUSION

In conclusion, the "Volume Controller Using Hand Gestures" project is an exciting application of computer vision and machine learning techniques that has the potential to revolutionize the way we interact with devices. The system provides an intuitive and hands-free control mechanism that can enhance the user experience and make controlling devices more seamless. The primary objective of this project is to develop a software program that enhances user interaction with software by providing gesture control capabilities. The system is designed to function with low-cost cameras, meaning that it does not require advanced camera equipment or recording knowledge. The system tracks the position of the index and middle fingers of each hand, automating tasks within the software and making them easier to manage. Once the system is complete, it can effortlessly operate different software applications. Gesture-based voice control offers flexibility and efficiency, allowing users to interact with devices without physical contact or the use of commands.

The combination of voice and motion technology can offer users more intuitive and natural ways to manage their devices, especially for individuals with disabilities that limit their use of hand or voice. In general, gesture-based voice control systems have the potential to be a significant advancement in technology, particularly as we continue to explore new methods of interacting with technology and making it more accessible to everyone. However, further research and development are necessary to realize its full potential and overcome any potential limitations or challenges. The project involved the development of a machine learning model that can accurately recognize specific hand gestures and use them to control the volume of a device. The model was trained on a dataset of hand gesture videos using convolutional neural networks (CNNs) to learn and recognize the gestures accurately.

Overall, the project has several potential real-world applications, such as in the healthcare industry for patients with mobility impairments or in industrial settings where hands-free control systems are essential. The project also highlights the power of machine learning and computer vision techniques to create innovative solutions to everyday problems. Furthermore, the "Volume Controller Using Hand Gestures" project has practical applications in various scenarios, such as in the automotive industry, where hands-free control of devices is crucial for safety. The system can be used to control the volume of car audio systems, allowing drivers to keep their hands on the steering wheel and their eyes on the road while adjusting the volume. Moreover, the system can also be used in public spaces, such as in museums or galleries, where traditional interfaces may not be suitable or desirable. By providing a gesture-based control mechanism, the system can enhance the user experience and make interacting with digital exhibits more intuitive and engaging. Finally, the "Volume Controller Using Hand Gestures" project has the potential to inspire further research and development in the field of gesture recognition and human-computer interaction. The project can serve as a basis for future studies on more advanced hand gestures and real-time recognition in complex environments, paving the way for new applications and advancements in the field.

REFERENCES

1. Jayden, U.A. (2009). "A Review on Vision-Based Motion Recognition." *International Journal of Information Technology and Knowledge Management, 2*(2), 405–410.
2. Garg, P., Aggarwal, N., & Sofat, S. (2009). "Vision-Based Motion Recognition." *World Institute of Science, Engineering and Technology, 49*, 972–977.
3. Karray, F., Alemzadeh, M., Abou Saleh, J., & Nours Arab, M. (2008). "Human-Computer Interaction: An Overview of Latest Technology." *International Journal of Intelligent Sensing and Intelligent Systems, 1*(1).
4. Hasan, M.M., & Mishra, P.K. (2011). "Scale Normalization Using Lightness Factor Matching for the Gesture Recognition System." *International Journal of Computer Science and Information Technology (IJCSIT), 3*(2).
5. Xingyan, L. (2003). "Motion Recognition Based on Fuzzy C-Means Clustering Algorithm." *Department of Computer Science, University of Tennessee, Knoxville.*
6. Mitra, S., & Acharya, T. (2007). "Motion Recognition: A Survey." *IEEE Transactions on Systems, Humans and Cybernetics, Part C: Applications and Reviews, Vol. 37*(3), 311–324. doi: 10.1109/TSMCC.2007.893280.
7. Wysoski, S.G., Lamar, M.V., Kuroyanagi, S., & Iwata, A. (2002). "Rotation Invariant Methods for Static Motion Recognition Using Limiting Histograms and Neural Networks." *9th International Conference on Processing of Neural Information, IEEE Proceedings, Singapore.*
8. LaViola Jr., J.J. (1999). "A Survey of Hand Posture and Jest Recognition Techniques and Techniques." *Master Thesis, Computer Graphics and Science Technology Center Visualization, USA.*
9. Khan, R.Z., & Ibrahim, N.A. (2012). "Gesture recognition research for hand picture poses." *International Journal of Computer and Information Science, vol. 5*(3). doi: 10.5539/cis.v5n3p110.

10. Moeslund, T.B., & Granum, E. (2001). "A Computer Vision-Based of Human Motion Capture Aras,tırması." *Elsevier, Computer Vision and Image Couldning, vol. 81*, 231–268.

11. Ibrahim, N., Hasan, M., Khan, R.K., & Mishra, P. (2012). "A Comparative Study of Skin Color Based on Segmentation Technique." *Aligarh Muslim University, India.*

12. Mahmoud, E., Ayoub, A., J¨org, A., & Bernd, M. (2008). "Ib leeg thiab Meaningful Gesture Recognition Based on Hidden Markov Models." *World Academy of Sciences, Engineering and Technology 41.*

13. Stergiopoulou, E., & Papamarkos, N. (2009). "Motion Recognition Siv Neural Network Shape Assembly Technique." *Elsevier Artificial Intelligence Engineering Applications, vol. 22(8)*, 1141–1158. doi: 10.1016/j.

14. Netinant, P., Tuaktao, Y., & Rukhiran, M. (2022). "Development of Real-Time Hand Gesture for Volume Control Application using Python on Raspberry Pi." *2022 The 5th International Conference on Software Engineering and Information Management (ICSIM).* https://doi.org/10.1145/3520084.3520085.

15. Shaik, D., PRIYA, E., Nikitha, G., KUMAR, K., & SHREE, N. (2022). "CONTROLLING VIDEOLAN CLIENT MEDIA USING LUCAS KANADE OPTIMAL FLOW ALGORITHM AND OPENCV." *YMER Digital.* https://doi.org/10.37896/ymer21.05/29.

16. Cibi, C., Vivekanandhan, G., Krishna, R., & Hemavathi, S. (2022). "Human Computer Interaction Using Arduino and Python." *2022 1st International Conference on Computational Science and Technology (ICCST)*, 1-3. https://doi.org/10.1109/ICCST55948.2022.10040307.

17. Dinh, D., Ngoan, P., Thang, N., & Kim, T. (2017). "A Single Depth Silhouette-Based Hand Gesture Recognition for Appliance Interfaces in Smart Home Environment.", 369–373.

18. Fels, S., & Hinton, G. (1997). "Glove-talk II - a neural-network interface which maps gestures to parallel formant speech synthesizer controls." *IEEE transactions on neural networks, 8(5)*, 977–984. https://doi.org/10.1109/72.623199.

Note: All the figures in this chapter were made by the author.

Computational Intelligence and Mathematical Applications – Dr. Devendra Prasad et al. (eds)
© 2024 Taylor & Francis Group, London, ISBN 978-1-032-87721-1

10 | Blockchain and Data Security: Building Trust in a Digital World

Dhananjay Batra[1], Shilpi Harnal[2], Paras Sachdeva[3]
Chitkara University Institute of Engineering and Technology, Chitkara University, Punjab, India

Rajeev Tiwari[4]
IILM University, Greater Noida, Uttar Pradesh, India

Raghav Bali[5]
Chitkara University Institute of Engineering and Technology, Chitkara University, Punjab, India

Shuchi Upadhyay[6]
Dehradun, Uttarakhand, India

Abstract: In today's digital world, building trust is essential for secure and seamless digital transactions. Protecting user data is essential given the recent rise in security breaches and digital surveillance. Through decentralized identification and other privacy methods, blockchain technology and distributed ledger technologies offer innovative potential for safeguarding user data, giving individuals authority over their data. Regulatory compliance, data portability, and monopolistic behavior are just a few of the ethical and political challenges that have been brought up by the rise of a few unscrupulous corporations managing trust relationships in the sharing economy. Blockchain technology's inherent decentralized and tamper-proof nature has recently helped it garner a lot of support and popularity. With an emphasis on digital identities, blockchain-based trust, programmable money, and marketplaces, a blockchain-powered idea for a shared and public programmable economy is presented. Through previous contact between strangers, trust is developed by estimating trustworthiness. The suggested technological stack is built with self-governance, autonomy, and shared ownership as its guiding principles, providing trust that goes beyond what is possible in the existing legal, political, and economic institutions. Thus, blockchain technology presents a significant opportunity to enhance data security and privacy, build trust in the digital world, and establish new principles and rules for a more secure and trustworthy digital ecosystem.

Keywords: Blockchain, Data security, Healthcare industry, IoT, System for smart industry

1. INTRODUCTION

Over 17 billion devices were linked to the Internet and in use worldwide in 2018, with about 7 billion of those being Internet of Things (IoT) devices, and by 2025, the estimate is for 35 billion people worldwide. All of these devices produce massive volumes of data in its raw form that are stored in data lakes and analyzed using Big data techniques such as batch and stream processing to produce new intelligence-focused apps and services. However, there is increased interest in data sharing and the acceptance of the collaborative economy model as data expands and changes into the new gold and IoT evolves into the new gold mine. The existing IoT operating and deployment paradigms, which are primarily focused on closed enterprise

[1]dh@gmaill.com, [2]shilpi13n@gmail.com, [3]parassachdeva0111@gmail.com, [4]errajeev.tiwari@gmail.com, [5]raghavbali790@gmail.com,
[6]shuchi.diet@gmail.com

DOI: 10.1201/9781003534112-10

environments, contrast with this. However, this horizontal approach, which involves the exchange of raw data between IoT devices or transactions on digital data markets, presents several significant challenges for authorization, authentication, and protecting the privacy and integrity of data, including:

1. Identification is the risk of associating an identifier such as a name or location with a real person along with the data about that individual.

2. The risk that localization and tracking bring depends on the ability to determine and record a person's position. Identification is presently the most significant threat during the information processing phase. Identification of the subject is necessary due to this danger. The most frequent privacy violations associated with this risk include GPS tracking and the disclosure of personal data, or simply having a general sense that you are being watched.

3. Blockchain technology is a revolutionary innovation that has overtaken the world in recent years. This technology has been applied across various sectors, including supply chain management, banking, and healthcare [1]. The Internet of Things (IoT) has become a significant technological trend in recent years and can completely change how we live and work. However, the IoT faces various challenges, including reliability and security issues. To address these challenges, two strategies have been proposed. The first strategy focuses on resource-constrained devices, While the second strategy focuses on cloud gateways and platforms. Utilizing a de-centralised model, such as blockchain, and the strength of smart contracts constitutes the second tactic. Blocks are the records, and cryptography is used to link them together. Distributed databases, peer-to-peer transmission, transparency, and irreversibility of records are among the foundational ideas of blockchain. In the Internet of Things (IoT), blockchain can be used to create a decentralized model, making it reliable and secure [2]. Users can create algorithms and rules that automatically initiate transactions between nodes by programming and connecting computational logic to blockchain transactions. Potential applications of blockchain technology are not limited to financial services, and it can be used to verify artwork, voting records, identity verification, transaction processing, and supply chain management. Blockchain technology has the potential to revolutionize many industries beyond cryptocurrencies, such as legal documents, healthcare, IoT, and governance. By providing a secure and transparent record of records, blockchain can help improve trust and security in a variety of applications. This paper believes that blockchain could mature within 2 to 5 years due to its widespread adoption in a variety of non-cryptocurrency applications. Blockchain has the potential to empower citizens in developing countries as it is widely applied in e-governance applications, healthcare, asset transfers, financial inclusion, and more. This paper discusses the principles of blockchain technology and its potential applications.

2. AN OVERVIEW OF THE BLOCKCHAIN TECHNOLOGY

A blockchain is a collection of sequentially ordered data blocks that are kept across all of the participating nodes in a decentralized form. Every transaction made on the blockchain generates a block, and the ledger of that transaction is distributed across all of the participating nodes, not kept centrally like in centralized or decentralised networks (as shown in Fig. 10.1), which is why it is known as Distributed Ledger Technology (DLT). [3]

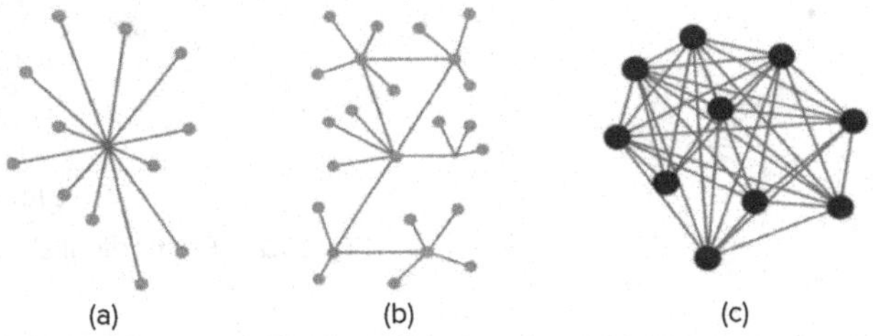

Fig. 10.1 (a) Centralized Networks, (b) Decentralized Network, (c) Distributed Ledger (adopted from Blockgeeks, 2016)

It is based on the following principles:

1. *Distributed Database:* Each party keeps a complete copy of the database and has access to the ledger. As a result, no one participant has ownership over the information or data, enabling each participant to independently verify the records of its transaction partners without the use of a middleman.

2. *Peer-to-Peer Transmission:* Blockchain is controlled by a peer-to-peer network that collectively follows a protocol for inter-node communication. Transactions are validated by consensus among the nodes in the network, using economic incentives like the consensus algorithm for proof of work.

3. *Transparency:* Anyone with access to the system can see transactions and the related data. [3] An individual alphanumeric address is used to identify each user and node. In a blockchain, transactions happen between addresses.

(a) *Irreversibility of Records:* A transaction that has been recorded in the ledger and saved cannot be changed. It is thus synchronized with each transaction record that was previously uploaded. The secure storage of information is enforced via a variety of machine techniques and methods.

In terms of the Internet of Things (IoT), there are two strategies for making it reliable and secure:

(b) A method that focuses on devices with limited resources, such as edge-layer devices and IoT sensors, actuators, and beacons. At the gateway or edge levels, a simple authentication and key agreement protocol can be created.

(c) The second approach focuses on cloud platforms and gateways, where data has been formatted, aggregated, pruned, and stored. The solution is to use a decentralized model like blockchain and the power of smart contracts. [4].

Because blockchain transactions may be written and connected to computational logic, users can create algorithms and rules that enable transactions between nodes to activate automatically.

3. BLOCKCHAIN AND DISTRIBUTED LEDGER BASED IMPROVED BIO-MEDICAL SECURITY SYSTEM

The Blockchain and Distributed Ledger Based Improved Bio-Medical Security system aims to make electronic shareable health records more trustworthy and private. An interpretation of a blockchain-based biomedical security system is shown in Fig. 10.2. The medical server is in charge of handling user queries and providing the necessary information in response. It uses blockchain technology to monitor user privacy and record-keeper trust. The system assesses trust and privacy standards using the medical server and the end-user blockchain. While trust is based on successful access and a high answer-to-request ratio, privacy is based on convergence and complexity. As a means of defense against attacks like man-in-the-middle and data manipulation, which it considers adversary models, the system introduces the trust model and concentric authentication. With appropriate protocols created in the server domain and the usage of suitable applications to receive data from the client side, the server-client-based blockchain technology is utilized to reduce man-in-the-middle attacks. [5]

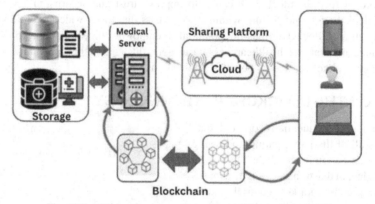

Fig. 10.2 Biomedical security system with blockchain [5]

4. BLOCKCHAIN-BASED OPEN CYBER THREAT INTELLIGENCE SYSTEM

To develop a trustworthy open cyber threat intelligence system, it is essential to ensure that the data can be traced and its contributors can be validated. One approach to achieve this is through the use of blockchain. The fundamental structure and operation of blockchain technology guarantees the trackability of the CTI data, while smart contracts can assess the data and its contributors' validity. Through the smart contract, the findings of these evaluations can be disclosed openly and transparently. Additionally, we talked about the privacy issues that could come up while sharing data on the blockchain. The proposed framework, known as BLOCIS (blockchain-based open cyber threat intelligence system) is a system for sharing open cyber threat intelligence that uses blockchain technology and is designed to withstand Sybil attacks [6]. The framework categorizes different layers based on the environment where the data is gathered and used, allowing for easy data interchange as shown in Fig. 10.3. There are three layers in the framework: the feed layer, the blockchain network layer, and the user layer.

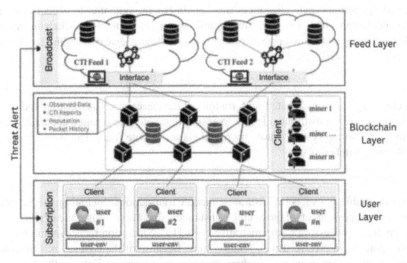

Fig. 10.3 Architecture of the blockchain-based open cyber threat intelligence system (BLOCIS)

5. APPLICATIONS OF BLOCKCHAIN TECHNOLOGY

Blockchain technology has the power to revolutionize many industries beyond cryptocurrencies. One of the main advantages of blockchain is its ability to enable secure and transparent records, making it ideal for applications where trust and security are key.

5.1 Healthcare

Blockchain in particular, a form of distributed ledger technology, has the potential to revolutionize the healthcare sector [7]. By offering a tamper-proof and visible method to track pharmaceuticals, it can be utilized to combat the problem of drug counterfeiting, thereby saving lives and minimizing financial losses [8]. Additionally, by providing a safe and functional platform for storing medical records, blockchain can enhance patient data management. The blockchain gives people authority over their medical records, allowing them to share them as needed with healthcare professionals [9]. By utilizing smart contracts, this technology can also assist in lowering treatment costs and do away with middlemen. In general, blockchain presents a wide range of options for enhancing the effectiveness and quality of healthcare services.

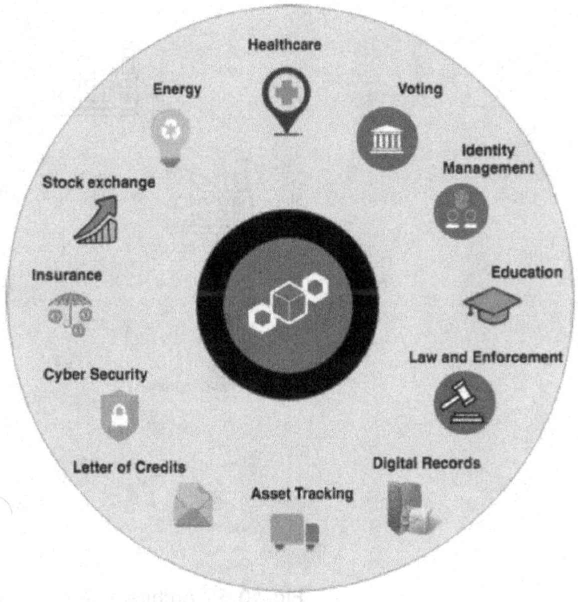

Fig. 10.4 Applications of blockchain technology

5.2 Energy Industry

The energy sector is utilizing blockchain technology to increase efficiency and dependability. Blockchain can make it easier for energy providers and consumers in microgrids to conduct transactions, enabling the purchasing and selling of extra energy [10]. Blockchain technology can be used by smart grids with bidirectional connectivity to conduct safe and private energy trade, with smart contracts maintaining power balance and flexibility. Energy trade in the Industrial Internet of Things (IIoT) may be made possible via blockchain [11]. Overall, the energy industry might save money and become more resilient if blockchain technology were to be used.

5.3 Stock Market

Blockchain technology has the potential to solve problems with transparency, trust, and interoperability in decentralized market systems like the stock exchange [12]. Transactions currently take longer than three days to complete due to middle-men, regulatory procedures, and operational trade clearance. Blockchain has the potential to decentralize and automate the stock exchange, do away with middlemen, and lower expenses. By utilizing smart contracts, blockchain can also serve as a regulator, reducing the need for third-party regulators in the post-trade process, trade and legal ownership transfer, and transaction clearing and settlement.

5.4 Voting

By using blockchain technology, issues associated with traditional databases, such as those in voting systems, can be solved. Blockchain technology, which offers a distributed ledger to ensure that votes are counted accurately, could help to resolve the lack of confidence in America's voting system caused by recent claims of vote fraud. The voter owns this ledger, which is also the one used to record the total number of votes.

5.5 Insurance

Blockchain technology can enable a variety of insurance market transactions, including the negotiating, purchasing, and registration of policies, the processing of claims, and reinsurance operations. Insurance policy automation using smart contracts can cut down on claim processing's administrative costs [13]. Smart contracts can automate the execution of terms by constructing insurance policies in precise if-then relationships. Because of the cost savings, insurance companies may be able to offer customers new automated insurance products and become more competitive. Furthermore, blockchain can help insurance companies go global.

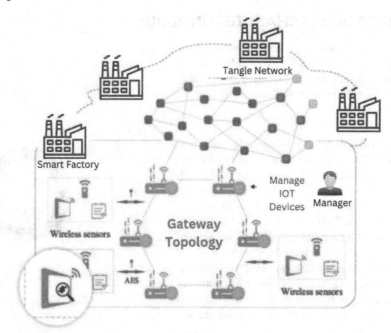

Fig. 10.5 Architecture of blockchain-based IoT system for smart industry

5.6 Identity Management

Blockchain technology can help solve the problem of the absence of a reliable method for protecting online identities. Blockchain can be used to create a secure identity that doesn't require a username or password by adding more security measures and controlling who has access to personal data. [14]. On blockchain-based identity management solutions, a digital ID can be created and allocated to each online transaction. Doing identification checks throughout each real-time transaction, will allow organizations to identify and eliminate the chance of fraud. Identity theft can be considerably reduced by using blockchain in conjunction with self-sovereign identity management because no central authority or outside party can access personal data without the user's agreement. To enhance decentralisation, transparency, and user control, various proposals have been made to use blockchain for identity management [15].

G. Trade Finance

Although the use of a letter of credit (LC) for the settlement of payments in trade finance is successful in reducing risk, low-value transactions are less likely to employ it due to its complicated process, high cost, and contractual delays. Blockchain technology can solve these problems by automating LC, lowering transaction fees, and simplifying business operations. Smart contracts can be designed to comply with all the requirements listed in the LC, which ensures payment once the trade item is received by the buyer. Blockchain can expedite the documentation process and guarantee trade finance security.

6. ADVANTAGES OF USING BLOCKCHAIN TECHNOLOGY

There are several possible advantages of using Blockchain technology across different businesses. In the healthcare industry, it can offer a safe and impenetrable way to share and preserve private patient data, as well as guard against unauthorized access and data manipulation [16]. It can be used to verify job applicant documentation, keep track of lifelong learning pursuits, and assist employers in making more informed hiring decisions in the educational sector. Blockchain technology may increase the effectiveness of replenishment, stop chemical theft and unlawful use, and trace product sourcing and movement in the commercial supply chain. Blockchain technology can give a solution to many of the problems these industries face as they develop and become more complicated by enhancing transparency, security, and efficiency.

7. FUTURE SCOPE OF BLOCKCHAIN TECHNOLOGY

Blockchain technology has the power to disrupt various industries by providing secure, decentralized, and transparent solutions. Here are a few unique examples of fields where blockchain can be used to improve in the future:

Gaming: Blockchain technology can enable decentralized gaming platforms that provide transparency and fairness in gaming transactions, secure asset ownership, and verify in-game item authenticity.

Education: Blockchain technology can help improve the education industry by enabling secure and tamper-proof records of academic achievements, reducing fraud in certificates and degrees, and facilitating the transfer of credits between institutions.

Charity and Non-Profit: Blockchain technology can help improve the transparency and accountability of charitable organizations by providing secure and transparent records of donations and ensuring that funds are used for their intended purpose.

Legal: Blockchain technology can help improve the legal industry by enabling smart contracts that can automate legal agreements, reducing the need for intermediaries, and providing secure and transparent records of legal transactions.

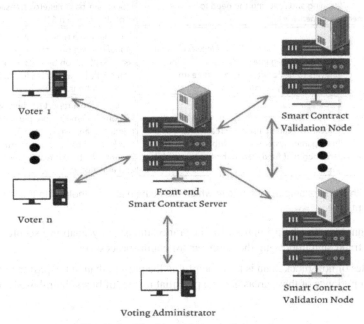

Fig. 10.6 The blockchain evoting system

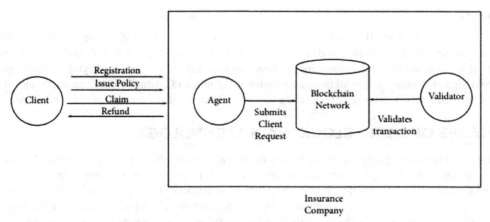

Fig. 10.7 Role of blockchain network in insurance company

Domain	How blockchain is more helpful in the present	Relevant Data
Supply Chain Management	Blockchain provides an immutable and transparent record of transactions, making it easier to track goods through the supply chain and ensure compliance with regulations.	According to a survey by Deloitte, 53% of companies plan to use blockchain for supply chain management.
Intellectual Property	Blockchain provides a secure and tamper-proof system for tracking ownership and transfer of intellectual property rights, reducing the risk of infringement and fraud.	According to a report by Accenture, 10% of the world's GDP is derived from intellectual property, making it a valuable area for blockchain-based solutions.
Energy Management	Blockchain can be used to create decentralized energy markets, enabling peer-to-peer energy trading and reducing the need for intermediaries.	According to a report by Navigant Research, the global market for blockchain-based energy transactions is expected to reach $19 billion by 2024.
Gaming	Blockchain can be used to create secure and transparent in-game economies, allowing players to trade virtual assets and earn cryptocurrency rewards.	According to a report by Newzoo, the global gaming market is expected to reach $200 billion by 2023, making it a potentially lucrative area for blockchain-based solutions.
Humanitarian Aid	Blockchain can be used to create secure and transparent systems for distributing aid and ensuring that funds reach those who need it most.	According to a report by the United Nations, 168 million people will need humanitarian assistance in 2021, making blockchain-based solutions crucial for ensuring that aid is distributed efficiently and effectively.

Fig. 10.8 Here is a table with some examples of fields in which blockchain is more helpful in the present than in the past, along with some relevant data

Insurance: Blockchain technology can help improve the insurance industry by enabling secure and transparent records of insurance claims, reducing fraud, and improving the efficiency of claims processing.

These are just a few examples of how blockchain is more helpful in the present than in the past in various domains, and the data provided underscores the importance of these areas and the potential impact of blockchain-based solutions.

8. CONCLUSION

In conclusion, the necessity for secure and open data storage has grown critical given the growing significance of data sharing and the adoption of a shared economy model. By offering safe and transparent records, blockchain technology has emerged as a solution to meet this demand. Blockchain technology has been investigated for usage in several industries, including voting, healthcare, energy, and the stock market. Biomedical security and open cyber threat intelligence have been improved by blockchain-based security systems. Beyond cryptocurrencies, blockchain technology applications have the power to completely transform a variety of industries. For example, they could improve the effectiveness and reliability of healthcare services, boost energy sector productivity and dependability, and address issues with decentralized market systems' transparency, trust, and interoperability.

REFERENCES

1. Kaur, G., Choudhary, P., Sahore, L., Gupta, S., & Kaur, V. (2023). "Healthcare: In the era of blockchain." In *AI and Blockchain in Healthcare*. Springer, pp. 45–55.
2. Bali, M.S., Gupta, K., & Malik, S. (2022). "Integrating iot with blockchain: A systematic review." In *IoT and Analytics for Sensor Networks: Proceedings of ICWSNUCA 2021*, pp. 355–369.
3. Catalini, C. (2017). "How blockchain applications will move beyond finance." *Harvard Business Review*, 2.
4. Boncea, R.M. (n.d.). "Rezumat teză de doctorat."
5. Liu, H., Crespo, R.G., & Martínez, O.S. (2020). "Enhancing privacy and data security across healthcare applications using blockchain and distributed ledger concepts." In *Healthcare*, 8(3). MDPI, p. 243.
6. Cha, J., Singh, S.K., Pan, Y., & Park, J.H. (2020). "Blockchain-based cyber threat intelligence system architecture for sustainable computing." *Sustainability*, 12(16), p. 6401.
7. McGhin, T., Choo, K.K.R., Liu, C.Z., & He, D. (2019). "Blockchain in healthcare applications: Research challenges and opportunities." *Journal of Network and Computer Applications*, 135, pp. 62–75.
8. Marfiuc, A.M. (2022). "Blockchain technologies."
9. Harnal, S., Sharma, G., Malik, S., Kaur, G., Simaiya, S., Khurana, S., & Bagga, D. (2023). "Current and future trends of cloud-based solutions for healthcare." In *Image Based Computing for Food and Health Analytics: Requirements, Challenges, Solutions and Practices: IBCFHA*, pp. 115–136.
10. Cohn, A., West, T., & Parker, C. (2016). "Smart after all: Blockchain, smart contracts, parametric insurance, and smart energy grids." *Geo. L. Tech. Rev.*, 1, p. 273.
11. Li, Z., Kang, J., Yu, R., Ye, D., Deng, Q., & Zhang, Y. (2017). "Consortium blockchain for secure energy trading in industrial internet of things." *IEEE Transactions on Industrial Informatics*, 14(8), pp. 3690–3700.
12. Lee, L. (2015). "New kids on the blockchain: How bitcoin's technology could reinvent the stock market." *Hastings Bus. LJ*, 12, p. 81.
13. Gatteschi, V., Lamberti, F., Demartini, C., Pranteda, C., & Santamaría, V. (2018). "Blockchain and smart contracts for insurance: Is the technology mature enough?" *Future Internet*, 10(2), p. 20.
14. Jacobovitz, O. (2016). "Blockchain for identity management." *The Lynne and William Frankel Center for Computer Science Department of Computer Science. Ben-Gurion University, Beer Sheva*, 1, p. 9.
15. Dunphy, P., & Petitcolas, F.A. (2018). "A first look at identity management schemes on the blockchain." *IEEE Security & Privacy*, 16(4), pp. 20–29.
16. Harnal, S., Jain, A., Rathore, A.S., Baggan, V., Kaur, G., Bala, R., et al. (2023). "Comparative approach for early diabetes detection with machine learning." In *2023 International Conference on Emerging Smart Computing and Informatics (ESCI)*. IEEE, pp. 1–6.

Computational Intelligence and Mathematical Applications – Dr. Devendra Prasad et al. (eds)
© 2024 Taylor & Francis Group, London, ISBN 978-1-032-87721-1

11

Fermatean Fuzzy Distance Measures with Pattern Recognition application

Tamanna Sethi*

Panipat Insititute of Engineering and Technology, Samalkha, Panipat, Haryana, India

Satish Kumarb, Megha Tanejaa and Rachna Khuranaa

Maharishi Markandeshwar (Deemed to be) University, Mullana, Ambala, Haryana, India

Abstract: When it comes to addressing different difficulties involving uncertainty, In comparison to fuzzy sets, intuitionistic fuzzy sets, and Pythagorean fuzzy sets, Fermatean fuzzy sets perform better. The fuzzy and non-standard fuzzy frameworks' distance measurements have applications in a number of fields like pattern analysis. Here, we suggest distance metrics for Fermatean fuzzy sets utilizing the idea of Jensen-Shannon divergence measure in this paper, We also go over all of their appealing qualities. We use numerical comparison to show that the proposed distance measure in the Fermatean fuzzy situation is superior to the current measurements. Finally, we show how the recommended measurements can be applied to pattern analysis.

Keywords: Entropy, Fermatean fuzzy set, Knowledge measure, Pattern recognition

1. INTRODUCTION

In real MADM (multi-attribute decision making) challenges, we are not always able to provide correct and fruitful information for the alternatives due to the uncertainty of experts and decision problems. Thus a unique fuzzy set (FS) developed by Zadeh (1965) was employed to address this problem. Fuzzy theory has been used in a variety of domains, including engineering, image processing, the sciences of the body and mind, social studies, and clustering. Fuzzy set theory is unable to show intuition because it only relies on measurement. Atanassov and Atanassov (1999) has produced a standpoint on the intuitionistic fuzzy set (IFSs) theory. The utilization of IFSs have been studied by numerous scholars. but its applicability is restricted in few circumstances. In order to address these problems, the "pythagorean fuzzy set (PyFS)" was developed by Yager (2013), who also explained the IFS theory. The PyFSs squares for the grades denoting membership (g) and non-membership (f) should be calculated, and the sum that results should be less than or equal to 1. Currently, it has been demonstrated that the PyFS theory is a fruitful method to tackle with ambivalent data in real-world circumstances. FFS is comparable to IFS and PyFS, but it offers more freedom in how misleading data can be expressed. It indicates that FFSs are advocating for membership grade with increased confidence and assertiveness. FFSs are better equipped and flexible to handle realistic information that is hazy or ambiguous. After the successful establishment of FFSs, numerous researchers began to conduct research on the idea of FFSs in many disciplines.

This paper contributes as:

1. We suggest a FF distance measures based on Jensen-Shannon divergence measure with their various properties.

---
*Corresponding author: tamannasethi.applied@piet.co.in

DOI: 10.1201/9781003534112-11

2. We contrast the known FF measures of compatibility with few recommended FF measures of distance.

3. The usefulness of the suggested distance measurements is then demonstrated using numbers, and they are contrasted with the current distance measures.

4. We provide examples to show how the suggested measurements may be used in pattern analysis.

The paper is structured as: Some definitions and fundamental ideas relating to Fermatean fuzzy sets are discussed in section 2. Section 3 gives the distance measure for fermatean fuzzy sets. In section 4, application of Fermatean distance measure has been presented. Section 5 contains the conclusion of the paper.

2. THE PRELIMINARY TASKS

This section discusses the fundamental definitions of Fermatean fuzzy sets (FFSs). The implementation of the entire work is then proposed using a few scoring functions.

Definition 1   Senapati and Yager (2019) A Fermatean fuzzy set ζ on L is explained as

$$\zeta = \{(p_i, g_\zeta(p_i), f_\zeta(p_i)) : p_i \in L\} \tag{1}$$

here a membership function is $g_\zeta : L \to [0, 1]$ and a non membership function $f_\zeta : L \to [0, 1]$ under the condition $0 \leq g_\zeta^3(p_i) + f_\zeta^3(p_i) \leq 1$, for all $p_i \in L$. Additionally, for all $p_i \in L$, Fermatean index is defined as $\Phi_\zeta(p_i) = \sqrt[3]{1 - g_\zeta^3(p_i) - f_\zeta^3(p_i)}$ $= \sqrt[3]{1 - g_\zeta^3(p_i) - f_\zeta^3(p_i)}$.

Definition 2   Senapati and Yager (2019) The score and accuracy value of any Fermatean fuzzy number are determined by $mu = (g, f)$:

$$Sco(\mu) = g^3 - f^3, \tag{2}$$

where $Sco(\mu) \in [-1, 1]$.

$$Acu(\mu) = g^3 + f^3, \tag{3}$$

where $Acu(\mu) \in [0, 1]$.

Definition 3   Senapati and Yager (2019) For any $\mu_1, \mu_2 \in FFN$, We have

- If $Sco(\mu_1) > Sco(\mu_2)$, then $\mu_1 > \mu_2$;
- If $Sco(\mu_1) = Sco(\mu_2)$, then
- if $Acu(\mu_1) > Acu(\mu_2)$, then $\mu_1 > \mu_2$;
- if $Acu(\mu_1) = Acu(\mu_2)$, then $\mu_1 = \mu_2$.

3. DIVERGENCE MEASURES FOR FERMATEAN FUZZY SETS

Definition 1   Suppose for $\zeta, \eta \in FFSs(L)$, a real-valued function $\Omega_{ff} : FFSs(L) \times FFSs(L) \to \Re$ is referred to be the FFS divergence measure if it satisfies below requirements:

1. $\Omega_{ff}(\zeta, \eta) \geq 0$; and $\Omega_{ff}(\zeta, \eta) = 0$ iff $\zeta = \eta$;
2. $\Omega_{ff}(\zeta, \eta) = \Omega_{ff}(\eta, \zeta)$;
3. $\Omega_{ff}(\zeta, \eta) \leq \Omega_{ff}(\zeta, R)$ and $\Omega_{ff}(\eta, R) \leq \Omega_{ff}(\zeta, R)$.

3.1 Divergence Measure for FFSs

Definition 3.2   Sethi and Kumar (2023a) Let $\zeta, \eta \in FFSs(L)$. Now we represent

$$\Omega_{MF}^i(\zeta, \eta) = \left(g_\zeta^3(p_i) + g_\eta^3(p_i)\right) \log\left(\frac{\left(g_\zeta^3(p_i) + g_\eta^3(p_i)\right)}{2}\right) - \left(g_\zeta^3(p_i)\log\left(g_\zeta^3(p_i)\right)\right) - \left(g_\eta^3(p_i)\log\left(g_\eta^3(p_i)\right)\right)$$

$$\Omega_{NMF}^i(\zeta, \eta) = \left(f_\zeta^3(p_i) + f_\eta^3(p_i)\right) \log\left(\frac{\left(f_\zeta^3(p_i) + f_\eta^3(p_i)\right)}{2}\right) - f_\zeta^3(p_i)\log\left(f_\zeta^3(p_i)\right) - f_\eta^3(p_i)\log\left(f_\eta^3(p_i)\right)$$

$$\Omega_{II}^i(\zeta, \eta) = \left(\phi_\zeta^3(p_i) + \phi_\eta^3(p_i)\right) \log\left(\frac{\left(\phi_\zeta^3(p_i) + \phi_\eta^3(p_i)\right)}{2}\right) - \phi_\zeta^3(p_i)\log\left(\phi_\zeta^3(p_i)\right) - \phi_\eta^3(p_i)\log\left(\phi_\eta^3(p_i)\right) \tag{4}$$

Now, using the three equations above, we provide the following equation above, we provide the following Jensen-Shannon fermatean fuzzy divergence measure:

$$\Omega_{ff}(\zeta,\eta) = -\frac{1}{n}\sum_{i=1}^{n}\left(\Omega_{MF}^{i}(\zeta,\eta) + \Omega_{NMF}^{i}(\zeta,\eta) + \Omega_{II}^{i}(\zeta,\eta)\right) \tag{5}$$

The proposed divergence measure's characteristics

1. $\Omega_{ff}(\zeta, \zeta \cup \eta) = \Omega_{ff}(\zeta \cap \eta, \eta) = \Omega_{ff}(\zeta, \eta)$
2. $\Omega_{ff}(\zeta \cup \eta, \zeta \cap \eta) = \Omega_{ff}(\zeta \cap \eta, \zeta \cup \eta) = \Omega_{ff}(\zeta, \eta)$
3. $\Omega_{ff}(\zeta, \zeta \cup \eta) + \Omega_{ff}(\zeta, \zeta \cap \eta) = 2\Omega_{ff}(\zeta, \eta)$
4. $\Omega_{ff}(\eta, \zeta \cup \eta) + \Omega_{ff}(\eta, \zeta \cap \eta) = 2\Omega_{ff}(\zeta, \eta)$
5. $\Omega_{ff}(\zeta \cup \eta, R) \leq \Omega_{ff}(\zeta, R) + \Omega_{ff}(\eta, R)$
6. $\Omega_{ff}(\zeta \cap \eta, R) \leq \Omega_{ff}(\zeta, R) + \Omega_{ff}(\eta, R)$
7. $\Omega_{ff}(\zeta \cup \eta, R) + \Omega_{ff}(\zeta \cap \eta, R) = \Omega_{ff}(\zeta, R) + \Omega_{ff}(\eta, R)$
8. $\Omega_{ff}(\zeta, \eta) = \Omega_{ff}(\zeta^c, \eta^c)$
9. $\Omega_{ff}(\zeta, \eta^c) = \Omega_{ff}(\zeta^c, \eta)$
10. $\Omega_{ff}(\zeta, \eta^c) + \Omega_{ff}(\zeta^c, \eta) = \Omega_{ff}(\zeta^c, \eta^c) + \Omega_{ff}(\zeta, \eta^c)$

4. APPLICATION OF THE FEMATEAN DIVERGENCE MEASURE

We present an example of the use of Fermatean divergence measurements in pattern analysis in this section.

4.1 Pattern Analysis

Now, we validate how the recommended FF distance measurements can be applied to the resolution of pattern classified matters. A non-recognized pattern in pattern analysis is changed into a recognized pattern by using several competency gauges, such as similarity, distance, etc. We shall contrast calculated results with currently available results. Now, here we solve several pattern analysis-related issues as discussed below: Jiang, Jin, Lee, and Yao (2019) Consider the Patterns F_1, F_2, F_3, F articulated in custom of FFFSs in W as:

$$F_1 = \{(pp_1, 0.34, 0.34), (pp_2, 0.19, 0.48), (pp_3, 0.02, 0.12)$$
$$F_2 = \{(pp_1, 0.35, 0.33), (pp_2, 0.20, 0.47), (pp_3, 0.00, 0.14)$$
$$F_3 = \{(pp_1, 0.33, 0.35), (pp_2, 0.21, 0.46), (pp_3, 0.01, 0.13)$$
$$F = \{pp_1, 0.37, 0.31), (pp_2, 0.23, 0.44), (pp_3, 0.04, 0.10)$$

The challenge is to determine which pattern (F) has the highest degree of similarity to pattern F_j; $j = 1, 2, 3$. In order to do this, we combine the current compatibility criteria with the suggested FF distance metrics. Following flowchart indicates the flow of work. Table 11.1 displays the calculated values. As can be seen from Table 11.1, F should be assigned to M2 as evidenced by the majority of FF distance measures, to include a specified FF distance measures.

As a result of the examples, we draw the conclusion that in analysis of pattern, the recommended FF measures of distance are stable with the current distance measurements.

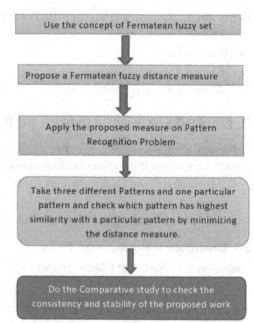

Fig. 11.1 A timeline diagram to show flow of work

Table 11.1 Estimated values of various FF compatibility measure related to the example

Divergence measures	(F_1, F)	(F_2, F)	(F_3, F)	Result
D_{PYY1} Peng, Yuan, and Yang (2017)	0.0208	0.0171	0.0176	F_2
D_{PYY2} Peng et al. (2017)	0.0208	0.0131	0.0149	F_2
D_{PYY3} Peng et al. (2017)	0.0167	0.0151	0.0163	F_2
D_{PYY4} Peng et al. (2017)	0.0187	0.0171	0.0176	F_2
D_{PYY5} Peng et al. (2017)	0.0208	0.0335	0.0347	F_2
D_{PYY6} Peng et al. (2017)	0.0405	0.0336	0.1569	F_2
D_{PYY6} Peng et al. (2017)	0.0405	0.0336	0.1569	F_2
D_{PYY7} Peng et al. (2017)	0.0408	0.1440	0.2763	F_2
D_{PYY8} Peng et al. (2017)	0.1758	0.2848	0.4361	F_2
D_{PYY9} Peng et al. (2017)	0.2506	0.4497	0.1652	F_2
D_{PYY10} Peng et al. (2017)	0.4077	0.1453	0.0306	F_1
D_{PYY11} Peng et al. (2017)	0.1812	0.0281	0.0310	F_2
D_{PYY12} Peng et al. (2017)	0.0346	0.0272	0.0146	F_2
D_{G1} Ganie (2022)	0.0173	0.0128	0.0146	F_2
D_{G2} Ganie (2022)	0.0174	0.0128	0.0146	F_2
D_{G3} Ganie (2022)	0.0175	0.0128	0.0147	F_2
D_{G4} Ganie (2022)	0.0175	0.0128	0.0147	F_2
Ω_{ff} (CurrentPaper)	0.00196	0.00156	0.00161	F_2

5. CONCLUSION

This paper has presented a distance measure for FFSs for Jensen-Shannon. The main-stream of the prevailing FF distance measures have generated not good enough results when calculating the distance between different FFSs. A number of approaches have also failed to explain every axiomatic requirement. However, the suggested FF distance parameters have consistently obtained fruitful results without running across any nonsensical situations. The usefulness of the proposed FF distance measures has been revealed in cataloging problems, and findings are compared to the ones already in use. Additionally, a multi-criteria decision-making problem was solved using the recommended measures, and the outcomes were consistent with those obtained using the existing measures. In the future, we'll demonstrate how the suggested distance measurements for FFSs might be used for clustering and medical diagnosis. Additionally, we will apply some recent expansions of FSs to the suggested technique for getting distance and knowledge measurements. Chen, Chang, and Pan (2012), picture fuzzy sets Cuong and Kreinovich (2013), spherical fuzzy sets Mahmood, Ullah, Khan, and Jan (2019), complex fuzzy sets Ramot, Milo, Friedman, and Kandel (2002) etc.

REFERENCES

1. Akram, M., Shahzadi, G., & Ahmadini, A. A. H. (2020). Decision-making framework for an effective sanitizer to reduce covid-19 under a fermatean fuzzy environment. Journal of Mathematics, 2020 .
2. Arya, V., & Kumar, S. (2020). Fuzzy entropy measure with an applications in decision making under bipolar fuzzy environment based on topsis method. International Journal of Information and Management Sciences, 31 (2), 99–121.
3. Atanassov, K. T., & Atanassov, K. T. (1999). Intuitionistic fuzzy sets. Springer.
4. Chen, S.-M., Chang, Y.-C., & Pan, J.-S. (2012). Fuzzy rules interpolation for sparse fuzzy rule-based systems based on interval type-2 gaussian fuzzy sets and genetic algorithms. IEEE transactions on fuzzy systems, 21 (3), 412–425.
5. Cuong, B. C., & Kreinovich, V. (2013). Picture fuzzy sets-a new concept for computational intelligence problems. In 2013 third world congress on information and communication technologies (wict 2013) (pp. 1–6).
6. Elie, S. (2013, 04). An overview of pattern recognition. Ganie, A. H. (2022). Multicriteria decision-making based on distance measures and knowledge measures of fermatean fuzzy sets. Granular Computing, 7 (4), 979– 998.

7. Garg, H., Shahzadi, G., & Akram, M. (2020). Decision-making analysis based on fermatean fuzzy yager aggregation operators with application in covid-19 testing facility. Mathematical Problems in Engineering, 2020 .

8. Gupta, R., & Kumar, S. (2022). A new similarity measure between picture fuzzy sets with applications to pattern recognition and clustering problems. Granular Computing, 7 (3), 561–576.

9. Jiang, Q., Jin, X., Lee, S.-J., & Yao, S. (2019). A new similarity/distance measure between intuitionistic fuzzy sets based on the transformed isosceles triangles and its applications to pattern recognition. Expert Systems with Applications, 116 , 439–453.

10. Mahmood, T., Ullah, K., Khan, Q., & Jan, N. (2019). An approach toward decision-making and medical diagnosis problems using the concept of spherical fuzzy sets. Neural Computing and Applications, 31 , 7041–7053.

11. Peng, X., Yuan, H., & Yang, Y. (2017). Pythagorean fuzzy information measures and their applications. International Journal of Intelligent Systems, 32 (10), 991–1029.

12. Ramot, D., Milo, R., Friedman, M., & Kandel, A. (2002). Complex fuzzy sets. IEEE Transactions on Fuzzy Systems, 10 (2), 171-186.

13. Senapati, T., & Yager, R. R. (2019). Some new operations over fermatean fuzzy numbers and application of fermatean fuzzy wpm in multiple criteria decision-making. Informatica, 30 (2), 391–412.

14. Sethi, T., & Kumar, S. (2023a). Multi-attribute group decision-making problem of medical consumption products based on extended todim-vikor approach with fermatean fuzzy information measure. Mathematical modeling and computing (10, Num. 1), 80–100.

15. Sethi, T., & Kumar, S. (2023b). Todim-vikor methods with pythagorean fuzzy information based emergency decision support model for economic growth factor selection. In 2023 13th international conference on cloud computing, data science & engineering (confluence) (pp. 340–345).

16. Shannon, C. E. (1948). A mathematical theory of communication. The Bell system technical journal, 27 (3), 379–423.

17. Tamanna, Kumar, S., Younis, J., & Hussain, A. (2023). A novel spherical fuzzy vikor approach based on todim for evaluating and ranking the opinion polls with Shannon entropy and jensen-shannon divergence measure. IEEE Access, 11 , 103242-103253.

18. Yager, R. R. (2013). Pythagorean fuzzy subsets. In 2013 joint ifsa world congress and nafips annual meeting (ifsa/nafips) (pp. 57–61).

19. Zadeh, L. A. (1965). Fuzzy sets. Information and control, 8 (3), 338–353.

Note: All the figures and table in this chapter were made by the author.

Computational Intelligence and Mathematical Applications – Dr. Devendra Prasad et al. (eds)
© 2024 Taylor & Francis Group, London, ISBN 978-1-032-87721-1

12 | Intelligent Threat Mitigation in Cyber-Physical Systems

Sandeep Singh Bindra[1]
Research Scholar, Chandigarh University, Mohali, India

Alankrita Aggarwal[2]
Associate Professor, Chandigarh University, Mohali, India

Abstract: IoT-based Cyber-Physical Systems (CPS) have become integral to modern industries, revolutionizing processes and functionalities. However, these systems face inherent vulnerabilities in their software, network, and hardware components, necessitating robust security measures. This paper introduces a novel approach to fortify CPS, presenting a secure data processing and transmission method. This research focuses on leveraging machine learning techniques to enhance the security of IoT-based CPS, specifically in industrial settings. The proposed Privacy-preserving CPS method demonstrates remarkable efficacy in detecting various cyber-attack types. Through rigorous experimentation and analysis, this method proves to be a pivotal advancement in safeguarding critical operations within industrial CPS environments. Incorporating advanced machine learning techniques empowers IoT-based CPS to bolster their defenses and mitigate potential threats effectively. This paper signifies a significant stride towards establishing a more secure and resilient foundation for the functioning of these complex systems.

Keywords: Attack, Cyber Security, IoT, Machine learning, Privacy policy, Threats

1. INTRODUCTION

Cyber-physical systems (CPS) and the Internet of Things (IoT) have created a new age of technological possibilities. A notable obstacle is the construction of manufacturing CPS, which is frequently characterized by geographic dispersion and operational complexity. However, the risk to the environment has been greatly increased by the computing technology's quick development and synthesis. This has caused vulnerabilities in various areas, including software applications, cloud services, endpoints, and networks. This ecosystem connection significantly affects ecological dynamics, affecting species numbers, biomass, and total productivity [1].

An intrusion detection system has been created to protect against potential threats. To detect deviations from the norm, this system uses anomaly-based approaches, which successfully identify possible threats that do not fit pre-existing threat profiles [2]. Additionally, by enabling continuous communication between control centers and field systems, the integration of industrial control systems networking assists in improving operational efficiency and flexibility [3]. The IoT is essential for creating a CPS that works together in the energy sector. This combination can potentially revolutionize resource optimization and energy management [4]. Focus has turned to defensive measures like authentication and encryption to strengthen security in this dynamic environment. These precautions are crucial given the wide range of linked data applications, which include speech recognition, picture classification, object identification, and more [6].

---
[1]sandeep.bindra@gmail.com, [2]alankrita.agg@gmail.com

DOI: 10.1201/9781003534112-12

Furthermore, the importance of protecting devices at the device level cannot be emphasized. Examining physical performance and accessibility rigorously protects the CPS's ability to operate as intended. Striking a balance is essential, though, since an excessive dependence on extra physical control devices might unintentionally introduce new points of failure within different CPS components [8]. Specification-based, anomaly-based, and signature-based strategies are divided into which misbehavior threat detection methods for IoT systems are divided under CPS [9]. This technique focuses on main code integrity and is specifically designed to meet the needs of embedded devices and computer software [10].

The CPS uses advances in neural network technology to improve fault diagnostics. In particular, neural networks support models performing breaker and relay operations, allowing precise diagnosis of defective parts [11]. Due to the existence of several crucial components, such as battery backups, power distribution units, surge protectors, and other power-related features, cyber power devices require attentive management and monitoring. To ensure flawless operation, integrating network device protocols and sensors necessitates particular attention during information processing [12]. Passive and active user inputs must be considered to get the data necessary for CPS to make well-informed decisions. Smart tools and biosensors are passive inputs that supplement the electronic data acquired through active user inputs and offer a reliable feedback mechanism, especially in inpatient data management [13]. The scientific community has paid great attention to this decision-making and data-collecting approach combination because it can potentially advance medical applications [14–15].

Digital substations are used in CPS to optimize infrastructure by lowering maintenance requirements and streamlining electrical systems. As a consequence, the operational framework becomes more effective and sustainable. This calls for a decrease in preservation supplies and convolution of cabling. Additionally, continuing research explores the critical analysis of machine learning-enabled CPS for IoT privacy protection, providing priceless insights after adoption [16]. This work provides the foundation for a future in Cyber-Physical Systems that is safer, more effective, and more sustainable by methodically tackling these challenges and utilizing cutting-edge technology.

2. RELATED WORKS

A wide range of approaches are presented in the literature reviewed in this area with the goal of strengthening security and privacy in IoT-based Cyber-Physical Systems (CPS). Here, we offer a thorough evaluation of each strategy, including its main goals, significant discoveries, and areas of application:

Collectively, these findings add to the discussion that is now being had about safeguarding IoT-based CPS. Each solution brings its ideas and approaches to the table by tackling distinct security and privacy issues.

3. PROBLEM STATEMENT

Cyber-Physical Systems (CPS) and Internet of Things (IoT) technology have recently been combined, opening up new business opportunities. These systems have revolutionized processes, but there are also drawbacks to this technical advance, notably in security. To guarantee the smooth functioning and security of these complex systems, the general challenges under CPS cover a variety of crucial difficulties that must be addressed:

- *Data Authentication Vulnerabilities:* The lack of reliable data authentication procedures is one of the main issues with CPS.
- *Failure of Breakers/Relays:* When a Breaker or Relay fails, determining the exact source of the problem can be challenging [16].
- *Misbehavior Detection in Noisy Environments:* Although the cyber-physical system functions in settings with high ambient noise levels, using lightweight misbehavior detection techniques does not effectively resist opportunistic attackers [20]. To protect against such dangers, more complex and flexible techniques must be developed.
- *Physical control devices prone to failure:* The extra physical control devices inside CPS are particularly prone to failure. These failures can potentially impair the system's smooth functioning, emphasizing the necessity for strong device-level security and backup procedures to lessen the effects of such failures.

Cyber-physical systems must resolve these issues to continue developing and functioning securely. Strengthening the resilience and reliability of CPS in complex and dynamic environments requires adopting robust authentication protocols, developing efficient fault diagnosis strategies, improving misbehavior detection techniques, and implementing proactive maintenance programs for physical control devices.

Table 12.1 Literature survey

Author(s)	Method	Advantage	Limitation	Performance metrics
Li. et al. [16]	The novel federated machine learning algorithm was used to detect cyber-attacks against cyber-physical systems (CPS).	The deep federated method detected many types of cyber-attacks on CPS.	This method detects cyber-attacks against cyber-physical systems. It is challenging to determine the cause of fault based on the position of the breaker and relays.	Accuracy, performance, and f1 score.
Rathore. S et al. [17]	The blockchain method provides a secure deep learning operation and provides low latency.	Here, the method was used to remove all control in centralized authority.	Some edge nodes attempted to prevent collaboration throughout the deep learning process.	Accuracy, precision, security and privacy analysis, latency delay.
Yizhou Shen et al. [18]	Development of a payoff matrix to track the communication between IoT devices and edge nodes.	They introduced a technique that prevents the leakage of sensitive data.	However, the introduced method does not consider the privacy preservation of the data sender.	Statistical probability, denying requests.
Tertytchny. G et al. [19]	With high accuracy, the CPS approach was utilized to distinguish between defective and attack classes.	In this, IoT IoT-based CPS method was used to increase the classification accuracy to detect attacks.	However, the method was unable to find the abnormality source.	Precision, accuracy, f-measure, and recall.
Choudhary. G et al. [20]	This method presented a lightweight description that identifies the attacks on IoT devices.	The accuracy of detecting the misbehavior confirms the requirement of protection against attacks on IoT devices.	However, this method was a lightweight description of identifying the attacks in IoT devices, which is unsuitable for the attackers when it takes place in a high-noise environment.	Precision, accuracy, f-measure, and recall.
Rajani Singh et al [21]	The newly proposed approach intends to replace their range-proof methods with the most effective range-proof solution.	The introduced method permits the combination of multiple in a single range, which enhances the efficiency of the ledger system	However, the prover's proof concept utilized in this research leaked data regarding the quadratic polynomial vector	Detection accuracy

4. RESEARCH AIM AND OBJECTIVES

A number of key goals are outlined in the suggested technique, all geared towards enhancing the security and resilience of cyber-physical systems (CPS). First, it aims to advance the system by enhancing its ability to withstand various security attacks. This requires strong safeguards and procedures to protect the CPS from potential intrusions and assaults, improving the operating environment's security. In order to detect security vulnerabilities inside the CPS framework, the approach also aims to use privacy preservation mechanisms. It tries to recognize and classify these security assaults by incorporating machine learning algorithms, offering a comprehensive threat detection method. This includes thoroughly evaluating the devices' operating capabilities to ensure they comply with privacy-preserving standards and fulfill performance criteria. The approach seeks to balance operational effectiveness and data safety by examining performance and accessibility, therefore fostering a strong and secure CPS environment.

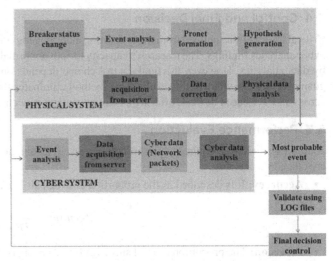

Fig. 12.1 Block diagram of the proposed method

The suggested study technique takes a thorough approach to dealing with the difficulties that failure analysis in cyber-physical systems (CPS) entails. This entails the seamless fusion of data-driven methodologies and advanced analytics, supplying a sturdy platform for in-depth research. Figure 12.1 offers a clear road map for the analytical process by visualizing the defined research technique.

5. PROPOSED METHODOLOGIES

5.1 Establishing the Data Analysis for Failure Assessment

A trigger-based scheme serves as the foundation for the failure analysis process, which is critical in assuring the accuracy of evaluating probable failures. Information is painstakingly acquired from physical and virtual arenas as the breaker's location changes. The rich dataset is then safely saved in cloud storage and available for in-depth examination. In parallel, a complex classification algorithm is used to analyze and condense important information acquired from the cyber network by utilizing the capability of ridge regression. At the same time, the deep auto-encoder algorithm carefully examines the physical data to ensure no helpful information is overlooked.

5.2 Formulation of Hypotheses

A hypothesis is put out in the context of network mistakes happening under typical circumstances; this hypothesis is predicated on the mutual dependency of each end of the parallel breakers. However, in instances when even a single breaker or relay malfunctions, this theory is put to the test. The difficulty is in tracing the origin of a malfunction using nothing more than the locations of the relays and breaker. Based on further rank and contextual knowledge, this sophisticated process of unwrapping breakers illuminates the present approach that must be examined.

5.3 Making Decisions Based on Data Analysis

An essential component of the study technique is the identification of probable underlying causes of node failure. Modern data science approaches are used for this purpose, allowing for a thorough and accurate study. This sub-chapter looks deeply into the complex identification process to unearth important insights.

- Convolutional neural networks (CNNs), a powerful subset of deep neural networks, excel in processing visual data, particularly spatial data like images. There are several levels in this algorithmic design, including hidden layers, output layers, and input layers. Fully linked, pooling, and convolutional layers are all contained inside these layers and work together to extract and interpret complex visual data.

- Recurrent Neural Networks (RNN) are a special artificial neural network type that sets themselves apart by creating connections between nodes in a graph that include temporal sequences. RNNs, which excel in binary classification and situations involving several classes and are tailored for sequential data, are crucial in intrusion detection. This method provides a considerable advancement in improving the precision of intrusion detection, which is essential for strengthening CPS.

5.4 Control and Final Decision

Critical information supplied from many sources, including log files and relays, is rigorously authenticated based on physical and cyber analytics findings. The research's capacity to spot abnormalities in security-compromise scenarios is further strengthened by data analytics. However, the final control choice depends on a labor-intensive but crucial manual validation process that guarantees the accuracy of the results. This method eliminates the need for time-consuming human checks of suspected fraud across all CPS-based devices, offering a clear advantage in accuracy and efficiency.

5.5 Performance Evaluation

A thorough review is required to confirm the suggested technique's effectiveness. The following important performance criteria are taken into account:

- **Accuracy:** It is described as the proportion between all the accurate forecasts and all the predictions.

$$Accuracy = \frac{TP + TN}{TP + TN + FP + FN}$$

- **Precision:** The proportion of total successfully detected positive identities to total correctly detected identities.

$$Precision = \frac{TP}{TP + FP}$$

- **Recall:** It is described as the proportion of total samples that are positive to the total samples.

$$Recall = \frac{TP}{TP + FN}$$

- **F-1 score:** The precision and recall values are taken into account while determining the F1 score's worth, and the result is as follows:

$$F-1\ score = 2 \times \frac{Precision \times Recall}{Precision + Recall}$$

6. SUMMARY

The Internet of Things (IoT) has become the foundation of Cyber-Physical Systems (CPS) in the modern world, using various fields, including virtual reality and data analysis. Fundamentally, CPS is a synthesis of computational and physical components that have been painstakingly planned to guarantee numerous operations' effective and secure operation. This concentrated effort aims to strengthen the defenses against possible assaults. We set out to protect these systems from cyber-attacks by utilizing the power of machine learning in CPS applications. This method starts with the initialization of the data, which prepares the ground for the following development of hypotheses. These theories play a crucial role in directing our research. Various cutting-edge deep learning methods are used to make well-informed judgments, offering the analytical capacity required for hypothesis confirmation. The results produced by the physical and cyber systems are closely examined as part of our methodology's final stages. This rigorous assessment aims to identify any indications of malignancy or compromise in CPS-based devices. An extensive collection of performance measures is used to assess the recommended technique's success thoroughly. Accuracy, precision, recall, and the F1 score are some measures that may be used to assess how reliable and robust our suggested strategy is. This exhaustive analysis validates our conclusions' accuracy and highlights our dedication to creating a defense system that can withstand the strictest testing.

7. CONCLUSION AND FUTURE SCOPE

In conclusion, this research aims to strengthen the security of cyber-physical systems (CPS) in the Internet of Things (IoT) age. To mitigate possible cyber dangers, the combination of cutting-edge data-driven methodologies and deep learning algorithms has shown encouraging results. We have shown the effectiveness of our technique in identifying and defending against possible system breaches through a comprehensive methodology combining hypothesis formulation and confirmation. The accuracy and precision of our threat detection techniques have been greatly improved by adding machine learning algorithms, notably Convolutional Neural Networks (CNNs) and Recurrent Neural Networks (RNNs). The suggested technique shows promise as a strong deterrent against cyberattacks aimed against CPS, as demonstrated by rigorous performance measurements. There are many future opportunities for additional research and improvement in this field.

- *Enhanced Deep Learning topologies:* Researching more intricate and adaptable neural network topologies, such as recurrent and convolutional models, may lead to the development of threat detection systems that are even more precise.
- *Real-time Threat Response:* The CPS framework might be strengthened by developing real-time threat response mechanisms that would allow quick action to be taken in the case of a detected cyber-attack.
- Integrating edge computing into CPS can increase the effectiveness and localization of threat detection while lowering dependency on centralized cloud resources.
- *Blockchain for protection:* How blockchain technology may secure CPS data transfers and keep an immutable trail of system occurrences might offer extra protection.
- *Multi-modal Data Fusion:* By combining information from several sources, such as sensory, visual, and aural inputs, a more complete picture of system behavior and possible dangers may be provided.
- *Human-in-the-loop Security:* Human decision-making and knowledge may improve the effectiveness of threat detection and response by being incorporated into the security process.
- *Applications across domains:* Extending the suggested technique to different fields outside CPS, including healthcare, smart cities, and autonomous cars, may provide insightful information on safeguarding networked systems.

The area of CPS security may move towards a more effective and flexible defense against the changing panorama of cyber threats by taking these risks.

REFERENCES

1. Li, B., Wu, Y., Song, J., Lu, R., Li, T. and Zhao, L. (2020). "DeepFed: Federated deep learning for intrusion detection in industrial cyber–physical systems." *IEEE Transactions on Industrial Informatics*, *17*(8), pp. 5615–5624.
2. Rathore, S. and Park, J.H. (2020). "A blockchain-based deep learning approach for cyber security in next generation industrial cyber-physical systems." *IEEE Transactions on Industrial Informatics*, *17*(8), pp. 5522–5532.
3. Jahromi, A.N., Karimipour, H., Dehghantanha, A. and Choo, K.K.R. (2021). "Toward Detection and Attribution of Cyber-Attacks in IoT-Enabled Cyber–Physical Systems." *IEEE Internet of Things Journal*, *8*(17), pp. 13712–13722.
4. Tertytchny, G., Nicolaou, N. and Michael, M.K. (2020). "Classifying network abnormalities into faults and attacks in IoT-based cyber physical systems using machine learning." *Microprocessors and Microsystems*, *77*, p. 103121.
5. Choudhary, G., Astillo, P.V., You, I., Yim, K., Chen, R. and Cho, J.H. (2020). "Lightweight misbehavior detection management of embedded IoT devices in medical cyber physical systems." *IEEE Transactions on Network and Service Management*, *17*(4), pp. 2496–2510.
6. Saha, T., Aaraj, N., Ajjarapu, N. and Jha, N.K. (2021). "SHARKS: Smart Hacking Approaches for RisK Scanning in Internet-of-Things and Cyber-Physical Systems based on Machine Learning." *IEEE Transactions on Emerging Topics in Computing*.
7. de Araujo-Filho, P.F., Kaddoum, G., Campelo, D.R., Santos, A.G., Macêdo, D. and Zanchettin, C. (2020). "Intrusion detection for cyber–physical systems using generative adversarial networks in fog environment." *IEEE Internet of Things Journal*, *8*(8), pp. 6247–6256.
8. Hao, W., Yang, T. and Yang, Q. (2021). "Hybrid statistical-machine learning for real-time anomaly detection in industrial cyber-physical systems." *IEEE Transactions on Automation Science and Engineering*.
9. Saad, A., Faddel, S. and Mohammed, O. (2020). "IoT-based digital twin for energy cyber-physical systems: design and implementation." *Energies*, *13*(18), p. 4762.
10. Li, P., Lu, Y., Yan, D., Xiao, J. and Wu, H. (2021). "Scientometric mapping of smart building research: Towards a framework of human-cyber-physical system (HCPS)." *Automation in Construction*, *129*, p. 103776.
11. Heartfield, R., Loukas, G., Bezemskij, A. and Panaousis, E. (2020). "Self-configurable cyber-physical intrusion detection for smart homes using reinforcement learning." *IEEE Transactions on Information Forensics and Security*, *16*, pp. 1720–1735.
12. Javed, A., Malik, K.M., Irtaza, A. and Malik, H. (2021). "Towards protecting cyber-physical and IoT systems from single-and multi-order voice spoofing attacks." *Applied Acoustics*, *183*, p. 108283.
13. Khan, Rabia, Amjad Mehmood, Zeeshan Iqbal, Carsten Maple, and Gregory Epiphaniou. (2023). "Security and Privacy in Connected Vehicle Cyber Physical System Using Zero Knowledge Succinct Non Interactive Argument of Knowledge over Blockchain." *Applied Sciences*, *13*(3), 1959.
14. Oppl, Sabrina, and Christian Stary. (2022). "Motivating Users to Manage Privacy Concerns in Cyber-Physical Settings—A Design Science Approach Considering Self-Determination Theory." *Sustainability*, *14*(2), 900.
15. Singh, S.K., Jeong, Y.S. and Park, J.H. (2020). "A deep learning-based IoT-oriented infrastructure for secure smart city." *Sustainable Cities and Society*, *60*, p. 102252.
16. Ahmed, A., Krishnan, V.V., Foroutan, S.A., Touhiduzzaman, M., Rublein, C., Srivastava, A., Wu, Y., Hahn, A. and Suresh, S. (2019). "Cyber physical security analytics for anomalies in transmission protection systems." *IEEE Transactions on Industry Applications*, *55*(6), pp. 6313–6323.
17. Keshk, M., Sitnikova, E., Moustafa, N., Hu, J. and Khalil, I. (2019). "An integrated framework for privacy-preserving based anomaly detection for cyber-physical systems." *IEEE Transactions on Sustainable Computing*, *6*(1), pp. 66–79.
18. Shen, Yizhou, Shigen Shen, Qi Li, Haiping Zhou, Zongda Wu, and Youyang Qu. (2022). "Evolutionary privacy-preserving learning strategies for edge-based IoT data sharing schemes." *Digital Communications and Networks*, 108290.
19. Nguyen, G.N., Le Viet, N.H., Elhoseny, M., Shankar, K., Gupta, B.B. and Abd El-Latif, A.A. (2021). "Secure blockchain enabled Cyber–physical systems in healthcare using deep belief network with ResNet model." *Journal of Parallel and Distributed Computing*, *153*, pp. 150–160.
20. Elsisi, M., Tran, M.Q., Mahmoud, K., Mansour, D.E.A., Lehtonen, M. and Darwish, M.M. (2021). "Towards secured online monitoring for digitalized GIS against cyber-attacks based on IoT and machine learning." *IEEE Access*, *9*, pp. 78415–78427.
21. Singh, Rajani, Ashutosh Dhar Dwivedi, Raghava Rao Mukkamala, and Waleed S. Alnumay. (2022). "Privacy-preserving ledger for blockchain and Internet of Things-enabled cyber-physical systems." *Computers and Electrical Engineering*, *103*, 108290.

Note: The figure and table in this chapter were made by the author.

Computational Intelligence and Mathematical Applications – Dr. Devendra Prasad et al. (eds)
© 2024 Taylor & Francis Group, London, ISBN 978-1-032-87721-1

13 | Lane Detection Under Lousy Light Conditions

Ritik Dwivedia[1]

Department of CSE, G L Bajaj Institute of Technology and Management, Gr. Noida

Kamal Kumar[2]

Department of CSE, National Institute of Technology, Uttarakhand, bDepartment of IT, IGDTUW, Delhi

Abstract: Lane detection plays a crucial role in autonomous driving systems because most accidents are caused by the fault of the driver. This paper aims to propose a Deep Learning or specifically a Fully Connected Neural Network based approach to predict road lanes in challenging scenarios such as adverse weather conditions, night, shadows, etc. Most of the lane prediction models are based on either a computer-vision based approach or a segmentation based deep learning approach. The proposed model evaluates based on regression-based method. The CULane dataset is used for the training, validation, and testing of the model. It finds/estimates the coordinates of the lanes in the image diffused with various natura0l/synthesized light conditions. Obtained coordinates are then used to draw the lanes in the road. The proposed model has shown to be effective and robust at predicting the lanes' coordinates with accuracy 85.57% and meagre loss 0.1232.

Keywords: Lane detection and Fully connected neural network

1. INTRODUCTION

According to the World Health Organization's reports(1) every year, deaths caused by road accidents or vehicle collisions are increasing and approximately 1.3 million deaths are caused due to accidents around the world, and roughly 20 to 50 million people individuals undergo non-fatal injuries as a result of these accidents. Most of the accidents are caused by the faulty driving skills, unavoidable obstacles, lack of attentions and careless driving that could be mitigated through the implementation of Advanced Driver Assistance Systems (ADAS)(2). The main significant role of ADAS in driving system is to reduce injuries and deaths by decreasing the count of road accidents. ADAS does not completely stops accidents, but it could protect against some human error. In passive safety measures, airbags, structure of the car body, seat belts, and head restraints are included whereas active safety measures include line keeping assistant (LKAs), electronic stability control (ESC), anti-lock braking systems (ABS), and other Advanced Driver Assistance Systems (ADAS) like intersection collision avoidance (ICAs) and Information and Communication Technology (ICT). Autonomous driving is a key feature needed for ADAS. It is a challenging and demanding area of research that has received lots of attention in recent years. The main problem in autonomous driving is detecting as well as predicting the lanes of the road. Computer vision methods and Deep learning-based(DL-based) methods are candidate methods for the detection and prediction of the lane.

Computer vision-based method for lane detection uses the hand-crafted low-level features like color, color-index, edges, texture information, etc for finding the lane(3) that works well under normal and limited weather conditions. When it comes to complex conditions or adverse conditions like rainy, cloudy, and lighting change, etc, CV based method can't handle with

[1]ritikdwivedi7777@gmail.com, [2]kamalkumar@nituk.ac.in

DOI: 10.1201/9781003534112-13

assurance and accuracy. Most of the algorithms use the deep learning-based method to reduce the overcome these lousy lighting conditions and complex weather conditions. Recently, some of research works employed deep neural networks to address the lane detection problem(4; 5). In the majority of cases, algorithms approach lane detection under classification techniques like semantic segmentation (4) wherein each pixel within an image is labeled to determine if it belongs to one of the lanes.

These techniques depend on the image segmentation maps of the lanes, which makes it a supervisory problem. The semantic segmentation-based method requires a large size of labeled data to achieve the desired output and requires heavy computation. To manage the challenges noted above, namely; Computation Intensive Approach, Loss Percentage, Size of DataSet, the work proposed in this paper contribute in the following ways:

- Firstly four lane-road images on the CULane dataset is used and performed some necessary pre-processing on the label of four-lane road images.
- Our approach relies on per-pixels labels which enable the detection of road lanes in images.
- A deep learning based approach is specifically designed to detect lane in complex scanarios.
- A performance accuracy of nearly 86% is achievable through proposed model given very low loss.

2. RELATED WORK

Before the widespread adoption of the deep learning method, traditional methods were in use and mostly utilized micro and macro features in images to detect lanes on road. For example, kluge et al.(7) separates the lane by considering the shape as a feature, Chiu(8) implemented color as a feature, whereas; the lanes normally appeared as straight and curved lines(9). Despite color and differential feature, ridge methods are used for lane detection such as Gaussian filter(10), hough transform(9), and Kalman filter(11). These methods fails to address varying weather conditions or when complex traffic conditions.

With the remarkable success of the deep learning-based methodologies in image noise classification, object detection, and medical image analysis(12), researchers applied similar methods to address the lane detection challenge. Huval et al.(13) proposes the first deep learning-based model in the domain of lane detection which is based on the overFeat and gave the output as a sort of segmentation-map and the subsequent step involved additional preprocessing through DBSCAN clustering to enhance the results. Authors utilized private dataset for training and testing purposes. VPGnet(4) presents a multi-tasking neural network designed to handle both lane detection and road marking detection tasks under the guidance of vanishing point. the author introduces Line-CNN, a method that builds upon the region proposal network (RPN) of Faster R-CNN. LaneNet(5) used instance segmentation for detecting the perspective transformation matrix for better fitting lanes.

Work in this paper provide a novel way by applying some pre-processing on the label of lane-road images. The evaluation of our approach is performed by using a method based on fully connected neural network which gives a good accuracy on CULane dataset.

3. PROPOSED MODEL

The general architecture of the proposed Lane Detection algorithm is as shown by Fig. 13.1.

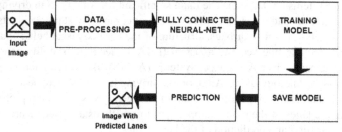

Fig. 13.1 Proposed model for lane detection

Source: Made by author

3.1 CULane(6) Datasets

The publicly available data-sets is used to evaluate proposed lane detection model which comprises a total of 133,235 images, with 88,880 images designated for training, 9,675 for validation, and 34,680 for testing. Additionally, the test set is categorized into nine distinct challenge categories, including normal, crowded, night, no line, shadow, arrow, dazzle light, curve, and crossroad. The primary emphasis of this dataset centers on the detection of four-lane markings, which hold greater practical significance in real-world scenarios. Other lane markings are not annotated. Each line in the dataset includes a (.txt) annotation file that provides the x and y coordinates of lane markings in the input image. Additionally, a per-pixel label is given, and four binary values (0 or 1) indicate the presence of four-lane markings from left to right. This annotation format is utilized in the 'train gt.txt' file for training purposes.

3.2 Pre-Processing

As per the discussion above, the CULane Dataset is considered for this work. This dataset consists of images containing 2 to 4 lanes. Research is done considering images containing 4 lanes only consisting of 35859 images. These images(no. of images: 35859) are further split into three sets for use in training(no. of images: 32265), testing and validation(no. of images: 3594). This work considers a total of 4 classes out of 9 classes present in CULane. These classes consists of curved lanes, nighttime, shadows (nearby cars, over-bridges, pedestrians jaywalking) and dazzled lighting condition. As discussed earlier, the annotations file contains the x and y coordinates of images containing four lanes. The y-coordinates being considered here from the dataset range from pixel value 250 to 590 with a pixel gap of 10, with x-coordinate corresponding to each y-coordinate. However, there are a few missing y-coordinates and their corresponding x-coordinates. To overcome a solution with data padding is proposed. Firstly, the y-coordinate is padded with missing data values. Further corresponding x-coordinate values is done via Alternate Padding. In this approach, the x-coordinate values of the missing y-coordinates are substituted with the adjacent available x-coordinates.

3.3 Model Description

The architecture for the proposed model used for prediction of the lane on the modified dataset model architecture is shown in Fig. 13.2. Here, a lane prediction model is developed based on a deep learning technique. The proposed architecture addresses the multi-point prediction problem and predicts the lane coordinates in the input image. The coordinates represent the lane line points of four lanes. The specification of the dataset utilized for the model training and testing after applying some pre-processing is discussed in pre-processing section.

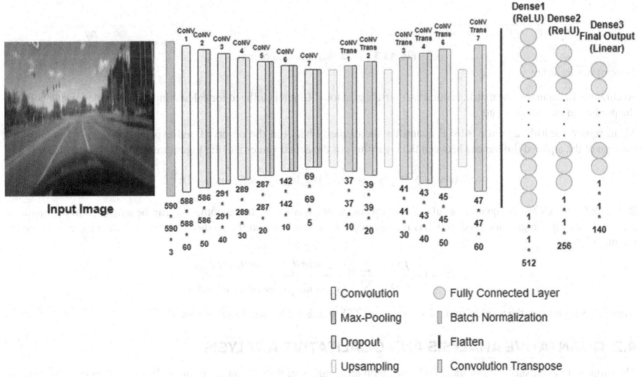

Fig. 13.2   Proposed neural network architecture for lane detection

Source: Made by author

Neural architecture has been used to develop the prediction model. Input images are resized to 590*590*3 then these images are fed to the batch normalization. Model consists of 7 convolution layers (Conv2D), four max-pooling layers, three upsampling, and seven convolution transpose layers (CoNV2dtranspose). Architecture also consists of 11 dropout layers. In the development process, Activation function used is ReLU and adam optimizer is used. The linear function is utilized in the final dense layer to predict the coordinates for each lane from images. Max-Pooling is utilized to subsample the features. Dropout is utilized to decrease the over-fitting of the model.

4. EXPERIMENTAL SETUP AND RESULT DISCUSSIONS

The lane detection model is constructed on a conventional dual-core 2.6 GHz hexa-core CPU paired with a graphical processing unit consisting of an NVIDIA GeForce RTX 3060 with 6 GB of memory. Dataset used for the experiment is CULane dataset which involved 35,869 images. These images were resized to dimensions of 590x590 pixels with three color channels for the purpose of training and testing. Subsequently, the data underwent normalization within the range of [0,1] before being fed into a CNN architecture for both training and testing.

4.1 Performance Parameter

For overall evaluation, the accuracy and loss parameter are implemented. They are elaborated as:

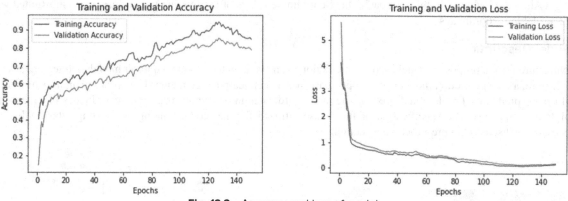

Fig. 13.3 Accuracy and loss of model

Source: Made by author

Accuracy—In accuracy, the matrix created two variables in total. Count is utilized for calculating the frequency match between the prediction and actual value.

Mean square logarithmic error- MSLE quantifies the disparity between the observed and expected values. It computes the average of the squared differences between the logarithms of the actual values and the logarithms of the predicted values.

$$loss = \frac{1}{N}\sum_{i=1}^{N}(\log(y\_actual + 1) - \log(y\_anticipated + 1))^2$$

R-squared score (R2)—R-squared calculates the percentage of variation in the outcome that can be attributed to the predictor factors. In multiple regression models, R2 is the squared correlation between the predicted values and the projected values of the model.

$$R^2 = 1 - \frac{RSS}{TSS} = \frac{\sum_{i=1}^{N}(y\_actual_i - y\_anticipated_i)^2}{\sum_{i=1}^{N}(y\_actual_i - mean(y\_actual))}$$

where RSS represents the residual sum of squares, and TSS stands for the total sum of squares.

4.2 QUANTATIVE ANALYSIS AND QUALITATIVE ANALYSIS

The proposed model successfully identified lanes with an accuracy of 94.87% on the training dataset, achieving a low loss of 0.0842. On the test dataset, the model maintained an accuracy of 85.57% with a loss score of 0.1232. Training was conducted using a batch size of 8 for up to 150 epochs. Figure 13.3 displays the model's accuracy and loss curve. R-squared score on test dataset is 0.8751. After the model's training phase, it was employed to predict lanes, as depicted in Fig. 13.4. The proposed model delivers lane predictions, with an average processing time of approximately 235 milliseconds per image.

Fig. 13.4 Output predicted by the model

Source: https://xingangpan.github.io/projects/CULane.html

5. CONCLUSION AND FUTURE WORK

The proposed lane detection model is based on fully connected neural architecture which predicts the lane's coordinates. After applying pre-processing on CULane dataset to give the complete input to fully connected neural network. The model is applied to various adverse conditions like lighting change, shadow, rainy, and night.

Further the model has achieved the accuracy of 85.57%. The work to be further expanded by transfer learning and attention-based mechanism.

REFERENCES

1. Road Safety. (n.d.). Retrieved April 4, 2022, from https://www.who.int/health-topics/road-safety
2. Ziebinski, A., Cupek, R., Grzechca, D., & Chruszczyk, L. (2017, November). Review of advanced driver assistance systems (ADAS). *AIP Conference Proceedings*, *1906*(1), 120002.
3. Cheng, H. Y., Jeng, B. S., Tseng, P. T., & Fan, K. C. (2006). Lane detection with moving vehicles in the traffic scenes. *IEEE Transactions on Intelligent Transportation Systems*, *7*(4), 571–582.
4. Lee, S., Kim, J., Shin Yoon, J., Shin, S., Bailo, O., Kim, N., ... & So Kweon, I. (2017). Vpgnet: Vanishing point guided network for lane and road marking detection and recognition. In *Proceedings of the IEEE international conference on computer vision* (pp. 1947–1955).
5. Neven, D., De Brabandere, B., Georgoulis, S., Proesmans, M., & Van Gool, L. (2018, June). Towards end-to-end lane detection: an instance segmentation approach. In *2018 IEEE intelligent vehicles symposium (IV)* (pp. 286–291). IEEE.
6. CULANE, CULANE DATASET. [Online]. Available: https://xingangpan.github.io/projects/CULane.html
7. Kluge, K., & Lakshmanan, S. (1995, September). A deformable-template approach to lane detection. In *Proceedings of the Intelligent Vehicles' 95. Symposium* (pp. 54–59). IEEE.
8. Chiu, K. Y., & Lin, S. F. (2005, June). Lane detection using color-based segmentation. In *IEEE Proceedings. Intelligent Vehicles Symposium, 2005.* (pp. 706–711). IEEE.
9. Jung, C. R., & Kelber, C. R. (2005). Lane following and lane departure using a linear-parabolic model. *Image and Vision Computing*, *23*(13), 1192–1202.
10. Aly, M. (2008, June). Real time detection of lane markers in urban streets. In *2008 IEEE Intelligent Vehicles Symposium* (pp. 7–12). IEEE.
11. Kim, Z. (2008). Robust lane detection and tracking in challenging scenarios. *IEEE Transactions on Intelligent Transportation Systems*, *9*(1), 16–26.
12. Joshi, D., & Singh, T. P. (2020). A survey of fracture detection techniques in bone X-ray images. *Artificial Intelligence Review*, *53*(6), 4475–4517.
13. Huval, B., Wang, T., Tandon, S., Kiske, J., Song, W., Pazhayampallil, J., ... & Ng, A. Y. (2015). An empirical evaluation of deep learning on highway driving. *arXiv preprint arXiv:1504.01716*.

Computational Intelligence and Mathematical Applications – Dr. Devendra Prasad et al. (eds)
© *2024 Taylor & Francis Group, London, ISBN 978-1-032-87721-1*

14

Data-Driven Precision Agriculture: Enhancing Crop Classification Accuracy and Agro-Environmental Analysis with Machine Learning

Aaskaran Bishnoi[1], Gurwinder Singh[2]
Department of AIT-CSE, Chandigarh University, Punjab, India

Ranjit Singh[3]
Department of Computer Science Engineering, Chandigarh University, Punjab, India

Abstract: This research paper presents a holistic approach for improving crop classification accuracy and conducting an extensive analysis of agro-environmental factors using machine learning techniques. The approach encompasses diverse data collection sources, rigorous data preprocessing for quality assurance, and precise crop classification employing feature selection and various machine learning algorithms. Additionally, it addresses the amalgamation of agro-environmental datasets, feature engineering, and the utilization of regression and clustering techniques for comprehensive analysis. In the results section, the paper provides a concise summary of average, minimum, and maximum prices for diverse agricultural commodities, offering valuable insights into pricing dynamics. Visual representations and statistical analyses elucidate trends and disparities in crop prices across distinct districts, equipping farmers with data-informed decision-making capabilities. Furthermore, the research delves into water quality metrics, furnishing a comprehensive overview of parameters such as pH, hardness, and turbidity. Visualizations and statistical synopses of water quality data facilitate a deeper understanding of the dataset's characteristics. The paper culminates with practical recommendations for farmers, emphasizing the prioritization of high-profit crops, consideration of geographic factors, and vigilant monitoring of price trends. Additionally, it underscores the critical role of water quality management in sustainable agriculture, highlighting key parameters that necessitate attention. In essence, this research contributes significantly to informed decision-making within precision agriculture and underscores the vital importance of data-driven approaches in fostering sustainable agro-environmental management.

Keywords: Progressive precision farming, Machine learning, Crop classification agro-environmental factors, Remote sensing, Sustainable agriculture, Data-driven decision-making

1. INTRODUCTION

Precision agriculture marks a transformative approach to modern farming, utilizing technological advancements to optimize practices and address challenges related to food security, resource constraints, and sustainability. At the forefront of this paradigm shift is "Data-Driven Precision Agriculture," a concept explored in this paper. Central to its success is the integration of machine learning, which plays a pivotal role in refining crop classification accuracy and conducting comprehensive agro-environmental analyses. The study delves into the complexity and variability inherent in crop management, highlighting the overwhelming volume and diversity of agricultural data, subjective human interpretation, and the dynamic nature of agricultural conditions. Machine learning is identified as the key solution to these challenges, excelling in pattern recognition, feature extraction, model generalization, ensemble learning, unbiased analysis, real-time monitoring, predictive analytics, and automated decision

[1]23YCS1020@cuchd.in, [2]singh1001maths@gmail.com, [3]ranjit.e10947@cumail.in

DOI: 10.1201/9781003534112-14

support. The research objectives focus on leveraging cutting-edge machine learning methods to significantly improve crop classification accuracy and thoroughly investigate various agro-environmental parameters. By enabling data-driven decision-making, the study aims to empower farmers with actionable insights for adaptive management strategies, contributing to more efficient and sustainable agricultural practices. Ultimately, the goal is to advance sustainable precision farming, promoting a shift towards technologically enabled methods that enhance productivity, reduce environmental impact, and ensure global food security.

2. RELATED WORK

The literature review provides a comprehensive overview of precision farming, spanning technological advancements, ethical considerations, policy implications, and sustainability concerns. Deng et al. [1] explore the potential of nanoenabled agriculture using machine learning, emphasizing its transformative role. Elsawah et al. [2] contribute a broader perspective on socio-environmental systems modeling, aligning with the goals of sustainable agricultural practices. Beyene et al. [3] address evolving soil challenges in Ethiopia, resonating with precision farming's focus on sustainability. The review includes studies on policy innovation by Tiikkainen et al. [4], the reuse of sludge by Mabrouk et al. [5], and ethical considerations by Hermann et al. [6]. Mueller et al. [7] contextualize challenges wthin the broader agricultural landscape, while Ali et al. [8] highlight the role of mobile technology. Shafi et al. [9] delve into precision agriculture methods, and Pathak et al. [10] assess factors influencing its spread. Liu et al. [11] explore ICT and blockchain in precision agriculture, and Gardezi and Bronson [12] examine psychological factors in the Midwest. Yarashynskaya and Prus [13] present a Polish case study, and Zhang et al. [14] integrate nanotechnology and AI for sustainable agriculture. Munz and Schuele [15] investigate economic success factors, while Hurst et al. [16] discuss augmented reality. Thakur et al. [17] review wireless sensor networks, and Larkin et al. [18] address perceived environmental improvements. Priya and Ramesh [19] focus on machine learning, Singh et al. [20] examine Indian precision agriculture, and Khattab et al. [21] propose an IoT strategy. Tey and Brindal [21] assess variables influencing adoption, and Cisternas et al. [22] review precision agricultural implementations. Bucci et al. [23] explore precision agriculture for sustainable farming, and Finger et al. [24] link precision farming with environmental impacts. Wolf and Wood [25] analyze precision farming's environmental legitimation, and Mandal and Maity [26] focus on small farms in India. Bongiovanni and Lowenberg-DeBoer [27] examine the relationship between precision agriculture and sustainability, and Bawa et al. [29] propose a support vector machine and image processing-based approach. Hakkim et al. [30] provide a comprehensive perspective on the future of Indian agriculture with precision farming, highlighting its transformative potential.

3. METHODOLOGY

Step	Description
1. Data Collection	Utilized OpenWeatherMap API [28] for historical weather data, including temperature, humidity, precipitation, wind speed, and other climate variables.
2. Data Preprocessing	Ensured data quality by collecting, integrating, and cleaning data for missing values, duplicates, and outliers. Applied feature engineering, scaling, normalization, time series decomposition, and lag features.
3. Feature Selection/Extraction	Identified and extracted features to capture unique qualities of each crop, considering both spatial and temporal elements.
4. Model Selection	Explored machine learning algorithms (e.g., decision trees, random forests, support vector machines) based on their suitability for handling data complexities and crop classification challenges.
5. Training and Validation	Systematically trained machine learning models on separate training and validation sets to identify characteristics and crop types. Evaluated model efficacy using the validation set.
6. Hyperparameter Tuning	Investigated hyperparameter tuning to improve the efficiency of machine learning models by systematically modifying hyperparameters and assessing their impact on model performance.

4. RESULTS AND DISCUSSION

An agricultural pricing analysis includes a summary table with average, minimum, and maximum prices for key commodities, followed by visual plots, box plots, heatmaps for data analysis, outlier detection, and correlation insights, concluding with an examination of price movements and actionable suggestions for farmers to optimize practices and financial returns. The box plot (1) visually illustrates regional price differentials for the same crop types, exemplified by median price variations of

Table 14.1 Crop summary

Commodity	Avg_Price	Min_Price	Max_Price
Amphophalus	2000	2000	2000
Apple	5500	5500	5500
Banana	1400	900	1800
Bhindi(Ladies Finger)	4250	3500	5000
Brinjal	1368.75	700	2000
Cabbage	420	300	500
Capsicum	2790	2050	4000
Carrot	631.875	300	900
Cauliflower	610	200	900
Chikoos(Sapota)	3000	3000	3000
Coriander(Leaves)	755	120	1500
Cotton	5445	5390	5500
Cucumbar(Kheera)	2250	1400	3500
Field Pea	1500	1500	1500
Garlic	3100	2200	4000
Ginger(Dry)	6500	6500	6500
Ginger(Green)	7046	5500	7730
Gram Raw(Chholia)	1000	1000	1000
Grapes	5416.667	3000	8500
Green Chilli	3104	1400	5000
Guava	2100	2000	2200
Kinnow	1214.286	500	2000
Leafy Vegetable	320	320	320
Methi(Leaves)	800	600	1000
Onion	965.4545	700	1500
Onion Green	1200	1200	1200
Papaya	1350	1200	1500
Peas Cod	1050	800	1200
Peas Wet	1133	700	1700
Pineapple	4000	4000	4000
Pomegranate	6500	6500	6500
Potato	453.6364	350	600
Raddish	415.4545	250	600
Spinach	614.2857	300	900
Squash(Chappal Kadoo)	1400	1400	1400
Tomato	1780	1200	2500
Turnip	550	400	700

carrots and cauliflower across districts (Fig. 14.2), enabling easy comparison and identification of regional disparities in crop prices. In Fig. 14.3, the heatmap displays a correlation matrix, visually representing relationships between numerical variables. Warmer colors indicate positive correlations, while cooler colors indicate negative correlations, aiding in the identification of patterns and relationships within the dataset. The subsequent box plot of 'modal_price' (Fig. 14.4) provides insights into the

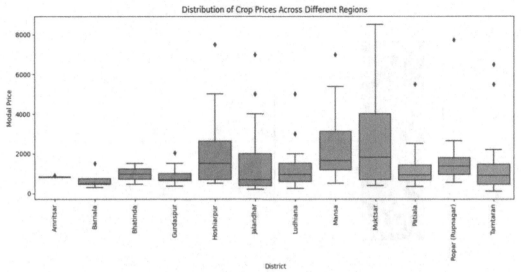

Fig. 14.1   Box plots of different crops

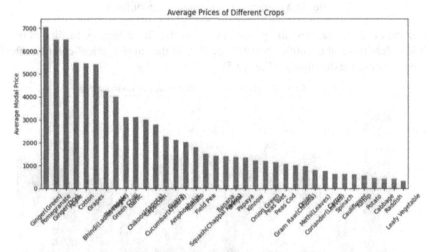

Fig. 14.2   Histograms of parameter distributions

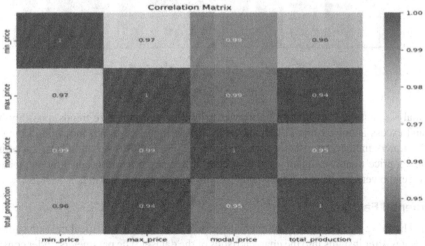

Fig. 14.3   Heat map

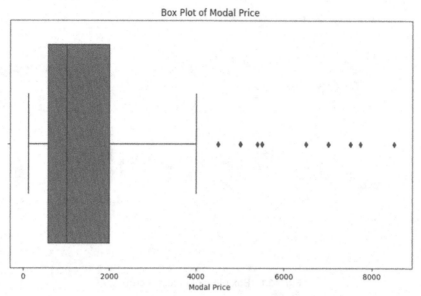

Fig. 14.4 Boxplot of parameter distributions

distribution, highlighting the median, quartiles, and potential outliers. Outlier detection, facilitated by the calculation of the Interquartile Range (IQR), identifies and quantifies potential outliers in the 'modal_price' column, offering a comprehensive understanding of extreme values in the distribution (Fig. 14.5).

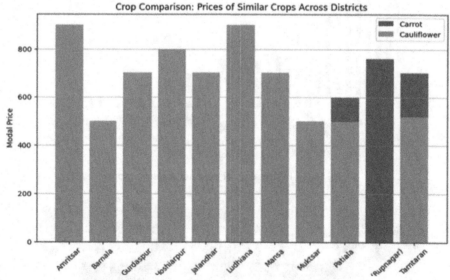

Fig. 14.5 Price comparison for two crops

In the Price Trends Analysis section, Carrot and Cauliflower exhibit decreasing price trends, possibly influenced by factors like falling demand and excess supply, while Potato shows an increasing price trend, indicating rising demand or scarcity. Recommendations for farmers include prioritizing high-profit crops like Cauliflower and Carrot, accounting for geographic factors, vigilantly tracking price trends, mitigating the impact of decreasing prices through strategic measures, engaging with local policymakers, and empowering farmers with insights for informed decision-making.

4.1 Agro-Environmental Factors

1) Water Composition: The Table 14.2 provides a comprehensive overview of basic statistical measures for various water quality parameters. These measures offer insights into the characteristics of the dataset. The parameters under consideration include pH,

Table 14.2 Insights from water quality metrics

	ph	Hardness	Solids	Chloramines	Sulfate	Conductivity	Organic carbon	Trihalomethanes	Turbidity
count	3276.000	3276.000	3276.000	3276.000	3276.000	3276.000	3276.000	3276.000	3276.000
mean	7.080	196.369	22014.093	7.122	333.776	426.205	14.285	66.396	3.967
std	1.470	32.880	8768.571	1.583	36.143	80.824	3.308	15.770	0.780
min	0.000	47.432	320.943	0.352	129.000	181.484	2.200	0.738	1.450
25%	6.278	176.851	15666.690	6.127	317.095	365.734	12.066	56.648	3.440
50%	7.081	196.968	20927.834	7.130	333.776	421.885	14.218	66.396	3.955
75%	7.870	216.667	27332.762	8.115	350.386	481.792	16.558	76.667	4.500
max	14.000	323.124	61227.196	13.127	481.031	753.343	28.300	124.000	6.739

Hardness, Solids, Chloramines, Sulfate, Conductivity, Organic Carbon, Trihalomethanes, and Turbidity. For each parameter, the table presents essential statistical information, such as the count (number of observations), mean (average value), standard deviation (measure of data spread), minimum and maximum values, as well as quartile values. These statistics collectively give a clear understanding of the central tendency, variability, and range of each parameter's values within the dataset.

As we know, correlation measures the strength and direction of a linear relationship between two variables. So a heatmap is generated (Fig. 14.6) using Seaborn's heatmap() function. The heatmap displays the correlation coefficients between pairs of variables in the dataset. The color intensity in the heatmap indicates the strength of the correlation, with cool colors (blues) indicating negative correlation and warm colors (reds) indicating positive correlation.

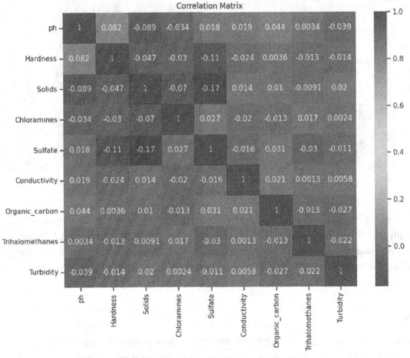

Fig. 14.6 Correlation heatmap

4.2 Water Composition Analysis

The figure depicts histograms with kernel density estimates to depict the distribution of key parameters, including the Water Quality Index (WQI), chloramines concentration, turbidity level, trihalomethanes concentration, organic carbon content, and pH level. These visualizations offer valuable insights into the variability and central tendencies of these factors within the dataset, aiding in the assessment of water quality and its implications for agricultural practices and environmental management.

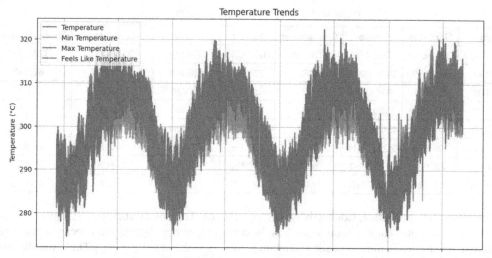

Fig. 14.7 Temperature variations

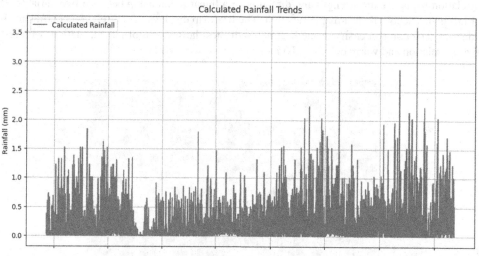

Fig. 14.8 Rainfall trends

As far the above discussion, we have provided statistical insights into various water quality parameters, including pH, hardness, and chloramines, through tables and histograms. These analyses offer a comprehensive understanding of water composition, vital for assessing its impact on agriculture and the environment. Finally, this section finishes with supplementary visualisations about temperature shifts and computed rainfall values, expanding the reader's understanding of the agro-environmental dynamics at play.

5. CONCLUSION

This paper presents a thorough examination of agricultural data, illuminating the interplay between price movements and agro-environmental parameters, two crucial elements of the agricultural landscape. First, the study breaks down the dynamics of prices for many different types of agricultural products in great detail, so that interested parties can make well-informed judgements regarding the crops they plant and the marketing techniques they employ. Farmers can better respond to regional market situations by adapting their practises in light of visual comparisons of crop prices across districts. Furthermore, specific pricing patterns for select commodities are recognised, such as the decline in the price of Carrot and Cauliflower or the rise in the price of Potato, providing farmers with concrete tips for overcoming obstacles and making the most of possibilities in the market.

The paper delves into the world of agro-environmental issues, with an emphasis on measures of water quality, in addition to the area of price. Insights into the central tendencies and variability of a wide range of water parameters are provided by the analysis. A better understanding of water quality is crucial for farming to be done in a sustainable way. To guarantee that water supplies sustain maximum crop development while minimising environmental impacts, it is crucial to maintain adequate water quality levels, as highlighted in the study, and to keep stakeholders apprised of potential areas of concern. In sum, this study makes important strides towards expanding our understanding of sustainable agricultural practises and data-driven decision making.

REFERENCES

1. Deng, P., Gao, Y., Mu, L., Hu, X., Yu, F., Jia, Y., Wang, Z. and Xing, B., 2023. Development potential of nanoenabled agriculture projected using machine learning. Proceedings of the National Academy of Sciences, 120(25), p.e2301885120.
2. Elsawah, S., Filatova, T., Jakeman, A.J., Kettner, A.J., Zellner, M.L., Athanasiadis, I.N., Hamilton, S.H., Axtell, R.L., Brown, D.G., Gilligan, J.M. and Janssen, M.A., 2019. Eight grand challenges in socio-environmental systems modeling. Socio-Environmental Systems Modelling, 2.
3. Beyene, S., Kibret, K. and Erkossa, T., 2023. Future/Emerging Soil Issues. The Soils of Ethiopia, pp.299-312.
4. Tiikkainen, O., Pihlajamaa, M. and A°kerman, M., 2022. Environmental impact bonds as a transformative policy innovation: Frames and frictions in the construction process of the Nutrient-EIB. Environmental Innovation and Societal Transitions, 45, pp.170-182.
5. Mabrouk, O., Hamdi, H., Sayadi, S., Al-Ghouti, M.A., Abu-Dieyeh, M.H. and Zouari, N., 2023. Reuse of Sludge as Organic Soil Amendment: Insights into the Current Situation and Potential Challenges. Sustainability, 15(8), p.6773.
6. Hermann, E., Hermann, G. and Tremblay, J.C., 2021. Ethical artificial intelligence in chemical research and development: a dual advantage for sustainability. Science and Engineering Ethics, 27, pp.1-16.
7. Mueller, L., Eulenstein, F., Dronin, N.M., Mirschel, W., McKenzie, B.M., Antrop, M., Jones, M., Dannowski, R., Schindler, U., Behrendt, A. and Rukhovich, O.V., 2021. Agricultural landscapes: history, status and challenges. Exploring and Optimizing Agricultural Landscapes, pp.3-54.
8. Ali, M., 2021. Mobile Technology Use by Rural Farmers and Herders.
9. Shafi, U., Mumtaz, R., Garc´ıa-Nieto, J., Hassan, S.A., Zaidi, S.A.R. and Iqbal, N., 2019. Precision agriculture techniques and practices: From considerations to applications. Sensors, 19(17), p.3796.
10. Pathak, H.S., Brown, P. and Best, T., 2019. A systematic literature review of the factors affecting the precision agriculture adoption process. Precision Agriculture, 20, pp.1292-1316.
11. Liu, W., Shao, X.F., Wu, C.H. and Qiao, P., 2021. A systematic literature review on applications of information and communication technologies and blockchain technologies for precision agriculture development. Journal of Cleaner Production, 298, p.126763.
12. Gardezi, M. and Bronson, K., 2020. Examining the social and biophysical determinants of US Midwestern corn farmers' adoption of precision agriculture. Precision Agriculture, 21(3), pp.549-568.
13. Yarashynskaya, A. and Prus, P., 2022. Precision Agriculture Implementation Factors and Adoption Potential: The Case Study of Polish Agriculture. Agronomy, 12(9), p.2226.
14. Zhang, P., Guo, Z., Ullah, S., Melagraki, G., Afantitis, A. and Lynch, I., 2021. Nanotechnology and artificial intelligence to enable sustainable and precision agriculture. Nature Plants, 7(7), pp.864-876.
15. Munz, J. and Schuele, H., 2022. Influencing the success of precision farming technology adoption—A model-based investigation of economic success factors in small-scale agriculture. Agriculture, 12(11), p.1773.
16. Hurst, W., Mendoza, F.R. and Tekinerdogan, B., 2021. Augmented reality in precision farming: Concepts and applications. Smart Cities, 4(4), pp.1454- 1468.
17. Thakur, D., Kumar, Y., Kumar, A. and Singh, P.K., 2019. Applicability of wireless sensor networks in precision agriculture: A review. Wireless Personal Communications, 107, pp.471-512.
18. Larkin, S.L., Perruso, L., Marra, M.C., Roberts, R.K., English, B.C., Larson, J.A., Cochran, R.L. and Martin, S.W., 2005. Factors affecting perceived improvements in environmental quality from precision farming. Journal of Agricultural and Applied Economics, 37(3), pp.577-588.
19. Priya, R. and Ramesh, D., 2020. ML based sustainable precision agriculture: A future generation perspective. Sustainable Computing: Informatics and Systems, 28, p.100439.
20. Singh, A.K., 2022. Precision agriculture in india–opportunities and challenges. Indian Journal of Fertilisers, 18(4), pp.308-331.
21. Khattab, A., Abdelgawad, A. and Yelmarthi, K., 2016, December. Design and implementation of a cloud-based IoT scheme for precision agriculture. In 2016 28th International Conference on Microelectronics (ICM) (pp. 201-204). IEEE.
22. Tey, Y.S. and Brindal, M., 2012. Factors influencing the adoption of precision agricultural technologies: a review for policy implications. Precision agriculture, 13, pp.713-730.
23. Cisternas, I., Vela´squez, I., Caro, A. and Rodr´ıguez, A., 2020. Systematic literature review of implementations of precision agriculture. Computers and Electronics in Agriculture, 176, p.105626.

24. Bucci, G., Bentivoglio, D. and Finco, A., 2018. Precision agriculture as a driver for sustainable farming systems: state of art in literature and research. Calitatea, 19(S1), pp.114-121.

25. Finger, R., Swinton, S.M., El Benni, N. and Walter, A., 2019. Precision farming at the nexus of agricultural production and the environment. Annual Review of Resource Economics, 11, pp.313-335.

26. Wolf, S.A. and Wood, S.D., 1997. Precision farming: Environmental legitimation, commodification of information, and industrial coordination 1. Rural sociology, 62(2), pp.180-206.

27. Mandal, S.K. and Maity, A., 2013. Precision farming for small agricultural farm: Indian scenario. American Journal of Experimental Agriculture, 3(1), p.200.

28. OpenWeatherMap. Historical Weather Data. https://home.openweathermap.org/history bulks/new. Accessed on [12/8/2023].

29. Bawa, A., Samanta, S., Himanshu, S.K., Singh, J., Kim, J., Zhang, T., Chang, A., Jung, J., DeLaune, P., Bordovsky, J. and Barnes, E., 2023. A support vector machine and image processing based approach for counting open cotton bolls and estimating lint yield from UAV imagery. Smart Agricultural Technology, 3, p.100140.

30. Hakkim, V.A., Joseph, E.A., Gokul, A.A. and Mufeedha, K., 2016. Precision farming: the future of Indian agriculture. Journal of Applied Biology and Biotechnology, 4(6), pp.068-072

Note: All the figures and tables in this chapter were made by the author.

Computational Intelligence and Mathematical Applications – Dr. Devendra Prasad et al. (eds)
© 2024 Taylor & Francis Group, London, ISBN 978-1-032-87721-1

15

The Digital Dilemma—Exploring the Impact of Social Media on Mental Health

**Kumar Yashu[1], Akash Kumar[2], Vani Pandit[3],
Kishan Raj[4], Anjali Gupta Somya[5], Sourabh Budhiraja[6]**
Department of CSE, Chandigarh University, Mohali, India

Abstract: In today's digital era, the widespread presence of social media platforms has revolutionized the manner in which people establish connections and engage in communication. This document explores the complex connection between the utilization of social media and mental well-being, with special attention given to the susceptibility of younger generations because of their extensive involvement with these online platforms. Based on extensive research and a descriptive study conducted among university students, our results highlight a strong connection between heightened social media engagement, the quantity of social networking platforms used, and mental health consequences, specifically the presence of anxiety and depression symptoms. This research emphasizes the critical necessity for additional inquiries into the intricate interaction between social media and the mental well-being of today's youth.

Keywords: Social media, Digital dilemma, Mental health, Adolescents, Online behavior

1. INTRODUCTION

In today's digitally connected era, social media platforms have become powerful tools for communication, information exchange, and social interactions, transcending geographical boundaries. While these platforms have transformed communication, there is a growing concern about the potential impact of digital technologies on mental health. This introduction sets the stage for an examination of the intricate connection between social media usage and mental well-being.[1] The digital revolution has reshaped how we connect and communicate, serving as virtual meeting places, information hubs, and spaces for self-expression. Despite the advantages, a digital dilemma emerges, as extensive social media use is associated with mental health challenges such as anxiety, loneliness, and feelings of inadequacy. This study focuses on the implications for mental health, particularly among the younger generation, aiming to uncover the mechanisms by which social media influences psychological well-being. As we navigate this digital landscape, understanding these effects becomes crucial for responsible usage and fostering holistic well-being.[2]

2. LITERATURE REVIEW

The pervasive presence of social media platforms in the modern world has transformed communication, information dissemination, and the formation of virtual communities. Research, exemplified by Huang (2017), emphasizes the nuanced impact of social media on psychological well-being, stressing the importance of considering personal characteristics and usage patterns. Studies by Primack et al. (2017) reveal a troubling association between regular social media use and increased social isolation and loneliness among young adults.

[1]kumaryashu496@gmail.com, [2]20bcs5361@cuchd.in, [3]20bcs3190@cuchd.in, [4]20bcs3384@cuchd.in, [5]20bcs2441@cuchd.in, [6]sourabh.e13134@cumail.in

DOI: 10.1201/9781003534112-15

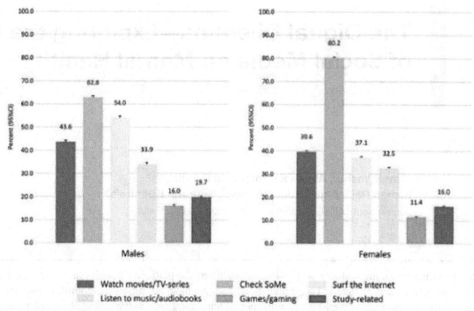

Fig. 15.1 Indulgence of people into different activities over internet[3]

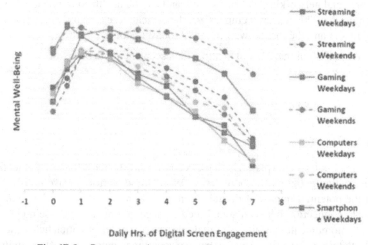

Fig. 15.2 Positive and negative effects of social networks[4]

Twenge's (2018) research explores the adverse correlation between screen time, including social media use, and the mental well-being of young individuals, indicating a negative link with overall well-being. Investigations into Facebook use, as conducted by Kross (2013), suggest a connection between heightened usage and a decline in subjective well-being among young adults. Hawi's (2017) study on social media addiction unveils an inverse relationship with life satisfaction, particularly among college students, with self-esteem serving as a potential mediator. The role of social comparison on platforms like Facebook is evident in Fardouly's (2015) research, linking increased usage to elevated levels of body dissatisfaction. Haferkamp et al. (2017) find that passive Facebook usage is associated with declines in subjective well-being. James (2020) delves into the Fear of Missing Out (FoMO) and its complex interplay with social media usage and overall well-being, suggesting both detrimental and beneficial effects.[5][6][7]

Naslund (2020) investigates the dual impact of social media on mental health, acknowledging its potential for empowerment and self-expression but also highlighting the risks of heightened anxiety, depression, and cyberbullying. This literature review establishes the intricate connection between social media and mental health, emphasizing the need for a comprehensive understanding of its multifaceted effects.[9]

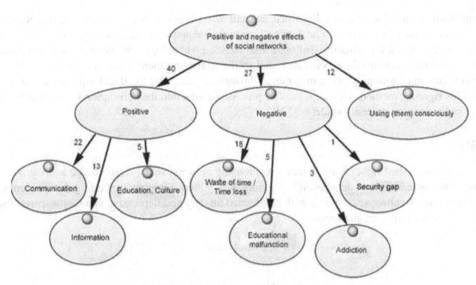

Fig. 15.3 Positive and negative effects of social networks[8]

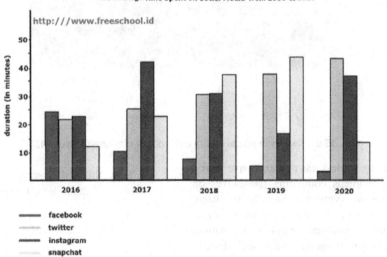

Fig. 15.4 Average time spent during the depicted years.[10]

3. METHODOLOGY

In this section, we present the methodology employed for a descriptive study among university students, emphasizing its significance for ensuring the validity and reliability of research findings.

The research design is a descriptive study, allowing systematic data collection and analysis for a comprehensive understanding. Participants were selected using a stratified random sampling technique, ensuring representation across various university student characteristics. The sample size, comprising many university students, provides sufficient statistical power. Data collection methods include structured survey questionnaires covering demographic information, social media usage patterns, and mental

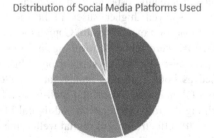

Distribution of Social Media Platforms Used

▪ Facebook ▪ Instagram ▪ Twitter ▪ Snapchat ▪ LinkedIn ▪ Pinterest

Fig. 15.5 Distribution of social media platforms used[11]

health assessment using standardized scales. Informed consent was obtained from participants, respecting their rights and providing comprehensive study information. The data collection period spanned multiple days to capture potential variations. Quantitative analysis, facilitated by statistical software like SPSS, involved descriptive statistics, correlation analysis, and regression analysis to explore relationships between social media usage and mental health outcomes. Ethical considerations, including confidentiality and anonymity, were maintained throughout, adhering to ethical guidelines and institutional review board approvals. This rigorous methodology aims to offer precise insights into the correlation between social media usage and mental health outcomes among university students.[12]

4. FINDINGS

The findings of our research shed light on the intricate relationship between social media usage and the mental well-being of young adults, emphasizing anxiety and depression symptoms. Key patterns emerged regarding the frequency and duration of social media engagement, revealing associations with heightened anxiety and depressive symptoms, particularly among those spending extended periods on specific platforms.

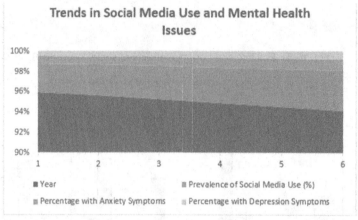

Fig. 15.6 Trends in social media use and mental health issues[13]

Anxiety symptoms, including restlessness and excessive worry, were prevalent among individuals comparing themselves on social media, while depressive symptoms, such as persistent sadness, were more common with frequent and prolonged use. Fear of Missing Out (FoMO) exhibited a multifaceted nature, correlating with increased social media usage and directly impacting mental health, leading to greater anxiety and feelings of inadequacy. Gender differences were noted, with females experiencing a higher prevalence of anxiety symptoms related to maintaining an idealized online image.

Younger participants, especially in their late teens and early twenties, displayed higher susceptibility to social media's adverse impacts on mental health. Coping mechanisms, such as digital detox, were reported, resulting in reduced anxiety and improved well-being. Social support on online platforms was identified as both a positive and negative factor, influencing

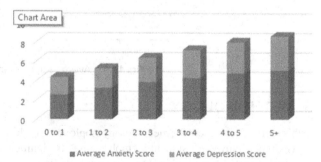

Fig. 15.7 Relationship between social media usage and anxiety/depression[14]

feelings of loneliness and mental well-being. These findings emphasize the importance of responsible social media use, highlighting the role of education and digital literacy in promoting healthier online behaviors. Further research is warranted to explore the effectiveness of digital well-being interventions and strategies for mitigating adverse outcomes associated with excessive social media utilization.

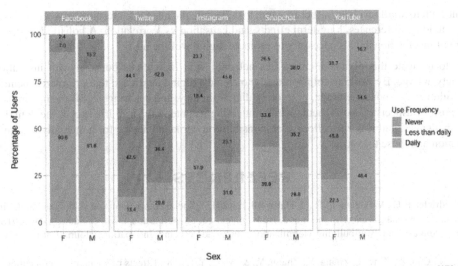

Fig. 15.8 The mental health and well-being profile of young adults using social media[15]

5. DISCUSSION

Our research has illuminated significant insights into the intricate relationship between social media usage and the mental health of young adults, prompting several noteworthy implications. Firstly, there is a need for further research to gain a nuanced understanding of how various platforms, usage patterns, and content types influence mental well-being. Longitudinal studies can offer valuable insights into the enduring effects of social media use, identifying trends and potential interventions over time. [16]

Recognizing the complexity of the relationship is crucial. While our study identifies correlations, establishing causation is challenging, as social media usage may both exacerbate existing mental health issues and serve as a coping mechanism or source of support. Individual differences play a role, with factors such as personality traits, coping strategies, and offline social support networks moderating the association between social media usage and mental health.

Addressing concerns about social comparison, social isolation, and screen time management is imperative. Encouraging media literacy and critical thinking can help users navigate social comparison, promoting digital spaces that foster genuine connections to combat feelings of isolation, and developing tools for healthy screen time management.[17]

The multifaceted nature of Fear of Missing Out (FoMO) and its impact on individuals call for increased awareness and education. Empowering individuals to recognize and manage FoMO and encouraging a healthy balance between online and offline experiences can alleviate the pressure to constantly engage with social media.[18]

Looking ahead, future interventions and support mechanisms are essential. Implementing digital well-being programs and features that promote responsible usage, self-reflection, and mental health support can contribute to a healthier online environment. Additionally, ensuring easy access to mental health resources within digital spaces can provide a safety net for individuals facing mental health challenges exacerbated by social media.[19]

6. CONCLUSION

In this comprehensive study, we have explored the intricate connection between social media usage and the mental well-being of university students, revealing the multifaceted dynamics inherent in this digital era. The digital dilemma is highlighted by the balancing act required to navigate the benefits and drawbacks of social media. While these platforms offer valuable opportunities for connection, information sharing, and self-expression, our findings emphasize potential drawbacks, including increased symptoms of anxiety and depression associated with excessive usage.

Encouraging responsible social media usage is crucial, and initiatives such as media literacy and digital well-being programs can empower individuals to navigate the digital landscape more adeptly. Recognizing the importance of mental health, ensuring accessible resources within digital spaces, and fostering a holistic approach are essential for promoting digital well-being.

As we conclude, the path forward involves further research to delve into the nuanced interplay between social media and mental health, exploring individual differences, long-term effects, and causation vs. correlation. A holistic approach, considering the multifaceted nature of this relationship, is necessary to promote digital well-being.

The responsibility to navigate this digital dilemma lies not only with individuals but also with institutions, social media platforms, and society at large. By working collaboratively to create a digital landscape that prioritizes mental health and well-being, fostering a culture of responsible usage, promoting digital literacy, and ensuring access to support, we can help young adults harness the positive aspects of social media while mitigating its potential harm. In the end, the digital age presents both opportunities and challenges, and it is our collective responsibility to strike a balance for a mentally healthy digital world for the younger generation and those to come.

REFERENCES

1. Fardouly, J., Diedrichs, P. C., Vartanian, L. R., & Halliwell, E. (2015). Social comparisons on social media: the impact of Facebook on young women's body image concerns and mood. Body image, 13, 38–45.Fardousi, M., & Sadeghi, H. (2018). The relationship between social media use and cyberbullying victimization among adolescents. International Journal of Cyber Criminology, 13(1), 1-16.

2. Gao, Y., Zhang, R., Cao, Z., Yang, J., Zhang, Y., Zhang, Y., & Wang, Y. (2018). Effects of social media on college students' anxiety: A social media usage questionnaire study. Journal of Addictions Nursing, 29(3), 183-189.

3. Hawi, N. S., & Samaha, M. (2017). The Relations Among Social Media Addiction, Self- Esteem, and Life Satisfaction in University Students. Social Science Computer Review, 35(5), 576–586.

4. Hunt, M. G., Marx, R., Lipson, C., & Young, J. (2018). No more FOMO: Limiting social media decreases loneliness and depression. Journal of Social and Clinical Psychology, 37(10), 751- 768.

5. Huang C. (2017). Time Spent on Social Network Sites and Psychological Well-Being: A Meta- Analysis. Cyberpsychology, behavior and social networking, 20(6), 346–354.

6. Kross, E., Verduyn, P., Demiralp, E., Park, J., Lee, D. S., Lin, N., Shablack, H., Jonides, J., & Ybarra, O. (2013). Facebook use predicts declines in subjective well-being in young adults. PloS one, 8(8), e69841.

7. Kroenke, K., Spitzer, R. L., & Williams, J. B. (2001). The PHQ-9: validity of a brief depression severity measure.

8. Kuss, D. J., Griffiths, M. D., Karila, L., Billieux, J., Pontes, H. M., & Van Rooij, A. J. (2014). Internet gaming disorder and social media addiction: A prospective analysis. Journal of Behavioral Addictions, 3(3), 112-118.

9. Primack, B. A., Shensa, A., Sidani, J. E., Whaite, E. O., Lin, L. Y., Rosen, D., ... & Miller, E. (2018). Social media use and depression in young adults: A longitudinal study. American Journal of Preventive Medicine, 54(2), 188-195.

10. Primack, B. A., Shensa, A., Sidani, J. E., Whaite, E. O., Lin, L. Y., Rosen, D., Colditz, J. B., Radovic, A., & Miller, E. (2017). Social Media Use and Perceived Social Isolation Among Young Adults in the U.S. American journal of preventive medicine, 53(1), 1–8.

11. Przybylski, A. K., Murayama, K., DeHaan, C. R., & Gladwell, V. (2013). Motivational, emotional, and behavioral correlates of fear of missing out. Computers in Human Behavior, 29(4), 1841-1848.

12. Spitzer, R. L., Kroenke, K., Williams, J. B., & Löwe, B. (2006). A brief measure for assessing generalized anxiety disorder: The GAD-7.

13. Steers, M. L., Wickham, R. E., & Messman, H. J. (2015). Social media and adolescent self- image: The mediating role of social comparison. Journal of Adolescent Research, 30(5), 663- 680.

14. Twenge, J. M. (2017). iGen: Why Today's Super-Connected Kids Are Growing Up Less Rebellious, More Tolerant, Less Happy— and Completely Unprepared for Adulthood—and What That Means for the Rest of Us. Atria Books.

15. Twenge, J. M., & Campbell, W. K. (2018). Associations between screen time and lower psychological well-being among children and adolescents: Evidence from a population-based study.

16. Twenge, J. M., Joiner, T. E., Rogers, M. L., & Martin, G. N. (2019). Increased depressive symptoms, suicide-related outcomes, and suicide rates among US adolescents after 2010 and links to increased new media screen time.

17. Wang, J. L., Wang, H. Z., Gaskin, J., & Hawk, S. (2017). The mediating roles of upward social comparison and self-esteem and the moderating role of social comparison orientation in the association between social networking site usage and subjective well-being.

Computational Intelligence and Mathematical Applications – Dr. Devendra Prasad et al. (eds)
© 2024 Taylor & Francis Group, London, ISBN 978-1-032-87721-1

16

IoT-Blockchain Integration for Optimizing the Performance of E-Health Systems: A Framework Design

Deepak Singla[1], Sanjeev Kumar[2]
Department of Computer Science and Engineering, Maharishi, Markandeshwar
(Deemed to be University), Mullana, Ambala (Haryana), India

Shipra Goel
Department of Computer and Application, DAV (PG) College, Karnal, India

Abstract: In the era of the Internet of Things (IoT), e-health has garnered significant attention among researchers. However, patient privacy has become a paramount concern due to the sensitivity of medical information. Presently, patient data predominantly resides in cloud-based healthcare systems, leaving users with limited control over their personal data. The escalating number of security breaches and privacy violations underscores the vulnerabilities of the traditional model, where third parties accumulate and manage vast troves of healthcare data. This study addresses these challenges by leveraging blockchain technology. We present a blockchain-based protocol designed specifically for e-health applications, which eliminates the requirement for third-party trust while providing an effective privacy-preserving access control mechanism. Unlike Bitcoin, our system's transactions are not solely financial, and we depart from conventional consensus methods like Proof of Work (PoW), recognizing their incompatibility with IoT device limitations. Our approach enhances network security, efficiency, and cost-effectiveness. Lastly, we provide a thorough security and privacy analysis, aiming to reshape the landscape of healthcare data management and security in the IoT-driven e-health domain.

Keywords: Internet of things, Blockchain technology, E-Healthcare, Privacy and security, Cryptography

1. INTRODUCTION

Numerous societies across the globe are grappling with a substantial surge in the number of patients seeking medical attention, leading to growing challenges in accessing primary healthcare professionals. The advent of the Internet of Things (IoT) has opened up the possibility for devices to seamlessly connect with each other and the internet, irrespective of time and location. Researchers anticipate that over 75 billion devices will join this interconnected ecosystem by 2025 [1]. Fortunately, within this landscape, e-health emerges as a prominent application of IoT, holding the promise of addressing these pressing healthcare challenges. IoT, in its broadest sense, describes a network of billions of interconnected, connected devices that can collect, store, and share data on a worldwide scale. Smartphones, vehicles, wearables, and other electronic devices with built-in sensors, CPUs, and network connectivity might all be considered among these gadgets [2]. Medical gadgets and electronic wearables, in particular, are often used IoT devices in the healthcare sector. IoT is used for a variety of purposes, including geriatric care, monitoring chronic patient care, and emergency situations [3]. E-health offers a potential solution by enabling medical practitioners to efficiently attend to a larger patient population while enhancing convenience for those in need. Patients

Corresponding authors: [1]deepaksingla.cse@piet.co.in, [2]dr.sanjeevrana@mmumullana.org

DOI: 10.1201/9781003534112-16

can maintain continuous connectivity with health-care providers as necessary, fostering a more patient-centered approach to healthcare delivery. Moreover, this innovative approach has the potential to curtail medical expenses while concurrently elevating the standard of care and treatment [4]. Recent advancements in IoT technology have paved the way for the integration of electronic wearable devices into e-health solutions[5]. Vulnerabilities abound, with potential security breaches enabling data tracking, unauthorized alterations, or theft by malicious actors, thus compromising patient information security.

2. BLOCKCHAIN

2.1 Blockchain Overview

Satoshi Nakamoto, the person who invented the first cryptocurrency known as Bitcoin, initially brings up blockchain technology [6]. Blockchain technology allows for the storing and sending of transactions. It keeps the information in a block-based ledger. A chain of blocks is created by each block attaching to the one before it. Peer-to-peer networks enable data transfer. Blockchain is a distributed ledger that may be shared in a secure, independent manner as a result. The banking and finance industries have been quite interested in blockchain during the past few years. It is presently utilized in a wide range of industries, including as insurance, energy, business, and healthcare. These characteristics help explain why blockchain is so well-liked across practically all businesses.

2.2 Blockchain Security

Blockchain security is supported by a number of factors, including:

Cryptography

Blockchains frequently leverage asymmetric cryptography, also commonly referred to as public-key cryptography. The security functionalities covered by this cryptographic technique include key creation, encryption, decryption, digital signatures, and identity verification [7]. Asymmetric cryptography is essential to the establishment of a key pair, which consists of a public key and a private key, in the context of blockchain technology. Using the private key, a personal digital signature is applied to confirm the sender's identity and the legitimacy of the account (See Fig. 16.1).

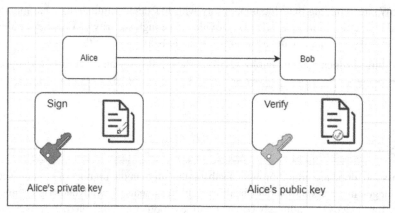

Fig. 16.1 Alice signs her mails with her private key using a public key digital signature mechanism. By utilizing Alice's public key, Bob may confirm that Alice signed the message

2.3 IPFS

InterPlanetary File System or IPFS for short [8], functions as a peer-to-peer file system over a network. The transition from location-based to content-based addressing is a key improvement in IPFS. We now utilize the data's hash to access it rather than the location's address. Each file kept in IPFS is given a distinct hash, assuring accurate retrieval. Consequently, all you need to do to locate this file is look up its hash (See Fig. 16.2).

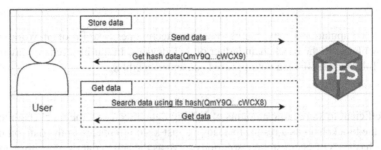

Fig. 16.2   Illustrate how to send/receive data in IPFS

3. RELATED WORKS

In order to address concerns about the privacy of personal data, researchers have looked into a variety of strategies [9]. One method for safeguarding personally identifiable in- formation is data anonymization. In anonymized datasets, the k-anonymity technique, for instance, ensures that each recorded data point cannot be distinguished from at least k-1 other recorded data points. Recent studies, however, have demonstrated that anonymized datasets have security flaws that may be exploited with very little information, therefore losing their anonymity. [10].

In situations where users must securely communicate data hosted on outside cloud servers, attribute-based encryption (ABE) is a useful method for maintaining data privacy and granular access control [11]. However, for system setup and secret key distribution to users, the majority of ABE schemes rely on a reliable private key generator (PKG). The disadvantage of such schemes is that the PKG has the power to decode any piece of information kept on the cloud server, which could result in problems like key misuse and privacy violations.

Furthermore, traditional cloud storage models use centralized storage, which creates a single point of failure that could endanger the stability of the entire system. Fully homomorphic encryption (FHE), for example, allows computations and queries on encrypted data [12], but it currently has practical drawbacks that prevent its widespread adoption.

Decentralised cryptocurrency systems have completely changed how transactions are conducted in recent years. As a leading example, Bitcoin uses blockchain technology to enable safe peer-to-peer exchanges and money transfers (bitcoins) without the need for middlemen. Since blocks of data are shared among all network nodes and act as an immutable ledger at their core, blockchain effectively does away with the necessity for centralised authorities. Blockchain technology's distributed nature has found use in a variety of sectors outside of finance.

Blockchain is now a key component of financial transactions [13] and has the potential to be used for decentralised Internet of Things (IoT) systems, identity-based Public Key Infrastructure (PKI), supply chain management, document existence verification, and decentralised storage solutions. The adaptability of this technology promises a paradigm change in many fields, offering improved security, trans- parency, and autonomy in a society that is becoming more interconnected.

In research referred to as [14], the development of a personal data access control and management system using blockchain technology with an emphasis on user privacy. Their original approach merged blockchain and off-chain storage methods, enabling the secure archiving of encrypted data outside the blockchain ledger. This technique saved pointers on the blockchain, providing a secure and reliable mechanism of data access.

By fusing IoT with blockchain, this research presents a cutting-edge framework for managing healthcare data while guaranteeing strong privacy and security protections.

4. OBSTACLES AND OUR RESOLUTIONS

In IoT-based e-health applications, it is crucial to guarantee the safe transfer and protection of patient privacy. As a result of its intrinsic decentralization and features like immutability and transparency, blockchain is ideally suited to address these issues. However, there are particular difficulties in integrating blockchain into IoT. To close the gap between blockchain technology and IoT-based e-health applications, we explore these issues in this context and explain our creative solutions.

4.1 Scalability

It is challenging to execute computationally expensive consensus techniques, like Proof-of-Work (PoW), owing to resource limitations, for adding new blocks to the blockchain ledger on IoT devices. In our blockchain network, we replace PoW with Practical Byzantine Fault Tolerance (PBFT), a voting-based consensus method.

4.2 Data Storage

It is neither practical nor efficient to store large amounts of IoT big data directly on the blockchain ledger. Our method avoids keeping the data itself on the blockchain as a result. Instead, we just keep references to the data on the blockchain, usually in the form of encrypted data hashes. Off-chain storage solutions serve as the actual data's repository and are securely protected.

4.3 Security of Data

Because patient health information is sensitive, we use symmetric key encryption as a strong security safeguard. Before being sent to the blockchain network, the data is first encrypted using a symmetric key encryption algorithm. This method guarantees that the data is safely saved in an encrypted manner even during off-chain storage, protecting patient confidentiality and data integrity.

4.4 Patients Privacy

This paper's primary focus is safeguarding patient privacy, given the high sensitivity of healthcare data. We operate under the assumption that medical personnel are honest yet inquisitive, adhering to the protocol. Our system allows patients to maintain a level of pseudonymous anonymity while storing medical staff profiles on the blockchain, enabling patients to verify the identities of medical personnel they interact with. Our proposed platform fulfills the following criteria:

- *Patient Health Information Ownership:* Patients exclusively retain ownership and control of their medical Information on our platform. In this setting, patients take on the crucial role of healthcare data owners while medical staff is classified as service providers and granted specific permissions.

- *Precise Access Control:* Patients have the ability to assign specific permissions to authorize medical personnel for accessing their healthcare data. They retain the flexibility to modify or revoke these access permissions at their discretion. These permissions are securely recorded as access-control policies on the blockchain ledger, with exclusive control residing in the hands of the patient for any alterations or revocations.

- *Data Auditability and Transparency:* Patients are afforded complete transparency and accuracy concerning their healthcare data, with a clear understanding of how medical personnel can access this information.

5. PROPOSED MODEL

Figure 16.3 show the main components of our system model, which are as follows:

- IoT clothing gadgets.
- The patient's mobile device.
- Medical personnel.
- The blockchain.
- Off-chain archiving.

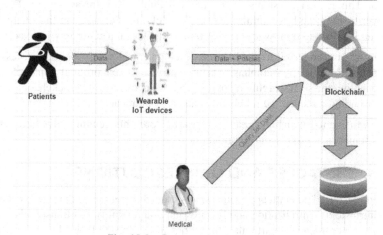

Fig. 16.3 Proposed system model

5.1 Wearable IoT Devices

The wearable IoT devices will gather all bodily healthcare data from the patient, such as blood pressure, body temperature, and heart rate [15]. These gadgets have limited energy, computation, and storage capabilities, among other resource deficiencies. Wearable IoT devices take these constraints into account and use short-range communication technologies like Bluetooth or Zigbee to deliver the healthcare data they have gathered to the patient's Smartphone.

5.2 Smartphone of the Patient

Smartphones outperform IoT devices due to their superior storage capacity, extended battery life, and enhanced processing capabilities. This added potency enables smartphones to handle intricate tasks, including computational and cryptographic operations. Moreover, smartphones facilitate data transmission through long-range communication channels, like cellular networks, effectively serving as gateways that empower patients to engage with the blockchain network seamlessly.

5.3 Medical Staff

Healthcare professionals, comprising doctors and nurses, must access patients' healthcare data to assess their medical condition and administer appropriate treatment based on the analysis of this information. This essential process ensures that patients receive the necessary care and interventions tailored to their specific health needs.

5.4 Blockchain

Our research eliminates the need for third-party intermediaries by using blockchain technology to store access policies. This strategy protects both the availability and integrity of patient data by fortifying the network against potential Denial of Service (DoS) assaults and single points of failure. Notably, our system refrains from storing the actual data on the blockchain. Instead, it retains only pointers to the data in the form of encrypted data hashes.

5.5 Off-Chain Storage

We place a high priority on patient privacy and use off-chain storage as a reliable technique for protecting encrypted patient data. The off-chain storage alternative we've selected is the InterPlanetary File System (IPFS), a peer-to- peer distributed file system designed to unite all computing devices into a single file system. IPFS makes use of content- addressed hyperlinks and high-throughput content-addressed block storage [21]. Our solution frees us from relying on a single central server and offers a definite advantage over traditional cloud storage systems by spreading and storing data over several computers throughout the world.

5.6 Privacy and Security Analysis

The efficiency of our suggested procedure in terms of security and privacy will be examined in this section. There are three essential security requirements that demand consideration when creating security measures for any model: the CIA paradigm, also known as confidentiality, integrity, and availability [22]. Only users who are authorized and have the necessary access permissions can access the system's communications thanks to confidentiality. We now present a summary of the evaluation of the primary security requirements in Table 16.1.

Table 16.1 Analysis of security requirements

Parameters	Recommended Solution
Confidentiality	Attained through the application of symmetric key encryption
Integrity	Hashing in blockchain serves the purpose of Ensuring data integrity.
Availability	Attained through the imposition of constraints on permissible transactions within the network.
Authorization	Utilizing digital signatures for the purpose of establishing authorization.

To protect privacy, we give data ownership top priority in our system. As a result, only users (patients) have the power to manage their data. Because blockchain technology is decentralized and digital signatures are used to authenticate network transactions, an adversary cannot compromise the system by pretending to be a user. By putting access-control rules on a blockchain ledger, where only the patient has the capacity to change or revoke them, we are able to implement fine-grained access control. In Fig. 16.4, we compare and contrast our suggested model with existing systems.

6. CONCLUSION

Healthcare data belonging to patients is highly sensitive in terms of both privacy and security. Consequently, it is unwise to rely on third-party entities for the management of such data, as these entities are susceptible to various forms of attacks and misuse. This study suggests a novel solution that makes use of blockchain and the Internet of Things (IoT). The urgent privacy and security issues relating to patient healthcare data in the context of e-health are what spurred our suggestion. By giving

Model Name & Property	Wang [4]	Young [11]	Zhao [12]	Bakar [17]	Moin [20]	Dorri [22]	Our Proposed Model
Authorization	T	T	T	T	T	T	T
Distributed Ledger-Based	F	T	T	T	F	T	T
Confidentiality Maintaining	T	T	T	T	T	T	T
IPFS Data Repository	F	F	F	F	F	F	T

Fig. 16.4 Contrasting various established methods

patients full ownership and control over the private health information collected by their IoT wearables, our platform would empower patients. Patients are given the tools to specify access control settings within a blockchain ledger in order to do this. Which members of the medical team have access to the patients' data is governed by these policies. In order to reduce the storage demand on the blockchain, we use off-blockchain storage in this platform. Additionally, in order to handle the resource constraints present in IoT environments, we deploy a suitable consensus method within the blockchain network.

REFERENCES

1. P. McGorry, T. Bates, and M. Birchwood, "Designing youth mental health services for the 21st century: examples from australia, ireland and the uk," The British Journal of Psychiatry, vol. 202, no. s54, pp. s30–s35, 2013.
2. A. Papa, M. Mital, P. Pisano, and M. Del Giudice, "E-health and wellbeing monitoring using smart healthcare devices: An empirical investigation," Technological Forecasting and Social Change, vol. 153, p. 119226, 2020.
3. D. Minoli and B. Occhiogrosso, "Internet of things applications for smart cities," Internet of things A to Z: technologies and applications, pp. 319– 358, 2018.
4. X. Wang, Z. Liu, and T. Zhang, "Flexible sensing electronics for wear- able/attachable health monitoring," Small, vol. 13, no. 25, p. 1602790, 2017.
5. B. Dai, C. Gao, and Y. Xie, "Flexible wearable devices for intelligent health monitoring," View, vol. 3, no. 5, p. 20220027, 2022.
6. A. Bora, M. R. Kothari, J. Chandhok, and S. P. Kurlekar, Introduction To Internet Of Things And Its Application. AG PUBLISHING HOUSE (AGPH Books).
7. D. Singla and S. K. Rana, "A systematic review on blockchain-based e- healthcare for collaborative environment," in International Conference on Emergent Converging Technologies and Biomedical Systems, pp. 361– 376, Springer, 2022.
8. D. Vujic̆ić, D. Jagodić, and S. Randić, "Blockchain technology, bitcoin, and ethereum: A brief overview," in 2018 17th international symposium infoteh-jahorina (infoteh), pp. 1–6, IEEE, 2018.
9. K. A. Aravind, B. R. Naik, and C. S. Chennarao, "Combined digital signature with sha hashing technique-based secure system: An application of blockchain using iot," Turkish Journal of Computer and Mathematics Education (TURCOMAT), vol. 13, no. 03, pp. 402–418, 2022.
10. T. V. Doan, Y. Psaras, J. Ott, and V. Bajpai, "Towards decentralised cloud storage with ipfs: Opportunities, challenges, and future directions," arXiv preprint arXiv:2202.06315, 2022.
11. A. L. Young and A. Quan-Haase, "Privacy protection strategies on face- book: The internet privacy paradox revisited," Information, Communica- tion & Society, vol. 16, no. 4, pp. 479–500, 2013.
12. Y. Zhao and J. Chen, "A survey on differential privacy for unstructured data content," ACM Computing Surveys (CSUR), vol. 54, no. 10s, pp. 1– 28, 2022.
13. P. K. Bhansali, D. Hiran, H. Kothari, and K. Gulati, "Cloud-based secure data storage and access control for internet of medical things using federated learning," International Journal of Pervasive Computing and Communications, no. ahead-of-print, 2022.
14. D. Singla and S. Rana, "Blockchain-based platform for smart tracking and tracing the pharmaceutical drug supply chain," in Examining Multimedia Forensics and Content Integrity, pp. 144–172, IGI Global, 2023.

Note: All the figures and table in this chapter were made by the author.

Computational Intelligence and Mathematical Applications – Dr. Devendra Prasad et al. (eds)
© 2024 Taylor & Francis Group, London, ISBN 978-1-032-87721-1

17

Issues Related to Accuracy and Performance during Classification Process in Sentiment Extraction from Image Description

Puneet Sharma[1], Sanjay Kumar Malik[2]

USIC&T, GGS Indraprastha University, New Delhi

Abstract: The task of sentiment extraction from visual descriptions encompasses several fields of study, including computer vision, natural language processing, and machine learning. The present work examines the significant concerns pertaining to the precision and efficacy in sentiment categorization while analyzing textual descriptions linked to photographs. The main aim of this study is to establish reliable approaches and models that can accurately perceive and comprehend the emotional tone or mood conveyed in picture descriptions. This study makes a valuable contribution to the field of sentiment analysis in the context of picture descriptions. The findings have the potential to enhance many applications such as social media analysis, e-commerce, and customer feedback analysis. The primary objective of this research is to enhance the proficiency of AI systems in comprehending and appropriately reacting touser emotions and feelings by addressing concerns related to accuracy and performance.

Keywords: Sentiment analysis, Image processing, Image classification, Machine learning

1. INTRODUCTION

The technique of extracting sentiment from picture descriptions is a multifaceted one that encompasses many crucial stages. Initially, it is necessary to extract the picture description, often in textual format, from the image (Tul *et al.* 2017). This task may be accomplished via the use of Optical Character Recognition (OCR) or other Natural Language Processing (NLP) approaches. After acquiring the text, the process of emotion extraction commences (Ahmed and Boudhir 2018). The subsequent stage is text pre-processing, including various activities such as tokenization, converting to lowercase, and eliminating stop words and punctuation marks. This helps clean the text data and prepare it for sentiment analysis. Sentiment analysis is the core of the process, where machine learning models or natural language processing techniques are employed to classify sentiment expressed in the text (Oliveira *et al.* 2020). This sentiment can typically be categorized into positive, negative, or neutral, or further refined into more nuanced emotions. The choice of sentiment analysis approach depends on the specific requirements of the task, such as whether fine-grained sentiment analysis is needed or a simple positive/negative classification suffices (Chandra *et al.* 2021). Accuracy, precision, recall, and F1-score are just few of the measures used to gauge how well a sentiment extraction model can detect emotions conveyed in picture captions (Narayanasamy *et al.* 2022). The classification process in sentiment extraction from image descriptions involves text extraction, pre- processing, sentiment analysis, and evaluation. It enables the interpretation of emotional content within textual descriptions associated with images (Hota *et al.* 2018), providing valuable insights into the sentiments and emotions expressed in visual content. Particularly important in the fields of computer vision and NLP (Zainuddin *et al.* 2018), is research on the precision and efficiency of sentiment extraction from visual descriptions. The process of extracting sentiment from image captions involves analyzing the linguistic expressions individuals use to describe

---

[1]puneetsharma2139@gmail.com, [2]skmalik@ipu.ac.in

DOI: 10.1201/9781003534112-17

visual content in order to determine their emotional states. The following are few fundamental elements of study within this particular field:

1. The evaluation of sentiment extraction algorithms is crucial in determining their accuracy and precision. The research may be oriented towards the development and enhancement of sentiment analysis models, with the objective of ensuring their ability to effectively capture the intended sentiment conveyed in picture descriptions. This encompasses the resolution of matters such as sarcasm, context, and ambiguity (Asghar *et al.* 2018).

2. Multimodal analysis is the integration of verbal and visual components inside image descriptions. The study may investigate methods for efficiently incorporating different modalities in order to improve the accuracy of sentiment extraction.

   The integration of textual and visual elements has the potential to enhance the comprehension of emotion, as shown by previous research (Alsaeedi and Khan 2019).

3. The exploration of sophisticated deep learning architectures, such as CNNs (Pandey *et al.* 2020) for image processing, as wellas RNNs (Liu and Cocea 2017), is being investigated. In the context of textual analysis, the present inquiry pertains to the examination and interpretation of written or spoken communication. The research might be directed towards the optimization of the model architecture in order to enhance the performance of sentiment analysis.

4. The investigation should focus on issues pertaining to labelled datasets. The development of sentiment analysis datasets that possess high quality, diversity, and representativeness is of utmost importance (Abo *et al.* 2018). In addition, it might be advantageous to investigate transfer learning approaches for the purpose of adapting models that have been trained on different sentiment analysis tasks (Bansal *et al.* 2019).

5. The analysis of sentiment is subject to substantial variation depending on the specific context in which it is articulated. The research may explore methodologies that take into account the contextual factors in order to enhance the precision of sentiment extraction from descriptions of images. The process may include a comprehensive analysis of the whole of the visual description, rather than focusing just on isolated lines or phrases (Ahmed *et al.* 2022).

The scope of sentiment analysis may be expanded to include the identification of emotions. The study aims to investigate methods for discerning and differentiating diverse emotions conveyed in visual descriptions, hence enhancing comprehension of sentiment in a more nuanced manner (Ahmed *et al.* 2022).

1. Real-time processing plays a critical role in applications such as social media or online reviews, where the timely analysis of sentiment is of utmost importance. Research can focus on developing efficient and fast sentiment extraction models that can process image descriptions in real-time, ensuring timely responses (Zhang *et al.* 2022).

2. *Ethical and Bias Considerations:* Address ethical issues related to sentiment extraction, such as bias and fairness. Ensure that the models do not perpetuate stereotypes or biases and are sensitive to the cultural and linguistic diversity of users (Reyero Lobo *et al.* 2023).

3. *Evaluation Metrics:* Develop and standardize evaluation metrics specific to sentiment extraction from image descriptions. Traditional metrics like accuracy, precision, recall, and F1 score may need to be adapted to this context.

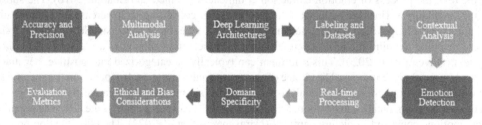

Fig. 17.1 Key aspects of research in domain (ahmed et al.2022)

Research in sentiment extraction from image descriptions that focuses on accuracy and performance is pivotal for improving the capabilities of AI systems (Hartmann *et al.* 2023) in understanding and responding to user emotions and sentiments, with potential applications in areas like social media analysis, e-commerce, and customer feedback analysis.

1.1 Challenges

Sentiment extraction from image descriptions using deep learning models presents several challenges related to accuracy and performance (Gaikwad *et al.* n.d., Cui *et al.* 2023). Some of the key issues include:

1. *Multimodal Fusion:* Integrating image and text information effectively is a complex problem. Deep learning models need to combine visual and textual features to make accurate sentiment predictions. The choice of fusion techniques and the model's ability to understand the relationships between these modalities can impact accuracy.

2. *Context and Ambiguity:* Understanding sentiment often requires context analysis. Sentiment can change based on the overall context of the image and description. Ambiguity in language can also lead to misinterpretations (Ray and Chakrabarti 2022). Deep learning models need to handle these nuances effectively.

3. *Training Time:* Deep learning models, especially large neural networks, can have extended training times. Reducing the time required for model training while maintaining accuracy is a critical performance challenge (Abdullah and Ahmet 2023).

4. *Overfitting:* Deep learning models are prone to overfitting, especially when the dataset is small or unrepresentative (Alzahrani *et al.* 2022). Addressing overfitting issues to improve model generalization is crucial for accurate sentiment extraction (Rodrigues *et al.* 2022).

To address these issues, ongoing research focuses on developing novel architectures, improved data collection and annotation techniques, advanced fusion methods, and model interpretability. Achieving a balance between accuracy and performance in deep learning-based sentiment extraction from image descriptions is a complex but essential endeavor for the successful implementation of such systems in real- world applications (Radiuk *et al.* 2022, Revathy *et al.* 2022).

2. PROBLEM STATEMENT

Sentiment extraction from image descriptions is a challenging task with several problems and issues related to both accuracy and performance.

- One of the primary concerns is the inherent complexity of the task itself.
- Unlike textual sentiment analysis, image description sentiment extraction combines visual and textual modalities, which can make it challenging for models to accurately capture & interpret the sentiment conveyed.
- Context, ambiguity, and nuances in language can lead to misinterpretations and affect the overall accuracy of sentiment analysis.
- Another significant issue is the scarcity of high-quality labeled datasets specific to image description sentiment analysis.

The choice of model architecture is a critical factor. Deep learning models, while powerful, can be computationally intensive and may require large volumes of training data to perform well Overfitting occurs when a model is too complicated, and as a result, it memorizes the training data instead of learning to generalize from it.

3. FACTORS INFLUENCING ACCURACY DURING IMAGE CLASSIFICATION IN DEEP LEARNING MODEL

Achieving high accuracy during image classification in deep learning models is a crucial goal. Several factors influence the accuracy of these models. Here are the key factors:

a. **Quality and Quantity of Data:** The quality and quantity of data are crucial factors influencing the accuracy of image classification in deep learning models. High-quality data ensures that the images used for training and evaluation is accurate, clearly labelled, and indicative of the kinds of situations the model may face in the actual world.

b. **Data Preprocessing:** Data preprocessing plays a pivotal role in influencing accuracy of image classification in deep learning models. It encompasses a series of essential steps to prepare raw image data for training. Proper data preprocessing can significantly enhance model performance.

c. **Model Architecture:** The choice of model architecture is a fundamental factor influencing the accuracy of image classification in deep learning models.

d. **Hyper-parameters Tuning:** Hyper-parameter tuning is a critical factor that significantly influences the accuracy of image classification in deep learning models. Hyperparameters are the settings and configurations that are not learned from the data but must be specified before training the model. They impact the model's learning process and its ability to generalize to new images.

e. **Transfer Learning:** Image categorization accuracy in DL models may be significantly impacted by transfer learning. It involves leveraging knowledge gained from pre-trained models on large datasets, typically from a different but related task, and applying this knowledge to a new image classification task. Transfer learning offers several advantages, impacting the model's performance positively. Leveraging pre-trained models and fine-tuning them for specific tasks can significantly enhance performance, particularly when working with limited data.

f. **Regularization Techniques:** It has a significant impact on how well deep learning models classify images. Overfitting is a significant issue in complicated models that may negatively affect the model's performance on unknown data, and these methods aim to mitigate this. Regularization approaches can enhance accuracy by limiting the model's potential and making it more robust to noise. Regularization methods, including L1 and L2 regularization, may be used to mitigate overfitting and improve precision.

g. **Data Balancing:** Image categorization accuracy in deep learning models may be greatly impacted by how well data is balanced. Data balance, when discussing picture classification, refers to the even distribution of class labels throughout the dataset used for training. Some classes may have much fewer samples than others, which may negatively impact accuracy when working with unbalanced datasets.

Poor results on minority classes may emerge from an imbalanced data set's tendency to favor the dominant class. As a result, the model may have trouble properly identifying and classifying the under- represented groups. Additionally, techniques like weighted loss functions or resampling strategies can assign higher importance to minority class samples during training. This ensures that the model pays more attention to these classes and results in a more balanced learning process as shown in Fig. 17.2.

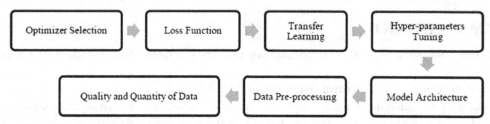

Fig. 17.2 Factors influencing accuracy during image classification in deep learning model

Source: Made by the author

Optimizing these factors requires careful experimentation and a deep understanding of the problem at hand. By systematically considering and fine-tuning these aspects, you can enhance the accuracy of image classification models in deep learning.

4. PERFORMANCE FACTORS INFLUENCE ON IMAGE CLASSIFICATION IN DEEP LEARNING MODEL

Performance in image classification using deep learning models is influenced by a wide range of factors. Understanding these factors and optimizing them is essential for achieving accurate and efficient image classification. Several performance factors play a significant role in influencing the success of image classification in deep learning models. The choice of model architecture is pivotal, with convolutional neural networks (CNNs) being a popular choice due to their ability to automatically extract relevant features from images. The quality and quantity of training data are equally important, as diverse, well-labeled data is essential for training a robust model. Hyperparameter tuning, including learning rates, batch sizes, and the number of training epochs, directly impacts the model's ability to converge effectively and avoid overfitting.

Selecting an appropriate loss function, regularization techniques like dropout and L2 regularization, and the optimizer are critical for guiding the learning process. Transfer learning, where pre-trained models are adapted for specific tasks, can significantly boost performance, especially when data is limited. Data balancing techniques address imbalanced datasets, ensuring that the model learns from all classes equally. Hardware and computer resources, such as GPUs or TPUs, can expedite training and inference.

5. PROCESS OF DATA COLLECTION

Data collection is made using two methods. First method is primary data collection where data is collected by any communication session at run time. Second method is to collect data from secondary data source such as twitter, Kaggle. In case of present research data has been obtained from secondary source.

6. EXPERIMENTAL SETUP

In present experiment, python environment has been configured in two ways. First is local platform where python is installed on system and Jupiter notebook is also set. But such configuration is complex for laymen thus second method has been used where Google Collaboratory is used that requires less technical skill. In present research Collaboratory has been used for simulation.

7. SCOPE OF RESEARCH

The scope of considering issues related to accuracy and performance during the classification process for sentiment extraction from image descriptions using deep learning models is vast and holds significant potential for various applications and research areas. Here are some aspects of the scope in this domain:

a. **Improved Sentiment Analysis Models:** Developing and fine-tuning deep learning models for sentiment extraction from image descriptions is an ongoing scope. Research can focus on creating more accurate and efficient models, leveraging advancements in neural network architectures and training techniques.

b. **Multimodal Sentiment Analysis:** The scope includes exploring innovative methods for effective fusion of textual and visual data, ensuring that deep learning models can leverage both modalities to enhance sentiment analysis accuracy.

8. CONCLUSION

Deep learning model picture categorization accuracy depends on data quality and quantity. A balanced mix of both is needed to train models that can categorize photos in real-world scenarios. To improve picture quality, preprocessing might involve histogram equalization and de noising. By eliminating distractions like borders and annotations, picture accuracy may increase. In conclusion, data pretreatment is crucial to deep learning picture categorization accuracy. Properly prepared data helps the model learn, generalize to new pictures, and classify more accurately. Another crucial model architecture element is transfer learning. Pretrained models like ImageNet can be fine-tuned for picture categorization. This method transfers knowledge from one area to another, improving accuracy when data is scarce.

REFERENCES

1. Abdullah, T. and Ahmet, A., 2023. Deep Learning in Sentiment Analysis: Recent Architectures. *ACM Computing Surveys*, 55 (8), 1–37.
2. Abo, M.E.M., Shah, N.A.K., Balakrishnan, V., and Abdelaziz, A., 2018. Sentiment analysis algorithms: Evaluation performance of the Arabic and English language. *In: 2018 International Conference on Computer, Control, Electrical, and Electronics Engineering, ICCCEEE 2018.* Institute of Electrical and Electronics Engineers Inc.
3. Ahmed, A.A.A., Agarwal, S., Kurniawan, Im.G.A., Anantadjaya, S.P.D., and Krishnan, C., 2022. Business boosting through sentiment analysis using Artificial Intelligence approach. *International Journal of System Assurance Engineering and Management*, 13, 699–709.
4. Ahmed, M. Ben and Boudhir, A.A., 2018. Innovations in Smart Cities and Applications. *Proceedings of the 2nd Mediterranean Symposium on Smart City Applications*, (January), 1–1046.
5. Alsaeedi, A. and Khan, M.Z., 2019. A study on sentiment analysis techniques of Twitter data. *International Journal of Advanced Computer Science and Applications*, 10 (2), 361–374.
6. Alzahrani, M.E., Aldhyani, T.H.H., Alsubari, S.N., Althobaiti, M.M., and Fahad, A., 2022. Developing an Intelligent System with Deep Learning Algorithms for Sentiment Analysis of E-Commerce Product Reviews. *Computational Intelligence and Neuroscience*.
7. Asghar, M.Z., Kundi, F.M., Ahmad, S., Khan, A., and Khan, F., 2018. T-SAF: Twitter sentiment analysis framework using a hybrid classification scheme. *Expert Systems*, 35 (1).
8. Bansal, A., Gupta, C.L., and Muralidhar, A., 2019. A sentimental analysis for youtube data using supervised learning approach. *International Journal of Engineering and Advanced Technology*, 8 (5), 2314–2318.
9. Chandra, S., Gourisaria, M.K., Harshvardhan, G.M., Rautaray, S.S., Pandey, M., and Mohanty, S.N., 2021. Semantic analysis of sentiments through web-mined twitter corpus. *In: CEUR Workshop Proceedings*. 122–135.
10. Cui, J., Wang, Z., Ho, S.B., and Cambria, E., 2023. Survey on sentiment analysis: evolution of research methods and topics. *Artificial Intelligence Review*, 56 (8), 8469–8510.
11. Gaikwad, R., ... S.G.I.S. and A. in, and 2023, undefined, n.d. MCNN: Visual Sentiment Analysis using Various Deep Learning Framework with Deep CNN. *ijisae.orgRS Gaikwad, SC GandageInternational Journal of Intelligent Systems and Applications in Engineering, 2023•ijisae.org*, 2023 (2s), 265–278.

12. Hartmann, J., Heitmann, M., Siebert, C., and Schamp, C., 2023. More than a Feeling: Accuracy and Application of Sentiment Analysis. *International Journal of Research in Marketing*, 40 (1), 75–87.

13. Hota, S., Technol, S.P.-I.J.E., and 2018, undefined, 2018. KNN classifier based approach for multi-class sentiment analysis of twitter data. *scholar.archive.orgS Hota, S PathakInt. J. Eng. Technol, 2018•scholar.archive.org*, 7 (3), 1372–1375.

14. Liu, H. and Cocea, M., 2017. Fuzzy rule based systems for interpretable sentiment analysis. *9th International Conference on Advanced Computational Intelligence, ICACI 2017*, 129–136.

15. Narayanasamy, S.K., Srinivasan, K., Hu, Y.C., Masilamani, S.K., and Huang, K.Y., 2022. A Contemporary Review on Utilizing Semantic Web Technologies in Healthcare, Virtual Communities, and Ontology-Based Information Processing Systems. *Electronics (Switzerland)*, 11 (3).

16. Oliveira, W.B. De, Dorini, L.B., Minetto, R., and Silva, T.H., 2020. OutdoorSent: Sentiment Analysis of Urban Outdoor Images by Using Semantic and Deep Features. *ACM Transactions on Information Systems*, 38 (3).

17. Pandey, S., Tekchandani, H., and Verma, S., 2020. A literature review on application of machine learning techniques in pancreas segmentation. *In: 2020 1st International Conference on Power, Control and Computing Technologies, ICPC2T 2020.* 401–405.

18. Radiuk, P., Pavlova, O., and Hrypynska, N., 2022. An Ensemble Machine Learning Approach for Twitter Sentiment Analysis. *In: CEUR Workshop Proceedings.* 387–397.

19. Ray, P. and Chakrabarti, A., 2022. A Mixed approach of Deep Learning method and Rule-Based method to improve Aspect Level Sentiment Analysis. *Applied Computing and Informatics*, 18 (1–2), 163–178.

20. Revathy, G., Alghamdi, S.A., Alahmari, S.M., Yonbawi, S.R., Kumar, A., and Anul Haq, M., 2022. Sentiment analysis using machine learning: Progress in the machine intelligence for data science. *Sustainable Energy Technologies and Assessments*, 53.

21. Reyero Lobo, P., Daga, E., Alani, H., and Fernandez, M., 2023. Semantic Web technologies and bias in artificial intelligence: A systematic literature review. *Semantic Web*, 14 (4), 745–770.

22. Rodrigues, A.P., Fernandes, R., Aakash, A., Abhishek, B., Shetty, A., Atul, K., Lakshmanna, K., and Shafi, R.M., 2022. Real-Time Twitter Spam Detection and Sentiment Analysis using Machine Learning and Deep Learning Techniques. *Computational Intelligence and Neuroscience*, 2022.

23. Tul, Q., Ali, M., Riaz, A., Noureen, A., Kamranz, M., Hayat, B., and Rehman, A., 2017. Sentiment Analysis Using Deep Learning Techniques: A Review. *International Journal of Advanced Computer Science and Applications*, 8 (6).

24. Zainuddin, N., Selamat, A., and Ibrahim, R., 2018. Hybrid sentiment classification on twitter aspect-based sentiment analysis. *Applied Intelligence*, 48 (5), 1218–1232.

25. Zhang, H., Shi, H., Hou, J., Xiong, Q., and Ji, D., 2022. Image Sentiment Analysis via Active Sample Refinement and Cluster Correlation Mining. *Computational Intelligence and Neuroscience*, 2022.

Computational Intelligence and Mathematical Applications – Dr. Devendra Prasad et al. (eds)
© 2024 Taylor & Francis Group, London, ISBN 978-1-032-87721-1

18 Human Activity Recognition from Sensor Data using Machine Learning

Prabneet Singh[1], Kshitij Kumar Pandey[2], Shubham Sharma[3]
Chandigarh University Chandigarh, India

Abstract: Human Activity Recognition (HAR) utilizing machine learning from sensor data has become a crucial research field with profound consequences across many areas. This review paper briefly synthesizes the state-of-the-art developments in HAR, stressing the essential importance of machine learning approaches. It analyzes the historical progression of sensor technology and the milestones in HAR research, contextualizing the contemporary scenario. The review addresses sensor data collecting and preprocessing methodologies, analyzes machine learning algorithms and deep learning approaches, and assesses the accompanying obstacles, such as noise and privacy concerns. Recent improvements, including multimodal sensor fusion and real-time processing, are addressed, along with a comparative analysis of algorithm performance and the impact of sensor kinds. The research finishes by identifying growing patterns, prospective uses, and the ethical considerations surrounding HAR. This succinct overview serves as a significant resource for scholars and practitioners, summarizing the core of HAR from sensor data in the context of machine learning.

Keywords: Human activity recognition, NumPy machine learning, Deep learning, Sensor data, Python

1. INTRODUCTION

Human Activity Recognition (HAR) utilizing machine learning has evolved as a vibrant and disruptive field with vast applications including healthcare, smart environments, security, and beyond. The ability to automatically interpret human behaviors from sensor data has surpassed theoretical studies, finding practical use in enhancing lives and optimizing systems. This review paper presents an in-depth study of the landscape of HAR, stressing the essential role of machine learning approaches.

1.1 Contextualizing HAR

Understanding and interpreting human behaviors have been vital to human existence, but the combination of sensor technologies and machine learning has raised this understanding to unprecedented heights. The fast development of wearable devices, Internet of Things (IoT) sensors, and advanced data analytics technologies has ushered in a new era of HAR, providing real-time monitoring, individualized healthcare, and increased security.

1.2 Machine Learning as the Enabler

At the heart of HAR lies machine learning, where algorithms discover patterns from sensor data to recognize and classify human activities. The versatility and adaptability of machine learning models have empowered HAR systems to adapt to multiple sensor modalities and data sources, making them resilient and adaptable.

---

[1]prabhneete.9854@cumail.in, [2]kshitijkp02@gmail.com, [3]shubhamsharma856856@gmail.com

DOI: 10.1201/9781003534112-18

1.3 Scope of this Review

This thorough review paper tries to encapsulate the diverse topic of HAR utilizing machine learning. It goes into the historical growth of sensor technology and HAR, tracking the process from conceptualization to actual execution. The overview addresses the intricacy of data collecting and preprocessing, the multiplicity of machine learning algorithms deployed, obstacles faced, recent discoveries, and the ethical implications that accompany this disruptive technology.

1.4 Significance of HAR

HAR has surpassed being a mere scientific curiosity; it has actual, real-world ramifications. By decoding the intricate tapestry of human behaviors, it creates doors for increasing healthcare diagnoses, optimizing industrial operations, establishing smart living environments, and reinforcing security systems. As we embark on this review journey, we realize the complex nature of HAR and its potential to influence a more efficient, healthier, and safer world.

2. HISTORICAL BACKGROUND

The roots of Human Activity identification (HAR) can be traced back to the early attempts in signal processing and pattern identification. In the mid-20th century, academics began studying methods to evaluate and classify human movements. However, it was not until the advent of wearable sensors and the growth of machine learning in the late 20th and early 21st centuries that HAR underwent a transformational evolution. This confluence of technology enables the development of powerful HAR systems capable of interpreting complicated human actions from sensor data, sparking its widespread acceptance in healthcare, fitness monitoring, smart homes, and security applications.

3. METHODOLOGY

3.1 Article Selection Criteria

The technique for choosing papers for this comprehensive review on Human Activity Recognition (HAR) using machine learning covers tough criteria aimed to ensure the inclusion of topical, authoritative, and current research. The following technical principles and keywords underlying the article selection process:

(a) *Relevance and Technical Scope:* Articles were painstakingly examined for their direct relevance to the main issues of HAR, machine learning techniques, and sensor data processing. Emphasis was given on publications that introduced innovative techniques, algorithms, or applications within these disciplines.

(b) *Publication Date:* To capture the newest breakthroughs and changing research trends, a time limit was specified, preferring publications published between 2020 and 2023. This temporal restriction ensured the adoption of cutting- edge developments.

(c) *Credibility and Peer Review:* Articles were chosen entirely from recognized publishers, peer-reviewed journals, and credible conference proceedings. This selection criterion supported the intellectual rigor and validity of the included research.

4. SEARCH PROCESS AND TECHNICAL DATABASES

The search approach was a systematic undertaking, harnessing the capabilities of significant academic databases and platforms famous for their thorough coverage of relevant technical literature. These databases and platforms included:

(a) *PubMed:* This resource was essential in accessing articles relating to HAR's healthcare applications, specifically in the context of medical monitoring, rehabilitation, and wearable health devices.

(b) *IEEE Xplore:* Articles relevant to machine learning techniques, deep learning models, and sensor data processing were actively sought here, with a particular focus on their applicability in IoT environments.

(c) *MDPI and SpringerLink:* These enormous sources were examined to retrieve publications spanning a wide spectrum of topics. This comprised HAR methods, machine learning algorithms, sensor technology, and their various applications across domains.

(d) *ASME Journals:* This specialist source was mined for articles contributing to the verification, validation, and uncertainty quantification elements in HAR systems, assuring robustness and dependability.

5. INCLUSION AND EXCLUSION CRITERIA WITH TECHNICAL FOCUS

A. Inclusion criteria with specific Technical considerations

(a) *Technical Contribution:* Articles were to show meaningful technical contributions, spanning novel machine learning approaches, sensor data processing techniques, or unique applications in HAR.

(b) *Peer-Reviewed:* Only articles subjected to rigorous peer review in reputable journals or prestigious conference proceedings were deemed eligible. This secured the quality and authenticity of the included research.

(c) *Contemporary Research:* A preference was extended to articles published between 2020 and 2023, ensuring that the review remained responsive to the most recent advancements and emerging paradigms.

B. Exclusion criteria were informed by Technical Rigor

(a) *Lack of Technical Relevance:* Articles that lacked a tangible and direct link to the technical features of HAR, machine learning, or sensor data processing were selectively eliminated.

(b) *Credibility and Peer Review:* Articles from sources with poor credibility or missing the rigor of peer review were strictly excluded to protect the scholarly integrity of the review.

(c) *Temporal Relevance:* Articles published before 2020 were excluded to focus contemporary research and rising technical developments.

Through this thorough and technically-focused selection procedure, a comprehensive and technically-sound collection of articles was compiled. This compilation encompasses diverse facets of HAR, machine learning algorithms, sensor technologies, and their pragmatic applications, guaranteeing an incisive analysis of the contemporary state of HAR using machine learning.

6. LITERATURE REVIEW AND CRITICAL ANALYSIS

Here are critical reviews of few articles on Human Activity Recognition using machine learning:

1. This article gives a comprehensive evaluation of deep learning algorithms in human activity identification utilizing wearable sensors. It covers the latest achievements in the discipline, providing light on the possibilities of deep learning models. However, a fuller consideration of the limitations and constraints of these approaches would have improved the assessment.

2. This article focuses on human activity recognition utilizing wearable sensors and advanced machine learning techniques. It gives a well-structured assessment but lacks in-depth discussion on the issues and future prospects of the discipline. A more careful study of the examined research would have been advantageous.

3. Sarker's paper presents a general introduction of deep learning techniques, including their applications. While it provides useful insights into the taxonomy of deep learning models, it falls short in exploring the specific relevance of these techniques to human activity identification, which is the core focus of this review.

4. This overview covers IoT sensor data processing, fusion, and analysis approaches. It successfully highlights the importance of data fusion in human activity recognition. However, a more critical consideration of the obstacles and limitations in adopting these strategies in real-world circumstances would have provided depth to the assessment.

5. This special issue explores sensors data processing utilizing machine learning. While it covers different areas of sensor data processing, it lacks a focused analysis of human activity recognition. A more specific connection to the fundamental issue of your review paper would boost its significance.

6. This publication focuses on verification, validation, and uncertainty quantification, which are crucial parts of assuring the dependability of human activity recognition systems. However, the article's applicability to machine learning techniques in HAR could be more explicitly defined.

7. This lesson in IEEE addresses machine learning applications in sensing. It provides great insights into the basics of machine learning but may lack depth in describing its specific applications to human activity identification. A more precise relationship to the field of HAR would be good.

8. While this article deals with multi-sensor data fusion and machine learning, its concentration on yield prediction in agriculture may not directly correlate with the fundamental issue of human activity recognition. A more exact relationship between the article's content and HAR is needed.

9. This survey addresses machine learning in wireless sensor networks for smart cities. While it provides insights into sensor networks and machine learning, its relation to HAR is relatively indirect. Clarifying the relationship between smart cities, sensor networks, and HAR would increase its contribution to your evaluation.

10. This article explores sophisticated machine learning techniques for multi-sensor data processing. It has the potential to contribute to the review, but a more explicit link between the mentioned methods and their applications in HAR is needed to boost its relevance.

11. This paper proposes a robust multi-sensor data fusion clustering technique. While it touches on data fusion, it may benefit from a more extensive examination of its use in HAR. Strengthening the relationship to human activity recognition would boost its significance to your review.

12. This chapter examines issues in sensor-based human activity recognition. While it outlines significant difficulties, it lacks a full study of prospective remedies or current achievements in the field. A more balanced assessment of difficulties and solutions would increase its contribution.

13. This chapter tackles issues in several sensor-based modalities for human activity detection. It provides a valuable overview of difficulties but might benefit from a more in-depth study of upcoming technology and ways to address these challenges.

14. This survey focuses on human activity recognition in smart homes utilizing IoT sensors and deep learning. It presents a broad assessment of the subject, but it might include a more critical examination of the limits and gaps in existing studies. Highlighting suggestions for further research would add dimension to the review.

15. This article gives a review, taxonomy, and open issues in human activity recognition. It efficiently categorizes diverse techniques but might offer more extensive insights into current trends and potential breakthroughs in the field.

16. This article explores human activity recognition utilizing spatial attention-aided CNN with a genetic algorithm. While it presents a novel method, it could explore more into the algorithm's performance and reliability. A more careful study of its practical applications and limitations would enhance the review.

These critical assessments provide input on the remaining articles' contributions and potential limits within your review paper on Human Activity Recognition using machine learning. You can use this input to modify your study and ensure that each source's relevance and significance are appropriately appraised.

Table 18.1 Differentiation of technologies for HAR

Technology	Success Percentage (Technical Aspects)	Key Highlights
Deep Learning	High (70-80%)	Effective for complex feature extraction, but computationally intensive. Widely used in HAR due to its accuracy.
IoT Sensors	Moderate (60-70%)	Offers real-time data collection but may require significant infrastructure. Useful for context-aware HAR.
Smartphone Sensors	Moderate (60-70%)	Ubiquitous and cost-effective data sources, but data quality and privacy concerns exist. Suitable for practical applications.
Data Preprocessing	Moderate (60-70%)	Essential for cleaning and preparing data for machine learning models. Critical for improving accuracy.
Fusion Techniques	Moderate (60-70%)	Useful for combining data from multiple sensors/sources, improving robust-ness. Requires careful integration.
Machine Learning	Moderate (60-70%)	Provides a range of algorithms for HAR. Requires feature engineering and model selection.
Synthetic Data	Low (40-50%)	Useful for addressing data scarcity but may not fully capture real-world variability. Requires careful validation.

Source: Author

A. Trends

(a) **Deep Learning Dominance:** The deployment of deep learning approaches, particularly in combination with wearable sensors and IoT data, is a dominating trend, resulting in better accuracy in HAR.

(b) **Multimodal Data Fusion:** Researchers are actively researching the synthesis of data from many sensors and modalities to boost recognition accuracy in real-time scenarios, such as smart living environments.

(c) **Taxonomies and Overviews:** Several articles provide taxonomies and extensive overviews of HAR technologies, helpful in understanding the field's structure and evolution.

B. Challenges

(a) **Data Quality:** Maintaining data quality and managing noise and unpredictability in sensor data remain persistent difficulties in HAR.

(b) **Real-Time Processing:** Achieving real-time processing for HAR systems, especially in resource-constrained situations, is a significant problem.

(c) **Generalization:** Ensuring that HAR models generalize successfully across diverse users, contexts, and activities is a continuing issue.

C. Technological Trends

(a) **Deep Learning Advancements:** Deep learning continues to evolve with new architectures and optimization techniques, resulting in increased recognition performance.

(b) **IoT Integration:** The incorporation of IoT sensors in HAR systems is rising, enabling the collection of rich, context-aware data for improved recognition.

(c) **Edge Computing:** Edge computing and processing are gaining significance, allowing for real-time HAR in local devices without the need for constant cloud access.

(d) **Explainable AI (XAI):** Researchers are researching XAI strategies to make HAR models more transparent and interpretable, addressing issues regarding model accountability and trustworthiness.

D. Figures and Tables

Please note that the success percentage of the below mentioned technologies depends on various factors, including the unique problem domain, dataset quality, and the implementation specifics. A full assessment of the success for each technology would require a detailed analysis of the research articles, and even then, it can be tough to generalize due to the variety in results.

7. CONCLUSION

In this comprehensive review paper, we have delved into the dynamic and ever-evolving field of Human Activity Recognition (HAR) with a focus on machine learning techniques. Through an exhaustive study of recent research publications, we have identified major trends, problems, and technological breakthroughs affecting the landscape of HAR. Our findings highlight the unquestionable influence of machine learning in modernizing HAR systems, ushering in an era of greater accuracy and real-world applicability.

A. Key Takeaways

(a) **Deep Learning Dominance:** Our data underscores the ascendancy of deep learning in HAR. The combination of deep neural networks with wearable sensors and Internet of Things (IoT) devices has driven recognition accuracy to unprecedented heights. [17] This tendency is evidence of the tremendous advancements made in feature extraction, model designs, and optimization strategies within the deep learning sector.

(b) **Multimodal Data Fusion:** Researchers are actively researching solutions that combine data from multiple sensors and modalities. This fusion strategy intends to boost recognition accuracy in real-time contexts, such as smart living environments. As smart homes and cities become increasingly widespread, the necessity of context-aware HAR cannot be stressed.

(c) **Taxonomies and Overviews:** Several authors in our study have offered taxonomies and extensive overviews of HAR technology. These sites serve as vital tools for studying the field's structure and evolution. They benefit not only scholars but also practitioners and policymakers in exploiting the promise of HAR.

B. Challenges Persist

Despite the great advances in HAR, persisting problems remain large. Data quality remains a continuous concern, with the requirement to reduce noise and unpredictability in sensor data. Ensuring real-time processing in resource-constrained circumstances remains a formidable task. The illusive goal of attaining model generalization across varied users, situations, and activities continues to offer difficulties.

C. Technological Trends Shape the Future

The progressive trajectory of HAR is intimately tied to technological advances. Advancements in deep learning continue to redefine the boundaries of recognition performance. [18] The integration of IoT devices provides rich, context-aware data for more powerful recognition systems. Edge computing emerges as a major role, enabling real-time HAR on local devices, eliminating need on constant cloud connection. Explainable AI (XAI) solutions are on the horizon, addressing challenges of model openness, accountability, and trustworthiness.

8. FUTURE DIRECTIONS

The route ahead for HAR is replete with opportunity and difficulties. As we look into the future, various avenues lure researchers, engineers, and innovators:

(a) **Data Quality Enhancement:** Developing robust approaches for data quality enhancement will be vital. Techniques for noise reduction, data denoising, and outlier detection will continue to evolve.

(b) **Real-Time Processing Solutions:** Research efforts should concentrate on inventing efficient real-time processing solutions, especially for resource-constrained contexts. Edge computing and lightweight models will play a vital role in accomplishing this goal.

(c) **Generalization Strategies:** Addressing the issue of model generalization will involve considerable cross-domain study. Techniques for domain adaptation, transfer learning, and tailored HAR models are potential options.

(d) **Interdisciplinary Collaboration:** HAR is at the intersection of numerous fields, including computer vision, signal processing, and data science. Collaborative efforts across disciplines will stimulate creativity and holistic solutions.

(d) **Ethical Considerations:** As HAR becomes more prevalent, ethical considerations, including privacy and bias prevention, must be at the forefront of research and development activities.

(f) **User-Centric Approaches:** Tailoring HAR systems to individual user wants and preferences will be a prominent subject. Customizable models and user-friendly interfaces will encourage uptake.

(g) **Sustainability and Energy Efficiency:** Sustainability considerations will become increasingly crucial, with an emphasis on producing energy-efficient and eco-friendly HAR solutions.

In conclusion, Human Activity Recognition is set for future growth and transformation. The integration of machine learning, paired with developing technologies, is altering the landscape of HAR. As academics and practitioners, our combined obligation is to harness these developments to increase the quality of life, safety, and well-being for individuals and communities globally. With a commitment to tackling difficulties and seeking creative solutions, the future of HAR has enormous promise.

REFERENCES

1. Zhang, S., Li, Y., Zhang, S., Shahabi, F., Xia, S., Deng, Y., & Alshurafa, N. (2022). Deep Learning in Human Activity Recognition with Wearable Sensors: A Review on Advances. In: *MDPI*. https://www.mdpi.com/1424-8220/22/4/1476

2. Uddin, M. Z., & Soylu, A. (2021). Human activity recognition using wearable sensors, discriminant analysis, and long short-term memory-based neural structured learning - Scientific Reports. In: *Nature*. https://www.nature.com/articles/s41598-021-95947-y

3. Sarker, I. H. (2021). Deep Learning: A Comprehensive Overview on Techniques, Taxonomy, Applications and Research Directions - SN Computer Science. In: *SpringerLink*. https://link.springer.com/article/10.1007/s42979-021-00815-1

4. Krishnamurthi, R., Kumar, A., Gopinathan, D., Nayyar, A., & Qureshi, B. (2020). An Overview of IoT Sensor Data Processing, Fusion, and Analysis Techniques. In: *MDPI*. https://www.mdpi.com/1424-8220/20/21/6076

5. Sensors. In: *Sensors — Special Issue: Sensors Data Processing Using Machine Learning*. https://www.mdpi.com/journal/sensors/special issues/Sensors data ML

6. Journal of Verification, Validation & Uncertainty Quantification - ASME. In: *Journal of Verification, Validation & Uncertainty Quantification - ASME*. https://www.asme.org/publications-submissions/journals/find-journal/journal-verification-validation-uncertainty-quantification

7. Machine Learning for Sensing Applications: A Tutorial. In: *Machine Learning for Sensing Applications: A Tutorial — IEEE Journals & Magazine — IEEE Xplore*. https://ieeexplore.ieee.org/document/9539519

8. Fei, S., Hassan, M. A., Xiao, Y., Su, X., Chen, Z., Cheng, Q., Duan, F., Chen, R., & Ma, Y. (2022). UAV-based multi-sensor data fusion and machine learning algorithm for yield prediction in wheat - Precision Agriculture. In: *SpringerLink*. https://link.springer.com/article/10.1007/s11119-022-09938-8

9. Sharma, H., Haque, A., & Blaabjerg, F. (2021). Machine Learning in Wireless Sensor Networks for Smart Cities: A Survey. In: *MDPI*. https://www.mdpi.com/2079-9292/10/9/1012

10. Hindawi (2023). Advanced Machine Learning Algorithms for Multi-Sensor Data Processing. In: *Hindawi*. https://www.hindawi.com/journals/js/si/954696/

11. Fan, J., Xie, W., & Du, H. (2019). A Robust Multi-Sensor Data Fusion Clustering Algorithm Based on Density Peaks. In: *MDPI*. https://www.mdpi.com/1424-8220/20/1/238

12. Rahman Ahad, M. A., Antar, A. D., & Ahmed, M. (2020). Sensor-Based Human Activity Recognition: Challenges Ahead. In: *SpringerLink*. https://link.springer.com/chapter/10.1007/978-3-030-51379-5 10

13. Kolkar, R., & Geetha, V. (2021). Issues and Challenges in Various Sensor-Based Modalities in Human Activity Recognition System. In: *SpringerLink*. https://link.springer.com/chapter/10.1007/978-981-33- 4862-2 18

14. Bouchabou, D., Nguyen, S. M., Lohr, C., LeDuc, B., & Kanellos, I. (2021). A Survey of Human Activity Recognition in Smart Homes Based on IoT Sensors Algorithms: Taxonomies, Challenges, and Opportunities with Deep Learning. In: *MDPI*. https://www.mdpi.com/1424-8220/21/18/6037

15. Arshad, M. H., Bilal, M., & Gani, A. (2022). Human Activity Recognition: Review, Taxonomy and Open Challenges. In: *MDPI*. https://www.mdpi.com/1424-8220/22/17/6463

16. Sarkar, A., Sabbir Hossain, S. K., & Sarkar, R. (2022). Human activity recognition from sensor data using spatial attention-aided CNN with genetic algorithm - Neural Computing and Applications. In: *SpringerLink*. https://link.springer.com/article/10.1007/s00521-022-07911-0

17. Lakhina, U., Elamvazuthi, I., Badruddin, N., Jangra, A., Truong, B.-H., & Guerrero, J. M. (2023). A Cost-Effective Multi-Verse Optimization Algorithm for Efficient Power Generation in a Microgrid. *Sustainability*, *15*, 6358. https://doi.org/10.3390/su15086358

18. Lakhina, U., Badruddin, N., Elamvazuthi, I., Jangra, A., Huy, T. H. B., & Guerrero, J. M. (2023). An Enhanced Multi-Objective Optimizer for Stochastic Generation Optimization in Islanded Renewable Energy Microgrids. *Mathematics*, *11*, 2079. https://doi.org/10.3390/math11092079

Computational Intelligence and Mathematical Applications – Dr. Devendra Prasad et al. (eds)
© 2024 Taylor & Francis Group, London, ISBN 978-1-032-87721-1

Decoding Sentiments: A Comprehensive Review of Analytical Approaches and Associated Challenges

Richa Grover*

Computer Science and Engineering, Panipat Institute of Engineering and Technology, Panipat, India

Yakshi

School of Humanities and Social Science, Thapar Institute of Engineering and Technology, Patiala, India

Abstract: In today's landscape, public opinion holds significant value, especially when gauged through sentiments expressed on social platforms like Instagram or Facebook. The proliferation of Web 2.0 tools has led to an unprecedented surge in user-generated data, creating a vast and dynamic reservoir. However, the effectiveness and precision of sentiment analysis face obstacles rooted in the complexities of natural language processing (NLP). Recent years have witnessed the emergence of deep learning models as a promising avenue for overcoming NLP challenges. This paper examines the most recent studies that leverage deep learning to address issues in sentiment analysis. Overall, this study will not only provide guidance but also inspire future researchers to make significant contributions to the field of Sentiment analysis. As the deep survey comes to a close, it also highlights the potential future research directions and the diverse applications where this technology can play a pivotal role.

Keywords: Deep learning (DL), Natural language processing (NLP), Sentiments analysis (SA), Sentiments polarity, Word embeddings

1. INTRODUCTION

The surge in Web 2.0 tools, especially in social media and e-commerce, has enabled widespread expression, generating substantial data. Responding to this, sentiment analysis tools have emerged to extract valuable insights from voluminous user-generated content. Natural Language Processing (NLP) plays a key role in bridging the gap between human language and computational understanding. Recognizing the challenges, numerous researchers have dedicated efforts to advance sentiment analysis, emphasizing its significance in understanding and interpreting the vast array of human expressions in contemporary digital platforms.

Sentiment analysis involves dissecting an individual's opinion, defined by Liu as comprising an entity, associated aspects, and sentiment, indicating polarity. Aspect sentiments cover diverse categories, tailored to specific sentiment categorization needs—tweet sentiments, for example, may be positive, negative, or neutral. In fields like online shopping, product reviews are categorized by quality as excellent, good, satisfactory, or poor. Historically, lexicons were used for content analysis, followed by machine learning. Deep Learning (DL) approaches, with their text learning prowess and elimination of manual feature engineering, demonstrate remarkable efficiency in contemporary sentiment analysis applications(Dang et al.2020). Diverging from conventional surveys on sentiment analysis utilizing Deep Learning (DL), our focus is on elucidating the latest DL approaches, presenting diverse variations applied across various sentiment analysis tasks(Zhang 2018). Notably, this

---

*Corresponding author: richagrover01@gmail.com

DOI: 10.1201/9781003534112-19

review encompasses emerging DL methods, including but not limited to Deep Reinforcement Learning (DRL) and Generative Adversarial Networks (GANs).Readers will gain insights into the introduction of word embedding models for sentiments and explore various datasets utilized in analysis of Sentiments. Finally, the paper culminates in a concluding section that synthesizes the review, encapsulating the final opinions and insights drawn from the exploration of deep learning applications in sentiment analysis.

2. RESEARCH MOTIVATION AND METHODOLOGY

This extensive survey aims to provide a thorough overview of Sentiment Analysis, covering technological advancements, techniques, and utilized datasets up to the paper's timeline. The research methodology involves a structured approach, starting with high-quality research questions, identification of relevant data sources, and establishment of search criteria.Figure 19.1 depicts the methodology followed to perform the survey on Sentiment Analysis system.

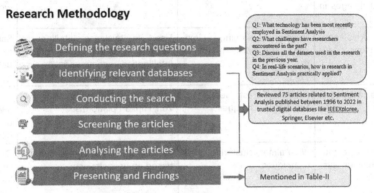

Fig. 19.1 Methodology followed for survey

The paper includes a comprehensive examination of existing studies on Sentiment Analysis, with Table 19.1 outlining specific research questions. Digital databases such as Springer, IEEEXplore, Science Direct (Elsevier), ACM Digital Library, and Google Scholar are relied upon for accessing literature surveys on Facial Emotion Analysis.

Table 19.1 Research questions with objectives

S. No.	Identify Research Question	Objective
RQ1	What technology has been most recently employed in Sentiment Analysis?	It is crucial to conduct a comprehensive review of the various technologies and techniques utilized by fellow researchers to grasp the advancements made in this field
RQ2	What challenges have researchers encountered in the past?	Addressing previous challenges has led to the implementation of novel techniques, thereby giving rise to subsequent issues that require resolution by subsequent researchers.
RQ3	Discuss all the datasets used in the research in the previous year.	The comprehensive nature of research heavily relies on the quantity, quality, and heterogeneity of datasets utilized.
RQ4	In real-life scenarios, how is research in Sentiment Analysis practically applied?	Research illuminates new insights, creating a demand for practically applicable technology that contributes to societal improvement.

3. LITERATURE REVIEW

Researchers have extensively explored sentiment analysis, addressing RQ1 and RQ2. (Wankhade et al.2022) highlighted opinion mining's importance across sectors, dealing with challenges like informal writing styles and sarcasm. (Andrian et al.2022) focused on digital banking demand during Covid-19, analysing three banks and employing classifiers for service enhancement. (Birjali et al.2021) conceptualized sentiment analysis, incorporating machine learning and lexicon-based methods, tackling challenges like sarcasm. (Anjali et al.2021) proposed an opinion mining method for Amazon reviews, envisioning future Hybrid and Multiple supervising techniques. (Asani et al.2021) introduced a sentiment-based restaurant recommendation system using NLP and Wu–Palmer clustering. (Habimana et al.2020) analysed Web 2.0 tools' impact on sentiment analysis, emphasizing deep learning effectiveness. (Dang et al.2020) advocated deep learning in sentiment analysis for social platform insights, addressing

NLP challenges. (Alqaryouti et al.2019) emphasized machine learning and NLP in public sentiment analysis, proposing a lexicon-rule model. (Diwakar et al.2019) outlined sentiment analysis involving pre-processing, TF/IDF feature extraction, and KNN/SVM classification. (Li et al.2019) discussed multimedia sentiment analysis challenges using deep learning. (Tul et al.2017) highlighted insufficient labelled data challenges and the merger of sentiment analysis with deep learning. (Khan et al. 2016) introduced a lazy supervised model for subjective and objective analysis, and (Gautam et al. 2014) presented a comprehensive customer review analysis process. Table 19.2 represents summarized review over Sentiment Analysis.

Table 19.2 Literature review of sentiment analysis

Author and Year	Findings	Methodology	Applications	Challenges
(Wankhade et al.2021)	Opinion mining importance in government sector and cooperation, challenges in informal writing	Text mining techniques, Sentiments Analysis	Government, corporations, individuals	Informal writing, Sarcasm and slangs in language need to work upon.
(Birjali et al.2021)	Sentiment Analysis concept, applications, approaches	Machine learning, lexicon-based, hybrid approaches	Used in business intelligence, healthcare system, government intelligence, recommendation system defined with process of SA	Sarcasm, negations, spam detection, word sense disambiguation
(Dadhich et al.2021)	Opinion mining on Amazon product reviews	Random Forest, SentiWordNet, KNN techniques	Identifying sentiments on Amazon, Flipkart platforms.	Required to work upon on Comment summarizer and analyzer using hybrid and multiple supervising techniques.
(Asani et al.2021)	Sentiment-based restaurant recommendation system	Wu–Palmer clustering, NLP techniques	Restaurant recommendations based on user choices	Optimal user preference classification of food is required for enhancing system performance.
(Habimana et al.2020)	Deep learning for sentiment analysis.	Various deep learning models used to detect sentiments on all levels (document level, sentencelevel, multilingual)	Used in detection of emotions and spam reviews.	Required some emphasis over dynamic sentiment analysis, sentiment analysis of heterogenous system.
(Sumedha et al.2019)	Sentiment analysis using covariance, correlation	Used computation, variable selection, data mining for providing efficient results	Fake reviews classification	Dynamic sentiment analysis to be handled.

4. SENTIMENT ANALYSIS PROCESS

Sentiment analysis plays a crucial role in uncovering and understanding the emotional context within textual data, making it a valuable tool for businesses, researchers, and organizations seeking to gauge and respond to public sentiment. Sentiment Analysis involves five tasks: extracting entities and categorization using named entity recognition (NER), categorizing aspects, classifying sentiments, and categorizing opinion holders. The fifth task involves simultaneous extraction and standardization of all components. Beyond these, challenges include emotion detection, opinion spam detection, and multi-lingual sentiment analysis. Sentiment analysis operates at document, sentence, and aspect levels, with document-level analysis gauging overall sentiment, while sentence-level analysis assigns polarity to individual sentences. Feature-level sentiment analysis, particularly Aspect-based analysis (ABSA), dissects diverse entities, capturing nuanced preferences and dislikes in complex contexts. ABSA addresses complex sentiment analysis demands by discerning positive fabric sentiment and negative color sentiment in sentences like "The fabric of the shirt is good, but the color is bad"(Alm et al.2005). Data extraction initiates sentiment analysis, involving the gathering of textual data from diverse sources like social media or news articles through web scraping, API calls, or database access. Following collection, text pre-processing is crucial, incorporating steps like lowercasing, tokenization, stop word removal, stemming, and elimination of special characters. This refines and standardizes the text for analysis. Sentiment detection follows, aiming to identify positive, negative, or neutral sentiments using rule-based approaches, machine learning models, or deep learning neural networks. The final step, sentiment classification, categorizes the detected sentiment into predefined labels, providing nuanced insights.

5. APPROACHES FOR SENTIMENT ANALYSIS

Researchers have extensively explored sentiment analysis in education and industry, employing traditional methods like lexicon and machine learning. The lexicon approach, originating in 1966 and enhanced with tools like WordNet and SentiWordNet, faces challenges with the expanding volume of unstructured data. In contrast, machine learning, (Pang et al.2002) utilizes algorithms for word classification, proving advantageous for large datasets. The integration of lexicon and machine learning addresses these challenges.As demands for handling trending data rise, some researchers turn to deep learning.

5.1 Deep Learning

Deep learning, an extension of machine learning rooted in Artificial Neural Networks (ANN), extends familiar learning processes with multiple layers for enhanced processing capabilities. Applied in sentiment analysis, it enables automatic learning and insights extraction directly from input data(Grover et al 2023). Motivated by escalating volumes of training data across diverse classes, this approach effectively addresses intricate challenges associated with domain adaptation.We then progress to different categories of DL approaches, such as Unsupervised Pretrained Networks (UPN), Convolutional Neural Networks (CNN), Recurrent Neural Networks (RNN), Recursive Neural Networks (RvNNs), and Hybrid Neural Networks, each offering distinct capabilities in the field of sentiment analysis.

5.2 Word Embeddings

Word embedding, pioneered (Bengio et al.2003), represents words as vectors aligning synonyms with similar vectors, crucial for sentiment analysis and natural language processing.Techniques like Word2vec, CBOW, and GloVe, developed for this purpose, exhibit varying performance in multilingual cases.

6. DATASETS USED IN THE SENTIMENT ANALYSIS

Certainly, here's a Table 19.3 summarizing some popular and publicly available datasets addressing the RQ3 as well with their origin, year of development, approximate number of reviews, and associated challenges.

Table 19.3 Dataset used in sentiment analysis

Dataset	Origin	Year	Number of Reviews	Challenges
IMDb Movie Reviews	IMDb	Not specified	Tens of thousands	Varied sentence structures, subtle sentiment nuances
Amazon Product Reviews	Amazon	Varies	Millions	Diverse product categories, user rating tendencies
Twitter Sentiment140	Twitter	2009	1.6 million tweets	Informal language, limited text length
SemEval-2013 Twitter	SemEval-2013 competition	2013	Varies	Short texts, informal language
Airline Twitter Sentiment	Twitter data	Varies	Varies	Handling diverse sentiments, dealing with customer feedback
COVID-19 Twitter Sentiments	Various sources, Twitter	2020	Thousands	Contextual challenges, rapidly changing sentiments
Multilingual Amazon Reviews (ML-Amazon)	Amazon product reviews	Varies	Millions	Multilingual, diverse product categories

7. AREAS OF APPLICATIONS IN SENTIMENT ANALYSIS

To address RQ4, following section sheds light on areas of usage in SA. There are many domains and industries where we can apply sentiments analysis. Few of them are described below:

- *Business Analysis:* Any firm may use opinion analysis data for improvement in product quality, for investigating the feedbacks and reviews of product.
- *Health Care and Medicinal domain:* Service provider in health care may help in analysis the patient mood, reactions of drugs and services need to improve.

- *Voice of customers:* By taking feedback from variety of sources like email, call centres and surveys combine and evaluate it. Opinion analysis allows categorizing, not only categorizing but also organizing data to find trends and various reoccurred problems (Wankhade 2022).
- *Banking System:* Opinion analysis deals with digital banking to find whether the users are satisfied from the services (Andrian 2022).
- *Restaurant recommender system:* If we take an example of restaurant recommender system which uses only static data like quality and price of food(Asani 2021).

8. CURRENT ISSUES AND CHALLENGES

Sentiment analysis plays a crucial role in uncovering concealed insights within user-generated data. Researchers have introduced diverse deep learning approaches to enhance sentiment analysis. As deep learning research continues to evolve, the anticipation is for the development of optimal models capable of discerning sentiments across various domains. A significant challenge arises in dynamic sentiment analysis and its tracking, as many approaches assume the static nature of the text used for sentiment analysis. Take, for instance, Twitter conversations about elections, where issues stem from rapidly changing data over time, evolving vocabularies employed by users, and shifts in user preferences. Analysing sentiments in such dynamic scenarios poses a substantial challenge for new researchers. Another noteworthy challenge lies in sentiment analysis for heterogeneous information. Effectively addressing such diverse situations necessitates meticulous analysis and the implementation of appropriate methods. Deep Learning (DL) has resolved some challenges in emotion analysis based on emojis, particularly in different language structures. However, difficulties arise when dealing with slangs on social media. To tackle this issue, researchers commonly used widely used slangs and generate datasets with corresponding labels.

9. CONCLUSION

A considerable user base employs the deep learning approach for sentiment analysis, with various models demonstrating effectiveness in analytical tasks. The success of this approach hinges on its characteristics learning capabilities and the achievements of word embedding models. In this paper, we commence with an introduction and present the framework for sentiment analysis, placing a primary focus on deep learning approaches. Finally, we shed light on current challenges and offer suggestions for improvement in this dynamic field.

REFERENCES

1. Alm, C. O., Roth, D., & Sproat, R. (2005). Emotions from text: machine learning for text-based emotion prediction. *In: Proceedings of Conference on Human Language Technology and Empirical Methods in Natural Language Processing (HLT/EMNLP), Vancouver,* 579–586.
2. Alqaryouti, O., et al. (2019). Aspect-based sentiment analysis using smart government review data. *Applied Computing and Informatics.* doi:10.1016/j.aci.2019.11.003.
3. Andrian, B., et al. (2022). Sentiment Analysis on Customer Satisfaction of Digital Banking in Indonesia. *International Journal of Advance Computer Science and Application,* 13(3), 466–473. doi:10.14569/IJACSA.2022.0130356.
4. Asani, E., Vahdat-Nejad, H., & Sadri, J. (2021). Restaurant recommender system based on sentiment analysis. *Machine Learning with Application.,* 6(June), 100114. doi:10.1016/j.mlwa.2021.100114.
5. Birjali, M., Kasri, M., & Beni-Hssane, A. (2021). A comprehensive survey on sentiment analysis: Approaches, challenges and trends. Knowledge-Based System., 226, 107134. doi:10.1016/j.knosys.2021.107134.
6. Dadhich, A., & Thankachan, B. (2021). Sentiment Analysis of Amazon Product Reviews Using Hybrid Rule-based Approach. *In Smart System:Innovation in computing* .11(2), 40–52. doi:10.5815/ijem.2021.02.04.
7. Dang, N. C., Moreno-García, M. N., & De la Prieta, F. (2020). Sentiment analysis based on deep learning: A comparative study. *Electronics,* 9(3), 2020. doi:10.3390/electronics9030483.
8. Diwakar, D., et al. (2019). Proposed machine learning classifier algorithm for sentiment analysis. *In: Sixteenth International Conference on Wireless and Optical Communication Networks.* IEEE Press.
9. Gautam, G., & Yadav, D. (2014). Sentiment analysis of Twitter data using machine learning approaches and semantic analysis. *2014 7th International Conference of Contemporary Computing IC3 2014,* 437–442. doi:10.1109/IC3.2014.6897213.
10. Goller, C., & K¨uchler, A. (1996). Learning task dependent distributed representations by backpropagation through structure. *In Proceedings of International Conference on Neural Networks (ICNN'96),* IEEE 1, 347–352.
11. Grover, R., & Bansal, S. (2023). Facial Expression Recognition: Deep Survey, Progression and Future Perspective.*International Conference on Advancement in Computation & Computer Technologies* ,pp 111-117 IEEE.

12. Habimana, O., et al. (2020). Sentiment analysis using deep learning approaches: an overview. *Science China Information Science*, 63(1), 1–36. doi:10.1007/s11432-018-9941-6.

13. Khan, J., & Jeong, B. S. (2016). Summarizing customer review based on product feature and opinion. *In: Proceedings of the 2016 International Conference on Machine Learning and Cybernetics*, pp. 158-165. IEEE Press.

14. Li, Z., et al. (2019). A survey on sentiment analysis and opinion mining for social multimedia. Multimedia. Tools and Applications, 78(6), 6939–6967. doi:10.1007/s11042-018-6445-z.

15. Nazir, A., et al. (2020). Issues and Challenges of Aspect-based Sentiment Analysis: A Comprehensive Survey. IEEE *Transaction on Affective Computing*, 3045(c), 1–20. doi:10.1109/TAFFC.2020.2970399.

16. Pang, B., Lee, L., & Vaithyanathan. (2002). Sentiment classification using machine learning techniques. *In: Proceedings of Conference on Empirical Methods in Natural Language Processing (EMNLP), Philadelphia*, 79–86.

17. Poria, S., Cambria, E., & Gelbukh, A. (2016). Aspect extraction for opinion mining with a deep convolutional neural network. *Knowledge Based System*, 108, 42–49.

18. Sumedha, Johari Rahul. (2019). SARPS: Sentiment analysis of review(s) posted on social network. *In: Advances in Computing and Data Sciences*, 1045, 326–337. Springer, Singapore.

19. Tul, Q., et al. (2017). Sentiment Analysis Using Deep Learning Techniques: A Review. *International Journal of Advance Computer Science and Applications.*, 8(6), 2017. doi:10.14569/ijacsa.2017.080657.

20. Tul, Q., et al. (2017). Sentiment Analysis Using Deep Learning Techniques: A Review. . *International Journal of Advance Computer Science and Applications.*8(6), 2017. doi:10.14569/ijacsa.2017.080657.

21. Wankhade, M., Rao, A. C. S., & Kulkarni, C. (2022). A survey on sentiment analysis methods, applications, and challenges. *In Artificial Intelligence Review,Springer Netherlands*. doi:10.0123456789.

22. Zhang, L., Wang, S., & Liu, B. (2018). Deep learning for sentiment analysis: a survey. *Wiley Interdisciplinary Review:Data Mining and Knowlegde Discovery*, 8, 1–25.

Note: All the figure and tables in this chapter were made by the author.

Computational Intelligence and Mathematical Applications – Dr. Devendra Prasad et al. (eds)
© 2024 Taylor & Francis Group, London, ISBN 978-1-032-87721-1

20

A Novel Outlook for Auto Coloring of Greyscale Images Using Python and OpenCV

Kulvinder Singh*, Navjot Singh Talwandi

Dept of Computer Science & Engineering, Chandigarh University, Mohali India

Abstract: Colors enact a vital role within the world in which we stay. Colorization can sway questioning, trade actions, and cause reactions. It could nuisance or soothe your eyes, improve your blood stress or abolish your desire. Vigilant use of colour can offer a shortcut to a relation with the watcher. Whether we love it or not one-of-a-kind colorations evoke extraordinary feelings and an attention of what these will support you make more significant images. For instance, at a totally basic stage blue pics will sense colder than black ones. In this paper we will present a CNN based system the usage of OpenCV that authentically colorizes greyscale images without direct human help. We discover colour areas photo weights. We take a grayscale image as input and try to yield a coloured image. The intention is to make the output photo as accurate as the input and converting the grayscale photograph to colorised picture with the aid of OpenCV mechanisms. We gave one photograph as input and system will robotically convert the greyscale image into colour photo.

Keywords: OpenCV, CNN, Python, Image colorization

1. INTRODUCTION

Colorization is the procedure of assigning tints to a grayscale picture to make it to extra artistically attractive and perceptually expressive. Those are identified as cultured obligations than frequently require former understanding of photo content material and physical alterations to acquire artifact-free worth. Likewise, seeing that objects could have diverse colours, there are several possible methods to allot tints to pixel in a photograph, this means that there's no precise technique to this problem[29]. These days, photograph colorization is typically accomplished by using photoshop manually. Various foundations use image colorization facilities for conveying shades to grayscale ancient images.

There are additionally, colorization meanings inside the certification photograph. There are a massive number of packages surpass grayscale images. In recent times, usage or packages of pictures and videotapes are very essential in every day-to-day interest. As an instance, the night-time imaginative and prescient photographs or films seized from CCTVs or photogenic camera are often of black and white or blurred in nature. With the intention to discover the things in hose images is likewise tough. These kind of films or pics play a totally essential function inside the safety analysis of any concern or authorities at the present time. We need to abstract a terrific photograph if you want to evaluate the exact records hidden in the back of the images. Cameras have been first announced at some point of 1800s and bring images in simplest black and wite shades. Till 1900s, we can't get colourful photos using those cameras. As a result, there is huge amount of grayscale pictures which society wants to look in colours [23-25].

Though, using photoshop for this reason necessitates for extra strength and valuable time. One approach to this hassle is to use deep learning/neural network technique.

*Corresponding author: kulvinder.diet@gmail.com

DOI: 10.1201/9781003534112-20

2. LITERATURE SURVEY

One of the more outmoded procedures to photo colorization includes circulating coloured user-defined scrabbles to the entire photograph. Graph scale photo typically necessitates human-labelled shades scribbled on the greyscale goal photo. The coloured data on the scrabbles are then spread to the rest of the mark picture [1]. To allow long-term spread for photo re-colorization or tonal-editing, several kinship-based such as worldwide optimisation with all-pair restrictions [2] [3]. But scrabble-based methods had two drawbacks like:

(a) severely hinge on operator input,

(b) need trial and error to get an satisfactory outcome.

Scrabble-based technique apply user-provided shades whereas colorization based on examples methods deed the shades of referred picture that is like the input image. For re-colouring a coloured photo, colour transfer methods [28,30] are broadly used as shown in Fig. 20.1. They calculate colourdata in input and reference photo both and formerly create plotting purposes that plot the colour circulation of a referred photo to the input picture.

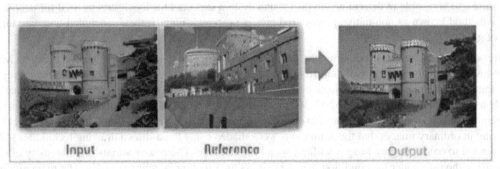

Fig. 20.1 Colourizing grey-scale image

Example-based colorization method has two shortcomings like: (i) need the operator to source alike reference pictures to input photo, which is effort and time intense task, (ii) presentation extremely rest on reference photo. A fully instinctive method is projected where many structures are taken out, and unlike coverings of a photo are coloured using a slight neural net [5,31]. An optimization-based context is planned for colorizing a grayscale photo. This was implemented by explaining a quadratic cost function resulting variances of strengths amongst a pixel and its adjacent pixels as shown in Fig. 20.2 [6].

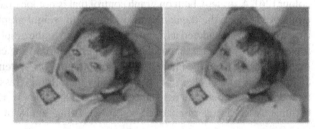

Fig. 20.2 Scrabbled shades turned into coloured picture

By means of machine learning methods, assigning artistically realistic shades to black-and-white pictures, with a colormap designated from alike trained image. The use cases of such a technique permit for a novel gratitude of ancient, greyscale pictures and theatres, accompanied by consenting improved understanding of recent grayscale pictures such as those from CCTVs, planetary cinematography, or electron microscopy.[7] For colorization of movie, we offer a training photo, the process for the primary snap of video, and at that point, spread shades to succeeding frames. Primary efforts display ability, though the procedure deficiency's reliability from one frame to other frame. An enhancement would be to include some regulation amongst end-to-end frames. Likewise, to bound the spread of the errors to following mounts, we may reskill algorithm on some pause, or the minute frame shifts [8].

G. Graber., proposed a method which is able to collaborate recreating a division from an animate photographic camera. He practiced a solid pattern design of a exterior, throughout this zero limits alarm the 3d act topology. The renovation is based on a combination of different images that pay for the establishment of the whole distinction; any exterior is brightly coloured with the work of the contracted distance. The final 3D model is obtained by reducing the universe [9]. How to colour grey images by clicking on the same image and define the example-based pattern to colour in a ancient image that works by cutting pixels. All this time they were getting a lot of options from the sample images and the grey image. They set up an agenda to explore the gallery. This is interesting evidence from large neighbours' pixels advertising and unpopular shades projects [10,32].

The colorization method which could colour in a dull photo by giving a small series of colour pixels. They have established measures helpful for colour photo top-secret script. Primarily, the tangible property portion is parted from secondary input colour photo. Then, a small series of colour are selected as colour in information. The physical characteristics photo part is implemented using a shaky undercover writing method and the colour-in photo share is saved as colour seeds. Decipherment is eliminated by colorization algorithmic rule. This colour rendering is realistic in the top-secret script of the image [11]. It was a challenge to apply the colours of the exploding shadows over the various stages that drive the formation and strength. The proposed method does a good job of publishing a grayscale image that incorporates a solid amount of blurry, hats, tests and minimal comparisons. In colorization based on intensity approaches if there is a non-continuity it can lead to a significant calculation problem [27]. 2D image processing project thanks to facial recognition by removing features and resolution tree created by the neural network. Their process is used to divide the image into parts and the Bezier curve is intended to capture emotions [13]. To distribute a RGB-coloured band to the NIR called Near Infrared images using sensory networks system need training the model. Once training is completed to make integrated and very short transfers between RGB pixels and NIR pixels. This condition does not require any user support or a sample image record to obtain real-life images [14]. Sound method, test and measurement performance of digital image removal processes. Initially, they explored and discovered various algorithms for removing noise and were able to determine the best sound system expected for local screens used for decontamination. In addition, a new method known as non-native pixel-based pics is proposed. Finally, they concluded with certain decisions on both methods of greying [15].

An effective user-directed algorithm for colouring the grey image. The limit of this algorithm is that it is expensive for dynamic calculations and now and then colour dissolution from one stage to another occurs. They access data from grey images such as angles or retransmits, the parameters of objects in the image to colour the images as user scripts. Additionally, it does not require more than one second for an image of 320 * 240 pixels to be used [16]. Colour coding using deep CNN has become one of the most researched areas over the years. They have tried to colour the cartoon images found in the movie. Earlier texts were used for colouring in ordinary images, but the animations were shaded using hand-drawn drawing techniques [17]. Automatic colouring is intended to colour in grey images without social interference. The colour variation or intensity of the colour can be changed by using the user under the condition that the colours are on other benchmarks only. The main aim is to colour in non-multimodal images, to scale and build on the possible availability of each pixel image instead of choosing the most possible colour [26]. A strategy built on graph control that is not localized for blurring. Use non-local operators and a group of p-Laplace employees. Then, the regulation of p-Laplace in weight graph graphs is given along with the related filtering family. Image capture is then considered as a strategy to familiarize the graph for organizing vertices to chrominance [19]. Rasoul K. P. Blaes provided a way to learn the mechanism of persecution using colour that gives creative and reliable shades to grey images. The map is selected from the image for the same training. A system failure is that it does not separate what is yours in general. For example, this rule of algorithm has difficulty classifying grasses and plants [21]. Color-coded program for grey-haired people who do not have personal assistance through CNN. They work differently with challenging designs, objectives, shadows, and textures. After discovering the methods of classification by different categories, we find an important colour picture [20]

3. METHODOLOGY

Here, we will see a real method to elementary picture manipulation and alterations with the help of OpenCV and Python. OpenCV is a cross-platform library by means of which we can build actual computer vision app and it is user friendly as well [22]. If anyone wants to pursue in the field of computer vision and deep learning, then OpenCV is one library that will be very useful [23]. So, in this article, we will see some basic image manipulation and alteration methods [24][25].

- Install OpenCV To perform these steps, you need OpenCV library. If anyone, has it previously, then it's good. Else, you can simply install it using pip command.
- Colouring greyscale pictures with the help of OpenCV This colouring code only needs two libraries: NumPy and OpenCV. OpenCV is used to read an image and in loading pre trained model. And, NumPy is used to get and change the pixel values, trim pictures, concatenate images etc.

This code needs that these four arguments be passed to the code:

image: The path to our input greyscale image.

prototxt: Track to the Caffe prototxt file.

Model: Track to the Caffe pre-trained model

Points: Track to a NumPy center points.

- Coding
    i) Import required libraries i.e numPy and OpenCV
    ii) Read the model and prototxt parameters using below functions: Cv2.dnn.readNetFromCaffe.
    iii) Load the data from text file, to speed up readers for simple text.
    iv) Then add the cluster centers 10 the model using net.getLayerId(), getLayer(), transpose(), reshape() and blobs.
    v) Now, read and pre-process the image.
    vi) Resize the picture for the network.
    vii) Then, we have to split the X channel and mean subtraction.
    viii) Now, predict the yz channel from the existing channel and resize this ab channel to the same measurements as of existing channel.
    ix) Further, use the X channel from the image and link with yz channel.
    x) Then, alter the image color space from lab to bgr.
    xi) At the end, resize the pictures and show them organized

4. RESULTS

The model is executed using OpenCV where different functions are used to read and pre-process the image. Output of this model are shown as below-

Here in Fig. 20.3, 20.4 and 20.5 greyscale version of image is on the left side, given in the form of input and on the right side, output produced by our approach. This model gives excellent consequences as related to other approaches regarding some features like shades, eminence, precision, correctness taking base pictures color into consideration. We can take the advantage of this model in many situations where we have only greyscale images or videos.

Fig. 20.3 Experiment input and output I

Fig. 20.4 Experiment input and output

Fig. 20.5 Experiment input and output III

5. CONCLUSION

In this research article, we have presented a novel approach to enhance the visual appeal of greyscale images by automating the coloring process using Python and the OpenCV library. The aim of this study was to explore an innovative solution for auto-coloring that could find applications in various domains, including image restoration, art, and entertainment. The results of our experiments and evaluations are promising, indicating that our method can accurately infer colors for a wide range of greyscale images. By training our model on extensive datasets, we have improved its ability to recognize objects, scenes, and patterns, leading to accurate color predictions. Hence, the auto-coloring of greyscale images using Python and OpenCV represents a compelling advancement in the field of image processing and computer vision. While our approach showcases significant potential, it is important to note that there is room for further refinement, such as improving the model's performance on complex images and expanding its training data.

REFERENCES

1. Iizuka, S., Simo-Serra, E., Ishikawa, H. (2016). Let there be Color!: Joint End-to-end Learning of Global and Local Image Priors for Automatic Image Colorization with SimultaneousClassification, 6.pdf, accessed on Jun 2018.
2. An, X., Pellacini, F. (2008). APPPROP: All-Pairs Appearance-Space Edit Propagation, *ACM Transactions on Graphics SIGGRAPH* 2008.
3. Xu, K., Li, Y., Ju, T., Hu, S.-M., Liu, T.-Q. (2009). Efficient affinity-based edit propagation using K-D tree, *ACM Transactions on Graphics*, 28(5).
4. Singh, K. (2021). Performance Comparison Of Wanet Protocols DSDV, DSR and AOMDV using Different Mobility Models and Varying Node Density, *2021 3rd International Conference on Advances in Computing, Communication Control and Networking (ICAC3N)*, pp. 1308- 1311. doi: 10.1109/ICAC3N53548.2021.9725587.
5. Cheng, Z., Yang, Q., Sheng, B. (2015). Deep Colorization, *Proceedings of ICCV 2015*, 29-43.
6. Yatziv, L., Sapiro, G. (2006). Fast image and video colorization using chrominance blending, *IEEE Transactions on Image Processing*, 15(5), 1120-1129.
7. Charpiat, G. (2009). Machine Learning Methods for Automatic Image Colorization, pages 1–27, October 2009.
8. Aggarwal, S., Vasukidevi, G., Selvakanmani, S., Pant, B., Kaur, K., Verma, A., Binegde, G. N. (2022). Audio Segmentation Techniques and Applications based on Deep Learning, *Journal of Scientific Programming*, 2022, Article ID 7994191, 9 pages, *Hindawi*.
9. Graber, G., Pock, T., Bischof, H. (2011). Online 3D reconstruction using convex optimization, *2011 IEEE International Conference on Computer Vision Workshops (ICCV Workshops)*, pp. 708–711.
10. Karlik, B., Sario¨z, M. (2009). Coloring gray-scale image using artificial neural networks, *2009 2nd International Conference on Adaptive Science Technology (ICAST)*, Accra, pp. 366-371.
11. Sharma, A., Abinashsingla, Boparai, R. S. (2012). Colourisation of black and white images-a survey, Department of Computer Science, BGIET, Sangnur, pp. 19-27.
12. Gairola, T., Singh, K. (2016). A review on dos and ddos attacks in cloud environment security solutions, *International Journal of Computer Science and Mobile Computing*, 5(7), pp. 136-141.
13. Babu, D. R., Shankar, R. S., Mahesh, G., Murthy, K. V. S. S. (2017). Facial expression recognition using bezier curves with hausdorff distance, *International Conference on IoT and Application (ICIOT)*, Nagapattinam, pp. 1-8.
14. Limmer, M., Lensch, H. P. A. (2016). Infrared colorization using deep convolutional neural networks, *ICMLA2016*, Anaheim, CA, USA, 18–20 December 2016, pp. 61–68.
15. Buades, A., Coll, B., Morel, J.-M. (2005). A review of image denoising algorithms, *A SIAM Interdisciplinary Journal, Society for Industrial and Applied Mathematics*, 4(2), pp. 490- 530.
16. Singh, K., Singh, A. (2018). Memcached DDoS exploits: operations, vulnerabilities, preventions and mitigations, *2018 IEEE 3rd International Conference on Computing, Communication and Security (ICCCS)*, pp. 171-179.
17. Futschik, D. (2018). Colorization using deep neural networks, *Czech Technical University*.
18. Lezoray, O., Ta, V. T., Elmoataz, A. (2008). Non-local graph regularization for image Colorization, *International Conference on Pattern Recognition*, France, pp. 26-38.
19. Hwang, J., Zhou, Y. (2016). Image Colorization with Deep Convolutional Neural Networks, *Stanford.edu*, pp. 01-07.
20. Sousa, A., Kabirzadeh, R., Blaes, P. (2013). Automatic Colorization of Grayscale Images, Department of Electrical Engineering, Stanford University.
21. Jakeera Begum, M., Venkata Rao, M. (2015). Collaborative Tagging Using CAPTCHA, *International Journal of Innovative Technology And Research*, Volume No.3, Issue No.5, pp. 2436 – 2439.
22. Sharma, R., Singh, K., Chand, L. (2016). Analysis of Image Processing Techniques for Road Anomalies Detection, *International Journal of Emerging Research in Management Technology*, 5(1).
23. Aggarwal, S., Singh, P. (2019). Cuckoo and Krill Herd Based K-Means++ Hybrid Algorithms for Clustering, *Expert Systems, Wiley Online Library*.
24. QuY. Qu, Y., Wong, T.-T., Heng, P.-A. (2006). Manga colorization, *ACM SIGGRAPH 2006 Papers, SIGGRAPH*, pp. 1214–1220.
25. Wu, F., Dong, W., Kong, Y., Mei, X., Paul J C., Zhang, X. (2013). Content-based colour transfer, *Computer Graphics Forum*, 32(1).
26. Lv, Z., Qiao, L. et al. (2021). AI-enabled IoT-Edge Data Analytics for Connected Living, *ACM Trans. Internet Technol.*, 21(4), Article 104.
27. Dhaka, V. S., Meena, S. V., Rani, G. et al. (2021). A Survey of Deep Convolutional Neural Networks Applied for Prediction of Plant Leaf Diseases, *Sensors*, 21(14), 4749.
28. Gaur, L., Singh, G., Solanki, A. et al. (2021). Disposition of Youth in Predicting Sustainable Development Goals Using the Neuro-fuzzy and Random Forest Algorithms, *HCIS, Springer*.

29. Arora, M., et al. (2020). A Systematic Literature Review of Machine Learning Estimation Approaches in Scrum Projects. In: Mallick, P., Balas, V., Bhoi, A., Chae, G. S. (eds) Cognitive Informatics and Soft Computing. *Advances in Intelligent Systems and Computing, vol 1040, Springer,* Singapore.

30. Sood, M. et al. (2019). Optimal Path Planning using Swarm Intelligence based Hybrid Techniques, *JCTN, Vol. 16 No. 9,* pp. 3717–3727.

31. Kumar, M. et al. (2022). Improved Deep Convolutional Neural Network based Malicious Node Detection and Energy-Efficient Data Transmission in Wireless Sensor Networks, *IEEE Transactions on Network Science and Engineering.*

32. Jhanjhi, N. Z., Brohi, S. N., Malik, N. A., Humayun, M. (2020). Proposing a hybrid RPL protocol for rank and wormhole attack mitigation using machine learning, *2020 2nd International Conference on Computer and Information Sciences (ICCIS),* pp. 1-6.

Note: All the figures in this chapter were made by the author.

Computational Intelligence and Mathematical Applications – Dr. Devendra Prasad et al. (eds)
© 2024 Taylor & Francis Group, London, ISBN 978-1-032-87721-1

21

Comparative Analysis of Machine Learning Algorithms used by Recommender Systems to Improve Marketing Strategies on E-commerce

Vikas Sethi* and Rajneesh Kumar

Maharishi Markandeshwar (Deemed to be University), Mullana, Ambala, Haryana (India)

Abstract: The ever-increasing number of online commodities and services makes it more difficult to extract precise information from the data pools. Through adaptive learning and recommendation systems, personalization recommender systems try to lessen this complexity. Recommender systems continue providing people with good or service suggestions in E-commerce. Now a days Machine learning algorithms has been applied to recommender systems from the field of artificial intelligence have increasingly been used in these systems and has become a major area of research field in artificial intelligence. The fundamental benefit of employing machine learning here is that if a model is trained what's in data, it can complete tasks on its own. Because of the large number of algorithms available, selecting an appropriate machine learning algorithm for a recommender system is very important. Here authors provide list of machine learning algorithms that is used by various recommender systems in E-commerce to improve the marketing strategies.

Keywords: Recommendation system, Naive bayes, Machine learning, Random forest, Decision forest, Deep learning, Clustering, Support vector machine

1. INTRODUCTION

The evolution of technological innovations has led to the digital transformation of almost all corporations, irrespective of field. With the availability and use of electronic materials growing so quickly, there may be a big data barrier that prevents easy finding of data over the Internet as a whole. Information retrieval systems like Google tried to solve the problem of quick access from abundant of data but not able to provide the prioritization and personalization. As consequently, recommender systems are nowadays more needed than anything. In accordance with the consumer's choices, desires, or consideration, recommender systems are screening strategies aimed at tackling the issue of overabundance of information by isolating meaningful data elements from a vast amount of user-supplied data. Both consumers and vendors of services benefit from recommender systems [1]. The main research directions for recommender systems are content-based, collaborative filtering, and hybrid techniques. The recommendation in content-based filtering is dependent on how comparable the contents are. If a customer buys an item, the system will suggest another item that is like the one they just bought. This algorithm's key benefit is that it doesn't need any data from additional users. By detecting the selections made by comparable users and subsequently proposing those things to a new user, collaborative filtering operates. This method will produce highly encouraging results when there is a sufficient volume of data. In order to improve system efficiency and get beyond some of the drawbacks and issues associated with traditional recommendation systems, hybrid filtering technique integrates many recommendation algorithms [2]. Machine learning algorithms are replacing regular algorithms as a result of progress in technology making it impossible for a typical suggestion engine to handle data that is constantly changing. Additionally, due to the abundance of stated approaches and types, the ML field lacks a defined categorization for its approaches. As a result, selecting an ML algorithm that satisfies individual's

---
*Corresponding author: vikassethi1987@gmail.com

DOI: 10.1201/9781003534112-21

needs when implementing a recommender system becomes complex and puzzling. The fact that recommender systems operate separately for all sorts of usage makes it difficult for academia to track patterns of ML algorithms in these systems [3].

This study presents a thorough assessment of how machine learning algorithms employed in RSs are explored and applied, as well as what the current trends in ML algorithm product development are. The main focus is on types of machine learning, mostly used machine learning algorithms in recommendation systems and also compare the recommender systems using machine learning.

Breakdown of the paper's structure: Section 2 explains the machine learning and its types, Section 3 explain mostly used machine learning algorithms in recommendation systems, Section 4 compares various recommender systems using different machine learning approach in E-commerce. And in the last conclusion and future work.

2. MACHINE LEARNING IN RECOMMENDER SYSTEMS

Machine learning is a subfield of computer science that enable machines to understand despite having to be directly instructed. One of the most fascinating concepts one has ever met. As the name implies, giving the computer the ability to learn, allowing it to become more like a living person. Machine learning algorithms are a set of instructions that can be used to solve any issue or difficulty. Algorithms are the building blocks of today's complex digital world, and they're created by programmers to tell the computer operating system how to complete jobs. However, in machine learning algorithms, it is the algorithms, not the computer programmers, that develop new rules, instructions, and recommendations, which is why they are termed machine learning algorithms. Instead of giving the computer instructions at every stage, the machine learning algorithm technique gives the computer instructions and guidelines that enable learning and understand from the data without the intervention of a computer programmer [3]. As a result, the computer may be utilized to accomplish complex tasks that a computer programmer would find impossible to complete manually. For example, for visually impaired people, we have a photo recognition tool that converts photographs into voice.

Machine learning is defined as the process of providing systems with training and a capacity for learning automatically and from their experiences with no making it apparent directed or designed. Machine learning focus mainly with the development of computer algorithms that can access any form of data and learn from it. The most basic operation in machine learning is delivering data for getting ready for the training procedure. After that, guidelines will be created by the system based on the outcomes. This is known as the creation of a new algorithm, which is sometimes referred to as the machine learning model. The same could be used to generate and develop many more designs using other dataset. The ability of a machine learning system to deduce new instructions from data provided or accessible is its main strength. In a machine learning model, data is extremely important; the more data provided for training, the more it learns. Machine learning types are basically divided into 4 parts and they are further subdivided as shown in Fig. 21.1 [4].

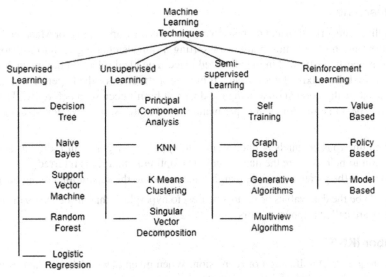

Fig. 21.1 Machine learning techniques

Source: Author

3. MACHINE LEARNING ALGORITHMS USED IN RECOMMENDATION SYSTEMS

3.1 Decision Tree

A Decision Tree can be defined as a tree-like graph with nodes signifying where we pick a feature and pose a question, edges signifying the question's responses, and leaves reflecting the output values or class value.

It's utilized in categorization where data is constantly segregated at each row according to criteria till the final result is produced. The outcome of this algorithm is an optimum result based on a tree structure with constraints or guidelines. Decision Nodes, Design Links, and Decision Leaves are three important components of the decision tree algorithm. Splitting, pruning, and tree selection are all part of the process. For big data sets with low time complexity, decision tree methods are very effective. This is mostly employed in client segmentation and the implemented for marketing strategies in the firm [4].

3.2 Random Forest

Random Forest, a technique which solve categorized data and analysis challenges. It is ensemble learning based method of integrating several categories to handle challenges and improve the effectiveness of the algorithm [5].

Random Forest, a classifier that takes the findings of numerous decision trees and aggregates them to enhance the dataset's predicted accuracy. Random forest is not dependent on individual decision tree; instead, it accumulates projections from every decision tree and uses the popular sovereignty of predictions to anticipate the final output.

The results will be more precise, as the number of trees increase and the issue of collinearity is addressed.

3.3 Naive Bayes

The Naive Bayes Algorithm is a Bayes Theorem based classification machine learning that aids in the categorization of data based on the computation of conditionally probability values. The Nave Bayes is made up of the keywords Nave and Bayes.

Its name, Naive, refers to the idea that the manifestation of a given aspect is unrelated to the description of any other aspects. If an item's size, hue, and aroma are used to identify it, then it can be said to be an apple if it is red, oval, and juicy. As a nutshell, each feature contributes to determining if it is an apple without depending on one another. It is based on the Bayes' Theorem, that's why called Bayes which is used to calculate the probability of a hypothesis previously learned [4].

It utilized class levels represented as feature values or vectors of determinants for classification and applied the Bayes theorem for computation.

It's a group of procedures that all work on the same principle: each set of classified features is distinct from the others. It's a probabilistic classifier, meaning it makes assumptions based on the likelihood of an object.

3.4 Support Vector Machine

A popular method for handling categorization and regression problems is the Support Vector Machine (SVM). Purpose of SVM algorithm is to find effective option or selection margin for sorting n-dimensional region into categories in the long haul, so that further datasets can be properly positioned in the allocated place. The ideal selection perimeter or boundary is known as a hyperplane. SVM is used to find the maximal nodes that aid in the construction of the hyperplane and support vectors are the prominent versions It is exceptionally good at organizing the data that hasn't been seen before [5]. This algorithm aims to find the best hyper plane with the most margin. An ideal hyper plane is used in the Support Vector Machine technique to categories data into multiple classes.

Hyperplane: There may be numerous channels boundary lines to split the categories in n-dimensional space, but we should choose the right decision limit to help analyze the data values. The optimum margin is referred to as the SVM hyperplane. We want to create a hyperplane with the greatest possible margin, that relates to the separation among data points.

Support Vectors: These would be the data values or vectors nearest to hyperplane that influence its location. Because they assist the hyperplane, such vectors are called Support vectors.

3.5 K Nearest Neighbor (KNN)

The KNN method is a technique for classification or regression. When given new unlabeled data, KNN trains a function that generates an appropriate result using designated data as inputs.

K is a positive integer value that determines the number of close by peers. The value of a neighbour is determined by the class set and the between points distance, which is described by Euclidean distance. The distance or length of a segment connecting two locations in the plane or three-dimensional space is known as the Euclidean distance.

The KNN algorithm presume that related units are next to adjacent ones. As per KNN, the new data and old cases are adjustable in each other, and it appoint the new data to the group is most similar to the original groups .The KNN technique saves all accessible information and identifies a novel piece of data relying on its relevance to the existing data. As a result, using the KNN approach, new data can be sorted into the well classification quickly. Due to the fact that it does not instantly benefit from testing, but rather saves the data and employs it to recognize it when the time comes, it is a slow learning mechanism. One use of KNN-search is the use of the KNN classifier in recommender system.

3.6 K-Means Clustering Algorithms

K-Means Clustering is a method for dealing with clustering issues. It's an iterative method for dividing a dataset into K non-overlapping subgroups or clusters depending on their attributes. It seeks to keep clusters as far apart as possible while keeping inter-cluster data points as uniform as possible. With K=2 for two groups, K=3 for three groups, and so on, it defines the number of clusters that need to be created within the process [5]. Each cluster is paired with a centroid in this centroid-based technique. Lowering the overall discrepancies between sampling points and the categories where it falls is the main goal of this method.

The technique separates an input of unorganised data into a number of groups, then repeats the procedure unless no better groupings are obtained. This method requires knowledge of the value of k. This method's main goals are to assign each piece of data to the nearest k coordinates or the centres and to progressively choose the best solution for K coordinates or centroids. A cluster is formed by the points of data that are close to a particular k-centre. Every group is thus separated from the others and contains points of data that are comparable.

3.7 Logistic Regression

In order to allocate values to a defined group of categories, logistic regression employs the logistic sigmoid function. The idea of possibility forms the basis of it. It is an approach to future prognostication. The sigmoid function, also known as a mapping from predictions to probabilities in machine learning, is used to translate anticipated values to probabilities. Using this, you can change any actual number into a number between 0 and 1.

When categorizing data into two or more classes, logistic regression is commonly utilized. The first is binary logistic regression, while the second is multi-class logistic regression. As the names indicate, the binary class has two classes: Yes/No, True/False, 0/1, and so on. There are more than 2 types for recognizing patterns in multi-class classification. In logistic regression, the likelihood which lies anywhere from 0 to 1 (including those of both) is normally determined. Likelihood can then be used to categorize the data.

3.8 Deep Learning

Deep learning is a method that teaches computers to learn by engaging in, just like people do. In order to perform classification tasks using photos, written material, or noise, a computational model must learn via something known as deep learning. Neural network architectures are used in the majority of deep learning approaches. An extensive collection of labelled data and multi-layered neural network designs that automatically learn features from the data are used to train models. The sheer number of neural network layers that are buried typically indicates how "deep" a network is. The feature harvesting and training processes in deep learning are autonomous. Relevant characteristics are extracted by themselves using a deep learning approach. Deep learning also does "end-to-end learning," in which the system is given unprocessed data and a job to complete, including classification, and it gradually understands the way to do it. Deep learning networks frequently get better as the volume of your data grows, which is a significant benefit [6]. Deep-learning architectures like deep neural networks, deep belief networks, reinforcement learning, recurrent neural networks and convolutional neural networks have been used in fields like artificial intelligence, recognition of speech, automated translation, biological informatics, development of drugs, image analysis for medicine, environmental science and gaming programs [7]. They have delivered outcomes on par with human expert performance, and in some cases even better.

4. COMPARISON OF DIFFERENT MACHINE LEARNING TECHNIQUES USED BY RECOMMENDATION SYSTEMS E-COMMERCE

Table 21.1 Comparing different machine learning techniques in recommendation systems

Author Name	Name of Recommender System (Year)	Machine Learning Approach used	Dataset
Asma Sattar et.al [8]	Novel Hybrid Approach (2017)	KNN	Movie Lens and Film Trust
Muyeed Ahmed et al [9]	Movie recommender system (2018)	K-means clustering	Movie Lens
Hafed Zarzour et al [10]	Collaborative filtering Recommendation System (2018)	K-means and SVD	Movie Lens
Antonio Jesús Fernández-Garcíaa et al [11]	Recommender system to component-based interfaces (2019)	Logistic regression, Decision trees, ANN and Decision Jungle	Data of ENIA interface
Baolin Y et al [12]	DMF (2019)	Deep learning	Movie Lens, Douban
Jiayu Han [13]	ADLFM(2020)	Deep learning with Latent Factor Model	Amazon (Electronics)
Suja Cherukullapurath Mana et al [14]	Hybrid recommendation system (2021)	(SVM)support vector machine and Naïve Bayesian algorithm	Twitter dataset and Movie lens
Arash Oshnoudi et al [15]	Clustering based Recommendation System (2021)	KNN and Auto MLP	Movie Lens
Rudraprasad Mondal [16]	DeCs(2022)	DNN	Movie Lens, Amazon and Douban
Vikas Sethi et al [17]	LCNA(2023)	LSTM, CNN	Amazon(electronics)

5. CONCLUSION AND FUTURE SCOPE

Recommender systems (RS) are now widely utilized in e-commerce, social media, and a variety of other fields. RS research has progressed a lot since its launch. Machine learning (ML), which are now paired with recommender systems, permit machines to understand basis of user knowledge and enormous datasets, design a recommendation model, and provide us an accurate recommendation, were the major milestone. The sharp rise of information generated by modern and automation devices needs the evolution of smarter approaches and solutions which can appropriately and effectively maintain, handle, analyze, and interpret information for optimal advantages. Machine learning-based recommender systems are one of the most effective answers to these problems, as they are excellent instruments for swiftly facilitating the relevant data process. This paper surveys various machine learning algorithms and provided most of the popular machine learning algorithms that is used by various recommender systems in E-commerce to improve the marketing strategies.

REFERENCES

1. Isinkaye, F.O., Folajimi Y.O., Ojokoh, B.A. (2015). Recommendation systems: Principles, methods and evaluation. *Egyptian Informatics Journal*, *16*(3), 261–273.
2. Sethi, V., Gujral, R. (2022). Survey of different recommendation systems to improve the marketing strategies on e-commerce. *Proceedings of the 2022 International Conference on Machine Learning, Big Data, Cloud and Parallel Computing (COM-IT-CON)*, 26–27 May 2022, Faridabad, India, pp. 119–125.
3. Portugal, I., Alencar, P., Cowan, D. (2017). The Use of Machine Learning Algorithms in Recommender Systems: A Systematic Review. *Expert Systems with Applications*, *97*, 205–227.
4. Batta, M. (2020). Machine Learning Algorithms -A Review. *International Journal of Science and Research (IJSR)*, *9*(1), 381-386.
5. Carvalho, T.P., Soares, F.A.A.D.M.N., Vita, R., Francisco, R.D.P., Basto, J.P.T.V., Alcalá, S.G.S. (2019). A systematic literature review of machine learning methods applied to predictive maintenance. *Computers & Industrial Engineering*, *137*, 1–14.
6. Mu, R. (2018). A Survey of Recommender Systems Based on Deep Learning. *IEEE Access*, *6*, 69009-69022.
7. Dargan, S., Kumar, M., Ayyagari, M.R., Kumar, G. (2020). A Survey of Deep Learning and Its Applications: A New Paradigm to Machine Learning. *Archives of Computational Methods in Engineering*, *27*, 1071–1092.

8. Sattar, A., Ghazanfar, M.A., Iqbal, M. (2017). Building Accurate and Practical Recommender System Algorithms Using Machine Learning Classifier and Collaborative Filtering. *Arabian Journal for Science and Engineering, 42*(8), 3229–3247.

9. Ahmed, M., Imtiaz, M.T., Khan, R. (2018). Movie Recommendation System Using Clustering and Pattern Recognition Network, *IEEE 8th Annual Computing and Communication Workshop and Conference (CCWC)*, 143–147.

10. Zarzour, H., Al-Sharif, Z., Al-Ayyoub, M., Jararweh, Y. (2018). A new collaborative filtering recommendation algorithm based on dimensionality reduction and clustering techniques. *IEEE 2018 9th International Conference on Information and Communication Systems (ICICS)*, 102–106.

11. Fernández-Garcíaa, A.J., Iribarne, L., Corral, A., Criado, J., Wang, J.Z. (2018). A recommender system for component-based applications using machine learning techniques. *Knowledge-Based Systems, 164*, 68–84.

12. Yi, B., Shen, X., Liu, H., Zhang, Z., Zhang, W., Liu, S., Xiong, N. (2018). Deep Matrix Factorization With Implicit Feedback Embedding for Recommendation System. *IEEE Transactions on Industrial Informatics, 15*(8), 4591–4601.

13. Han, J., Zheng, L., Xu, Y., Zhang, B., Zhuang, F., Yu, P.S., Zuo, W. (2020). Adaptive Deep Modeling of Users and Items Using Side Information for Recommendation. *IEEE Transactions on Neural Networks and Learning Systems, 31*(3), 737–748.

14. Mana, S.C., Sasipraba, T. (2021). A Machine Learning Based Implementation of Product and Service Recommendation Models. *IEEE 2021 7th International Conference on Electrical Energy Systems (ICEES)*, 543–547.

15. Oshnoudi, A., Neysiani, B.S., Aminoroaya, Z., Nematbakhsh, N. (2021). Improving Recommender Systems Performances Using User Dimension Expansion by Movies' Genres and Voting-Based Ensemble Machine Learning Technique. *IEEE 2021 7th International Conference on Web Research (ICWR)*, 175–181.

16. Mondal, R., Bhowmik, B. (2022). DeCS: A Deep Neural Network Framework for Cold Start Problem in Recommender Systems. *2022 IEEE Region 10 Symposium (TENSYMP)*, 1–6.

17. Sethi, V., Kumar, R., Mehla, S., Bhandari, A., Nagpal, S., Rana, S. (2023). LCNA-LSTM CNN based attention model for recommendation system to improve marketing strategies on e-commerce. *Journal of Autonomous Intelligence (2024), 7*(1).

Note: All the figure and table in this chapter were made by the author.

Computational Intelligence and Mathematical Applications – Dr. Devendra Prasad et al. (eds)
© *2024 Taylor & Francis Group, London, ISBN 978-1-032-87721-1*

22

Evolution of Recommender System: Conventional to Latest Techniques, Application Domains and Evaluation Metrices

Vikas Sethi[1]
Panipat Institute of Engineering and Technology, Samalkha, Haryana (India)

Rajneesh Kumar[2]
Maharishi Markandeshwar (Deemed to be University), Mullana, Ambala, Haryana (India)

Suresh Chand Gupta, Sunny Kuhar, Sumit Kumar Rana
Panipat Institute of Engineering and Technology, Samalkha, Haryana (India)

Abstract: Today's world needs recommender systems because of rising Internet availability, rising personalization tendencies, and shifting computer usage patterns. Overloaded with information is a problem that is typically solved by recommendation systems (RS). Nowadays, there is a lot of information being produced, which makes it challenging for users to locate the pertinent information about goods and services that suit their tastes and interests. Although recommender systems have become prevalent in e-commerce, multimedia, tourism, and many other industries and are effective at producing good suggestions, they continue to face difficulties including cold start, sparsity, scalability, and many more. Scholars have gotten more and more engaged in RSs in the past few years, and numerous analyses of literature have been completed in order to look into the attributes, difficulties, and methodologies of various RSs. But neither of the preceding audits offered an in-depth investigation of every one of RSs. For the purpose of educating and directing novice researchers who are interested in the topic, the authors of this paper offer a thorough evaluation of the recommendation system. In order to act as a guide for future research and practise in this area, this paper examines the fundamental methodologies and widely used approaches, including machine and deep learning techniques in recommender systems with assessment metrices as well as the application areas of recommender systems.

Keywords: Recommender system, Machine learning, Deep learning, Collaborative filtering, Cold-start, Sparsity and trust

1. INTRODUCTION

The exponential rise in data production from electronic and automated equipment is unfathomably large. It could be challenging for people to sift across every bit of such information and identify those that are crucial components. Many online commerce companies make product recommendations to their customers because there are lakhs of items accessible through a single site. The overwhelming number of choices that the typical user has can result in an overload of information [1]. Recommender systems aim to reduce overwhelm with data while also personalizing an individual's encounter by offering individuals accurate tailored recommendations of products or services that suit what they want. A recommendation system (RS) uses the information given to decide whether something will prove relevant to a consumer. A recommendation engine (RE) is another name for a recommender system (RS).

Corresponding authors: [1]vikassethi1987@gmail.com, [2]drrajneeshgujral@mmumullana.org

DOI: 10.1201/9781003534112-22

The key distinction between an information retrieval (IR) system and a recommender system (RS) is that the latter takes into account both past user preferences for products as well as the past sales history of those products across all users. Therefore, for a RS, both the user and the item are significant entities, while for a system for retrieving information, just the object was a significant entity [1]. Internet-based shops and merchants use these types of platforms, and throughout the past several years, the prevalence of their use continues to be swiftly increasing. Every business has a recommendation system; thus, businesses select the recommender mechanism as the primary method of identifying products for customers based on customer preference to product matches. It assists users in daily tasks such as locating their favorite tunes or video or means of transportation [2]. The user interaction and overall organization earnings or strategy are both positively impacted by top-notch RSs. Both vendors and consumers are benefited by recommender systems. In an online buying environment, they lower the entire transaction costs associated with searching and choosing products. It has been demonstrated that recommendation systems enhance the way decisions are made and the efficacy of decision-making. Considering they serve as a successful approach to sell additional goods in an e-commerce environment, recommender systems increase revenues [3].

Extensive studies have already been made to take a peek at the traits, complications, and methodologies that affect different RSs in recent years as RSs have attracted the consideration of a broad spectrum of experts. But none of these audits gave an exhaustive review across all RSs. People frequently engage with a variety of recommender systems in areas other than e-commerce, such as recommending books, jobs on LinkedIn, YouTube videos, news articles, and friends on Facebook, as well as recommending research papers, courses to students [3] [4].

The author of this publication provides a thorough introduction to RSs. The authors emphasize many RSs classifications and examine contemporary RSs' problems, such as cold start, data sparsity, scalability, and trust. How to assess the effectiveness of the RSs using several metrics, including MAE, RMSE, recall, precision, and F-measures. Determine the way a variety of sectors and different fields have used RSs. Following is an outline for this piece of writing: The overview of suggestion and its transition from traditional to new methodologies is presented in Section 2. The practical areas of RSs are outlined in Section 3, and the various evaluation metrics are described in Section 4. And in the last conclusion.

2. RECOMMENDER SYSTEM

The role of recommendations has grown in significance and altered how users and websites interact. The recommender system is thought of as a method of supporting and strengthening the overall decision-making process when particular knowledge and understanding are inadequate. Recommender systems help users deal with the issue of overabundance of information by offering them a range of options, unique content, and customized content [4]. A strong filtering system known as a recommender system predicts client opinions and buying tendencies. In the area of electronic commerce, the problem of contradicting information could be helped by a recommender system. The procedure for making decisions as well as quality have both been enhanced by recommendation systems. Recommendation systems increase profits in the field of online commerce because they are a wonderful way to promote a variety of goods. People can do much more with the assistance of recommendation systems than just seek for information [5].

Finding the client's needs—whether they be for information, a product, a service, music, videos, or other content—is done using the recommender framework (RS). Numerous industries, including e-commerce, education, tourism, scientific research, and many more, use recommender systems on a regular basis.

Recommendation system algorithm consists of mainly 3 steps, for creating an individual's profiling and portraiture to support the suggestion, firstly accumulate every piece of vital and significant details pertaining to the user's behaviors and the internet pages or materials they frequented. The user's behaviours are separated identified in the second stage utilizing the knowledge and understanding generated in the preceding stage, and the consumer receives suggestions or proposed commodities, goods, and amenities that they might like [2]. Since the advent of recommendation systems, many different recommendation algorithms have developed, and as their accuracy continues to increase, the techniques of nowadays are both far more accurate and efficient in their ability to provide recommendations.

2.1 Conventional Techniques of Recommender Systems

Since the recommender systems are introduced, many recommendation algorithms have emerged. The recommendation techniques can be divided into three main categories: Content-Based Filtering (CBF), Collaborative Filtering (CF) and Hybrid Filtering.

Content-based filtering (CBF)

CBF is a pretty straightforward conventional recommendation technique. The items that the CBF approach suggests are comparable to the things that attract consumers. The most important stage is information comparison between things and users. Content-based filtering separates items based on how similar the content of previously observed items is to that of newly discovered ones. While a specific user has watched a motion picture, CBF suggests further films that are similar to the first. Movie genre, cast, humor, drama, director, or plot can all influence how similar two films are. The cold start issue that CF has is not present in CBF. From the very beginning, something can be suggested to a new user based on his background data, which includes his current age and place of residence [6].

Another benefit of CBF is that it may be explained to the user, so he knows exactly anything has been suggested to him. In contrast, CF lacks the CBF's potential to be explained. The bubble effect is a drawback of CBF since it may suggest a movie that a user may not enjoy because it is quite exactly like the previous one, they watched. For instance, if a person has seen ToyStory-1, ToyStory-2 will be suggested to him, and so forth.

Collaborative filtering

In collaborative filtering, once an intended customer's preferences have previously matched those of another user, things which the other individual strongly suggests will be suggested to the intended recipient going forward. It is of the utmost importance to understand what extent a user has previously liked an item for collaborative filtering. Someone who uses it has the option of either directly or indirectly expressing his preference for a product. The collaborative filtering method creates a record of individual user preferences about products (user-item matrix) [6].

It then pairs users with comparable tastes and inclinations to come up with solutions through reviewing their profile information. These users create a community of their own. Users can get suggestions for goods they haven't yet evaluated but have heard good things about from other users in their neighborhood. The generation of models by collaborative filtering additionally assists customers in exploring novel interests, which is another benefit. Memory-Based Collaborative Filtering and Model-Based Collaborative Filtering are two categories under which this model falls. Whenever an individual is novel and their preferences is undetermined or whenever the item is novel and hasn't been tried by any individual, CF has cold start issues.

Hybrid filtering

To improve performance, hybrid systems combine two or more strategies. Due to the dependence of the Content-Based Filtering model on an individual's overview of the item's and the Collaborative Filtering model on an individual's item grading information, both filtering methods have constraints. To overcome the shortcomings of both recommendation filtering techniques and to enhance suggestion effectiveness, a hybrid recommendation model been presented [2].

The fundamental goal of hybrid filtering is to deliver suggestions that are more accurate and profitable than a particular recommendation filtering techniques because, when two distinct approaches are combined, the advantages of one particular approach outweigh the drawbacks of the other. In a combination framework, it is also possible to integrate various recommendation strategies to mask the shortcomings of one algorithm, therefore a hybrid method is employed to benefit from both approaches. According to the method that integrates filtering techniques, the hybrid recommendation model is classified into seven categories: weighted hybridization, switching hybridization, cascade hybridization, mixed hybridization, feature combination, feature-augmentation, and meta-level.

2.2 Issues in Recommender System

Numerous academics have developed and explained various recommendation algorithms in response to needs, however cold start, sparsity, scalability, trust and serendipity are some of the unresolved problems and challenges that are handled by current recommender systems.

Cold start

It is often viewed as a trouble whenever someone who recommends something does not have adequate knowledge concerning the individual or something to make reliable recommendations in this situation. This is one of the key problems affecting the performance of the recommendation system. The technology cannot become updated with the newly added customer's choices because they haven't done assessed any of them, therefore their online profile remains useless [4].

Sparsity

When the client numbers are comparable to or beyond a certain number of available data elements, recommender systems function successfully. However, there are always fewer users than there are available items. Therefore, a dismantled profile matrix results in less accurate suggestions when there is a high number of data components in the collection and user resistance to rating products [4].

Scalability

It is characterized as a characteristic of recommender systems wherein the framework is capable of handling suggestions despite evolving circumstances especially when there exist numerous people and merchandise and when the amount of information expands regularly as fresh individuals and goods get added to the collection. A recommender system finds it very challenging to manage such large and dynamic datasets [4].

Trust and privacy

Along with the advancement of internet access and technologies around the globe, purchasing goods online or websites for e-commerce are growing fairly quickly. Although the rise in users, electronic commerce continues to encounter a number of difficulties, including one and this is gaining customers' trust while making purchases. Customers fear that they will be taken advantage of and fear receiving a good or service that differs from the information on the website. It is believed that customer reviews, whether they be comments or ratings, have the power to change how purchasers behave while making purchases. So, trust should be established among users [7].

Serendipity

It can be characterised as a chance event or consequence that is advantageous. While serendipitous results or things can occasionally be beneficial and useful for users, if the recommender system just offers these results or items without also recommending similar items, consumers may come away with the feeling that the system is unreliable. The collaborative filtering strategy has the potential to provide unexpected results because this recommendation algorithm only employs "neighbours" to promote items rather than taking the item's content into account [8] [9].

2.3 Latest Techniques

Machine learning algorithms have begun to replace traditional algorithms as a result of advancements in new technologies that make it impossible for standard recommendation system strategies to handle data in real-time. Deep learning and machine learning are currently undergoing another shift. There is a lot of potential in the field of study of recommendation systems employing machine and deep learning, which is also a machine learning subgroup, as these learning methods have prospective applications. Recommender systems employing machine and deep learning techniques are used in order to enhance the efficiency of precision of predictions, deal with difficulties with data scarcity as well as cold start, and more. This results in a better user experience with better recommendations.

Machine learning techniques

Computer science's subject of machine learning gives computers the ability to comprehend without being explicitly trained. Giving the computer the capacity to learn will, as the name suggests, enable it to resemble someone actually a living being. A collection of guidelines known as machine learning algorithms can be utilised to address any problem or challenge. This is why they are called machine learning algorithms: in machine learning algorithms, fresh regulations, directions, and suggestions are created by the algorithms themselves rather than by programming experts, enabling the computer to learn from the data and make inferences on its own [10]. Retrieving crucial data and learning from previous customer behaviour are two functions of machine learning in recommender systems. Machine learning employs a variety of strategies, including supervised learning, semi-supervised learning, unsupervised learning, and reinforcement learning, to tackle data difficulties. Decision trees, random forests, naive bayes, support vector machines, clustering, k-means, and k closest neighbours are a few of the machine learning techniques that are most frequently utilized [11].

Deep learning tehcniqes

The study of deep learning is a new area in machine learning. Artificial neural networks are the cornerstone of the "deep learning" subfield of machine learning. It possesses capacity to discover complicated patterns and connections inside data. Not every algorithm in deep learning needs to be preprogrammed. due to the simple reason this system has its foundation on deep neural networks (DNNs), which are additionally referred to as artificial neural networks (ANNs). Deep neural networks, that employ a multitude of nodes that are interconnected to interpret and transform statistics, are the essential component of deep learning [12]. The primary idea behind it is to eliminate the need for manually designing features in conventional machine learning by extracting characteristics from data through the combination of lower-level characteristics to create stronger the highest-level lexical abstracts. Deep learning has significantly altered recommendation architectures and increased options to boost recommender performance. Considering these algorithms were capable to get around the shortcomings of conventional algorithms and deliver recommendations of high calibre, new developments in deep learning-based recommender systems (DLRS) have sparked plenty of curiosity. [13]

In addition to enabling the codification of more complicated Abstractions that are as a means to describe data organization in the upper stages, deep learning is able to efficiently capture the non-stationary and complicated user-item associations. For recommender systems, deep neural networks, deep belief networks, reinforcement learning, auto-encoders, recurrent neural networks, and convolutional neural networks are the deep learning methods most frequently utilized [14] [15].

3. APPLICATION DOMAINS FOR RECOMMENDER SYSTEM

Many application categories in current RS research are identified in this publication. This section goes through the different application domains and contexts that are used to make recommendations. Due to the increasing growth of data, current RS researchers are focusing on the designing of recommender systems.

3.1 E-commerce

It is undeniable that the Internet increases market rivalry, resulting in the emergence of numerous perspectives aimed at developing successful marketing techniques. Many recommender systems for e-commerce applications have been created in recent years. E-shopping is a specialty and prominent e-commerce field. The majority of some of the most prominent online stores, such as eBay and Amazon, now use recommender services to manage their customers find items to shop. On these B2C online marketplaces, commodities are able to be offered through aggregate best- performing, user profiles, or an analysis of the their prior spending habits for an estimate for subsequent spending habits. Research additionally offered some sophisticated models for a variety of online shopping environment aspects [3].

3.2 E-learning, New and Documents Recommendation

Because of the increasing expansion of resources, recommender systems have traditionally suggested webpages, papers, and news. Several recommendation systems systems are designed around keyword analysis, and textual material, including webpages and news is typically characterized as a set of keywords that can be obtained from historical data, URLs, and online services. The migration of learning recommenders to mobile platforms is a common theme in the reviewed literature. Owing to the pervasive utilisation of mobile phones and their convenience of use, educational RS has been included into mobile platforms. These recommender systems are typically designed to assist learners with identifying programmes, areas, as well as supplies which they are interested in passionate about [5] [8].

3.3 Multimedia Domain

Customization of multimedia content using social platforms is an essential consideration in the multimedia realm. The vast array of social media resources and services has overwhelmed users. With the popularity of mobile phones in recent times, there seems to be an especially rapid increase of movie, video, and music resources. Users, on the other hand, on mobile devices, they got irritated when browsing for material that delights them. Many movie and music recommendation algorithms have been created to address the issue [3].

3.4 Travel and Tourism

In the tourism industry, recommender systems play an important role. There are numerous tourism websites accessible that give visitors with timely information about the various services available in the area. This is critical since the internet has revolutionized the tourist industry to the point where most travelers no longer utilize travel agents. Tourists have a lot of access to tourism information thanks to the internet and smart devices. Tourists can use e-tourism recommender systems to get recommendations. Certain systems focus on certain sights and locations, whereas others offer tour packages that also include housing, meals and transportation [5].

4. EVALATION METRICES

Finding the qualities that create an excellent RS and figuring out how to quantify them are necessary when assessing a RS. The metrics that are used vary according to the method that is used. Accuracy is evaluated using statistics and decision support accuracy metrics. The structure of the set of data and the variety of operations that the recommender system will carry out define the practicality of each statistic [1].

Metrics for statistical accuracy assess by comparison of anticipated ratings with the user's actual rating. Accuracy, Mean Absolute Error (MAE), Root Mean Square Error (RMSE), and Correlation are frequently employed.

The computation for MAE looks like this: $\text{MAE} = \dfrac{1}{N}\sum_{i=1}^{M}\left(V_i - \hat{V}_i\right)^2$

N is the number of datapoints, V_i the i-th assessment and $\hat{V}_i$ the related assessment. The lower the MAE, better the forecasting.

Greater absolute mistake is prioritized by Root Mean Square mistake (RMSE), and the lower the RMSE, better the accuracy for recommendations are. $RMSE = \sqrt{\dfrac{\left\|\sum_{i=1}^{N} V_i - \hat{V}_i\right\|^2}{N}}$

Reversal rate, weighted errors, receiver operating characteristics (ROC), precision recall curve (PRC), precision, recall, and F-measure are commonly used measures for decision support accuracy [1] [3]. These measures assist users in choosing from among the available things those that are of extremely high quality. The metrics see the prediction process as a binary operation that separates the good things from the bad things. They are computed as

Precision = Perfectly suggested things / Total suggested things

Recall = Perfectly suggested things / Total useful suggested things

Precision and recall can be combined into a single statistic using the F-measure, which is defined as: F-measure = 2PR/(P + R)

5. CONCLUSION

People are discovering recommender systems to have become a useful tool to develop options that reflect their specific requirements. Additionally, it gives consumers the opportunity to use items and amenities that are not always accessible via the platform, especially contributes to solving the challenge of content overabundance which has become a fairly common occurrence with information retrieval systems. In order to increase the accuracy of recommendations, this study explained how recommender systems evolved from traditional methodologies to machine and deep learning-based techniques. The problems with recommendation systems were also covered. The authors of this paper additionally offer illustrations of how recommendation systems are used in media, e-health, transportation, and other sectors. The most common evaluation metrics for recommendation systems were also covered. Investigators are going to gain from such information, which also serves as a manual for improving innovative recommendation methodologies.

REFERENCES

1. Isinkaye, F.O., Folajimi Y.O., Ojokoh, B.A. (2015). Recommendation systems: Principles, methods and evaluation. *Egyptian Informatics Journal*, 16(3), 261–273.
2. Ko, H., Lee, S., Park, S., Choi, A. (2022). A Survey of Recommendation Systems: Recommendation Models, Techniques, and Application Fields. *Electronics*, 11(1).
3. Fayyaz, Z., Ebrahimian, M., Nawara, D., Ibrahim, A., Kashef, R. (2020). Recommendation Systems: Algorithms, Challenges, Metrics, and Business Opportunities. *Appl. Sci.*, 10(21), 7748.
4. Sethi, V., Gujral, R. (2022). Survey of different recommendation systems to improve the marketing strategies on e-commerce. *Proceedings of the 2022 International Conference on Machine Learning, Big Data, Cloud and Parallel Computing (COM-IT-CON)*, 26–27 May 2022, Faridabad, India, pp. 119–125.
5. Haruna, K., Ismail, M.A., Suhendroyono, S., Damiasih, D., Pierewan, A.C., Chiroma, H., Herawan, T. (2017). Context-Aware Recommender System: A Review of Recent Developmental Process and Future Research Direction. *Appl. Sci.*, 7(12), 1211.
6. Guha, R. (2021). Improving the performance of an artificial intelligence recommendation engine with deep learning neural nets. *Proceedings of the IEEE 2021 6th International Conference for Convergence in Technology (I2CT)*, 2–4 April 2021, Maharashtra, India, 1–7.

7. Selmi, A., Brahmi, Z., Gammoudi, M.M. (2016). Trust-based Recommender Systems: An overview. *27th IBIMA Conference*, Milan (Italy).

8. Bai, X., Wang, M., Lee, I., Yang, Z., Kong, X., Xia, F. (2019). Scientific Paper Recommendation: A Survey. *IEEE Access*, *7*, 9324-9339.

9. Wang, C., Deng, Z., Lai, J., Yu, P.S. (2019). Serendipitous Recommendation in E-Commerce Using Innovator-Based Collaborative Filtering. *IEEE Transactions on Cybernetics*, *49*(7), 2678-2692, July 2019.

10. Portugal, I., Alencar, P., Cowan, D. (2017). The Use of Machine Learning Algorithms in Recommender Systems: A Systematic Review. *Expert Systems with Applications*, *97*, 205-227.

11. Mahesh, B. (2020). Machine Learning Algorithms - A Review. *IJSR*, *9*(1), 381-386.

12. Khanal, S.S., Prasad, P.W.C., Alsadoon, A., Maag, A. (2020). A systematic review: machine learning based recommendation systems for e-learning. *Education and Information Technologies*, *25*, 2635-2664.

13. Mu, R. (2018). A Survey of Recommender Systems Based on Deep Learning. *IEEE Access*, *6*, 69009–69022.

14. Batmaz, Z., Yurekli, A., Bilge, A., Kaleli, C. (2019). A review on deep learning for recommender systems: challenges and remedies. *Artificial Intelligence Review*, *52*, 1–37.

15. Sethi, V., Kumar, R., Mehla, S., Bhandari, A., Nagpal, S., Rana, S. (2023). LCNA-LSTM CNN based attention model for recommendation system to improve marketing strategies on e-commerce. *Journal of Autonomous Intelligence (2024)*, *7*(1).

Computational Intelligence and Mathematical Applications – Dr. Devendra Prasad et al. (eds)
© 2024 Taylor & Francis Group, London, ISBN 978-1-032-87721-1

23 | Securing Cloud Data: A Comprehensive Analysis of Encryption Techniques and Recent Advancements

Juvi Bharti[1] and Sarpreet Singh[2]

Department of Computer Science, Sri Guru Granth Sahib World University Punjab, India

Abstract: This paper provides an in-depth analysis of cloud vulnerabilities and explores the role of encryption techniques in addressing security challenges within cloud services. We compare a range of symmetric and asymmetric encryption algorithms and discuss recent advancements in modified encryption models. The study highlights the importance of understanding the encryption landscape, selecting appropriate techniques based on specific requirements and threat models, and keeping up-to-date with the latest developments in cryptography to ensure data security in the cloud.

Keywords: Security, Cloud computing, Encryption technology and algorithms, Privacy, Cipher

1. INTRODUCTION

The rapid adoption of cloud computing has revolutionized the way businesses and individuals store, process, and manage data. Cloud services offer scalability, cost-effectiveness, and flexibility, making them an attractive choice for organizations of all sizes. However, this widespread adoption has also brought forth a myriad of security concerns, as sensitive information is now being stored outside traditional, on-premises infrastructure. As cloud services continue to evolve, it is crucial to understand the vulnerabilities inherent to these environments and the various encryption techniques that can be employed to mitigate the associated risks. Cloud vulnerabilities can be broadly classified into two categories: those arising from the cloud infrastructure itself and those resulting from user negligence or misconfiguration. Infrastructure-related vulnerabilities include issues like virtual machine (VM) escape, side-channel attacks, and denial of service (DoS) attacks [1]. These attacks can result in unauthorized access to data, data leakage, or loss of service availability. On the other hand, user-related vulnerabilities stem from improper management of access controls, weak authentication mechanisms, and misconfigured security settings [2]. Both types of vulnerabilities can have severe consequences, and it is essential for organizations to proactively assess and address the associated risks.

Encryption is a key component of any robust cloud security strategy. It is the process of converting data into a form that is unreadable to unauthorized users, ensuring that sensitive information remains confidential, even if an attacker gains access to the cloud environment [3]. Encryption techniques can be categorized into two primary types: symmetric and asymmetric encryption.

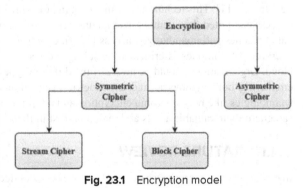

Fig. 23.1 Encryption model

[1]juvibharti@yahoo.com, [2]ersarpreetvirk@gmail.com

DOI: 10.1201/9781003534112-23

2. SYMMETRIC ENCRYPTION

It is also known as secret key encryption, employs a single key for both encryption and decryption [4]. This means that the same key is used to transform the plaintext (unencrypted data) into cipher text (encrypted data) and to convert the cipher text back into the plaintext. Symmetric encryption algorithms, such as the Advanced Encryption Standard (AES), are generally faster and more efficient than their asymmetric counterparts, making them suitable for encrypting large volumes of data [5]. However, symmetric encryption presents a key management challenge, as the secret key must be securely shared between the communicating parties. If the key is intercepted or compromised, the encrypted data is at risk of being exposed [6].

Asymmetric encryption

It is also known as public key encryption, addresses the key management issue by utilizing a pair of mathematically related keys: a public key and a private key [7]. The public key is used for encryption, while the private key is used for decryption. As the name suggests, the public key can be freely shared, while the private key must remain confidential. Common asymmetric encryption algorithms include RSA, Elliptic Curve Cryptography (ECC), and the Diffie-Hellman key exchange [8]. Asymmetric encryption is generally slower and more computationally intensive than symmetric encryption but offers a higher level of security due to the separation of the encryption and decryption keys [9].

IMPORTANCE OF CLOUD ENCRYPTION

Fig. 23.2 Need of cloud encryption

In addition to these fundamental encryption techniques, there are various hybrid approaches that combine the strengths of both symmetric and asymmetric encryption. One such approach is the use of secure key exchange protocols, such as the Transport Layer Security (TLS) protocol, which enables the secure transmission of symmetric keys using asymmetric encryption [10]. This hybrid approach facilitates efficient and secure communication between parties, leveraging the speed of symmetric encryption while maintaining the security benefits of asymmetric encryption. Another emerging area of research is homomorphic encryption, which allows for computations to be performed directly on encrypted data without the need for decryption [11]. This technique has the potential to significantly enhance cloud security by enabling cloud service providers to process encrypted data without ever having access to the underlying plaintext, effectively mitigating the risks associated with data breaches and unauthorized access [12].In conclusion, understanding cloud vulnerabilities and implementing appropriate encryption techniques is critical to ensuring the security and privacy of sensitive information stored and processed in the cloud. Organizations should be aware of the different types of vulnerabilities and assess their cloud environments accordingly, employing both symmetric and asymmetric encryption techniques to protect their data. By staying abreast of emerging technologies like homomorphic encryption and adopting a proactive approach to cloud security, organizations can effectively safeguard their valuable assets and maintain trust in their cloud-based services.

3. LITERATURE REVIEW

In recent years, numerous research studies have been conducted to explore various aspects of cloud vulnerabilities and encryption techniques. This literature review provides a brief overview discussing the key findings and contributions to the field.

In [13], the authors propose an enhanced security framework for cloud data storage using block chain technology. They discuss the limitations of existing cloud security solutions and highlight the potential benefits of utilizing block chain for data integrity, access control, and auditing purposes. The proposed framework combines encryption, data fragmentation, and block chain to ensure data confidentiality, integrity, and availability in cloud environments. This study contributes to the growing body of literature on the application of block chain technology in cloud security.

A study by [14] presents a comprehensive survey on the security challenges and vulnerabilities associated with cloud computing. The authors systematically review the literature, identifying and categorizing various threats, attacks, and vulnerabilities in the cloud environment. They also discuss different security mechanisms and countermeasures, including encryption, to address these challenges. This work provides a valuable reference for researchers and practitioners seeking to understand the current state of cloud security research.

In [15], the authors propose a new method for secure data sharing in the cloud using attribute-based encryption (ABE) and block chain technology. They introduce a decentralized access control mechanism that leverages ABE to ensure fine-grained access control over encrypted data and utilizes block chain for secure and transparent auditing. The proposed approach provides enhanced security, scalability, and privacy compared to traditional centralized access control mechanisms.

A study by [16] investigates the performance of different cryptographic algorithms in the context of cloud storage. The authors analyse the efficiency, security, and resource consumption of various symmetric, asymmetric, and hash-based encryption algorithms. Their findings suggest that the choice of encryption algorithm significantly impacts the overall performance and security of cloud storage systems, highlighting the importance of selecting appropriate encryption techniques based on specific use cases and requirements.

In [17], the authors present a novel privacy-preserving scheme for secure data storage and retrieval in cloud environments. Their approach combines homomorphic encryption with a secure indexing mechanism to enable search and retrieval operations on encrypted data without revealing sensitive information. The proposed scheme addresses both data confidentiality and privacy concerns, making it a promising solution for securing sensitive data in the cloud.

A study by [18] explores the use of machine learning techniques to detect and mitigate cloud vulnerabilities. The authors propose a framework that employs machine learning algorithms to analyse cloud system logs and identify potential vulnerabilities in real-time. This proactive approach to cloud security management allows organizations to detect and address vulnerabilities before they can be exploited by attackers, improving overall security and resilience.

In [19], the authors present a comprehensive analysis of the security and privacy challenges associated with the adoption of multi-cloud environments. They discuss various security risks and vulnerabilities in multi-cloud settings and propose a set of best practices and countermeasures to mitigate these risks, including the use of encryption and secure access control mechanisms. This study contributes to a better understanding of the unique security challenges associated with multi-cloud deployments and provides practical guidance for organizations implementing multi-cloud strategies.

4. COMPARISON OF EXISTING METHODS

In this section the goal is to compare various existing encryption methods to better understand their performance, security levels, and applicability in different scenarios. Through a detailed analysis, we will evaluate the unique advantages and limitations of different encryption techniques, such as symmetric and asymmetric encryption, as well as their modifications. This comparison will enable us to identify the most appropriate encryption methods for specific use cases and environments, thereby providing insights on how to effectively enhance data security in the cloud.

Table 23.1 provides an overview of various encryption algorithms, highlighting their key size, security level, speed, memory usage, key management complexity, and use cases. From the table, it is evident that there is no one-size-fits-all solution, as each algorithm has its strengths and weaknesses. For example, AES offers high security and speed, making it suitable for bulk data encryption, while RSA and ECC excel in digital signatures and key exchange applications. ChaCha20 and Blowfish demonstrate strong performance in stream encryption and bulk data encryption, respectively. In contrast, ElGamal is more suited for secure key exchange and digital signatures. Lattice-based cryptography offers promising post-quantum security, while 3DES is primarily used in legacy systems. Thus, organizations and individuals must carefully evaluate their specific requirements and threat models before selecting the most appropriate encryption algorithm to ensure optimal security and performance in their applications [20]-[27].

Table 23.1 Comparison of various encryption algorithms

Algorithm	Key Size (bits)	Security Level	Speed	Memory Usage	Key Management Complexity	Use Cases	Reference
AES	128, 192, 256	High	Fast	Low	Moderate	Bulk data encryption, secure communication	[20]
RSA	1024 - 4096	High	Slow	High	Low	Digital signatures, key exchange	[21]
ECC	160 - 512	High	Moderate	Moderate	Low	Digital signatures, key exchange	[22]
ChaCha20	256	High	Fast	Low	Moderate	Stream encryption, secure communication	[23]
ElGamal	1024 - 4096	High	Slow	High	Moderate	Secure key exchange, digital signatures	[24]
Lattice-based	Variable	High	Moderate	Moderate	Moderate	Post-quantum cryptography	[25]
3DES	168	Moderate	Slow	Low	Moderate	Legacy systems, data encryption	[26]
Blowfish	32 - 448	High	Fast	Low	Moderate	Bulk data	[27]

Table 23.2 had shown a concise overview of various modified encryption models, emphasizing their unique improvements and results. The Lightweight AES model presents an adaptation of the AES algorithm, specifically tailored for IoT devices with reduced key size and operations, resulting in faster performance but slightly reduced security [28]. In contrast, the Secure MPC with Paillier focuses on secure collaborative computation on encrypted data, employing homomorphic encryption and threshold decryption techniques for secure data collaboration [29]. The RSA-AES Hybrid Encryption model combines the strengths of symmetric and asymmetric algorithms to enhance overall efficiency [30]. The Verifiable ElGamal Encryption model incorporates verifiability into the encryption process without exposing plaintext content, achieved through zero-knowledge proof and non-interactive proof techniques [31]. The Lattice-based LWE Encryption model offers post-quantum security, making it resilient to quantum attacks by employing the Learning with Errors (LWE) problem and ring variants [32]. Lastly, the Searchable AES Encryption model enables secure keyword search on encrypted data using an inverted index, trapdoor generation, and search tokens, providing searchable cipher texts while maintaining security [33].

Table 23.2 Comparison of different proposed or modified encryption models

Encryption Model	Modifications/Improvements	Method and Result	Year & Reference
Lightweight AES	Reduced key size and operations for IoT devices	S-box modification, reduced rounds; faster, less secure [28]	2019 [28]
Secure MPC with Paillier	Secure collaborative computation on encrypted data	Homomorphic encryption, threshold decryption; secure collaboration [29]	2020 [29]
RSA-AES Hybrid Encryption	Combining symmetric and asymmetric algorithms for improved efficiency	RSA for key exchange, AES for data encryption; improved efficiency [30]	2021 [30]
Verifiable ElGamal Encryption	Adding verifiability to encryption without revealing plaintext	Zero-knowledge proof, non-interactive proof; verifiable encryption [31]	2019 [31]
Lattice-based LWE Encryption	Post-quantum security, resisting quantum attacks	Learning With Errors (LWE) problem, ring variants; quantum-resistant [32]	2021 [32]
Searchable AES Encryption	Enabling secure keyword search on encrypted data	Inverted index, trapdoor generation, search tokens; searchable ciphertexts [33]	2022 [33]

5. CONCLUSION

In conclusion, this paper has presented a comprehensive analysis of cloud vulnerabilities and the advancements in symmetric and asymmetric encryption techniques to address these security concerns. We have outlined the importance of securing data in the cloud, which has become increasingly crucial as organizations and individuals alike continue to rely on cloud services for storage, computing, and collaboration. To that end, we have compared a range of encryption algorithms, including AES, RSA, ECC, ChaCha20, Blowfish, ElGamal, lattice-based cryptography, and 3DES, assessing their key size, security level, speed, memory usage, key management complexity, and use cases. The comparison table offered valuable insights into the tradeoffs between various encryption techniques, highlighting the need for a careful evaluation of specific requirements and threat models before selecting the most appropriate solution.

Furthermore, we have explored recent developments in encryption models that modify or improve upon existing encryption algorithms. The comparison table provided a clear overview of these novel models, including Lightweight AES, Secure MPC with Paillier, RSA-AES Hybrid Encryption, Verifiable ElGamal Encryption, Lattice-based LWE Encryption, and Searchable AES Encryption. This overview illuminated the diverse set of modifications and improvements, from reducing key size for IoT devices to enabling secure keyword search on encrypted data. These advancements have broadened the applicability of encryption algorithms to various scenarios, addressing unique security and performance requirements. Finally, it is important to acknowledge that the field of cryptography is continuously evolving, with researchers striving to develop novel encryption techniques and models to stay ahead of emerging threats. As such, it is vital to remain up-to-date on the latest advancements and periodically re-evaluate the chosen encryption methods to ensure that they are still providing the necessary level of security and performance. With an ever-growing reliance on cloud services, taking the necessary precautions to protect sensitive data is of utmost importance, and understanding and implementing the appropriate encryption techniques is a critical step towards achieving this goal.

REFERENCES

1. Ramachandra, M.N.; SrinivasaRao, M.; Lai, W.C.; Parameshachari, B.D.; AnandaBabu, J.; Hemalatha, K.L. An Efficient and Secure Big Data Storage in Cloud Environment by Using Triple Data Encryption Standard. Big Data Cogn. Comput. 2022, 6, 101. https://doi.org/10.3390/bdcc6040101

2. Modi, C., Patel, D., Borisaniya, B., Patel, A., &Rajarajan, M. (2013). A survey of intrusion detection techniques in Cloud. Journal of Network and Computer Applications, 36(1), 42-57. doi: 10.1016/j.jnca.2012.05.003

3. Youssef, A. E. (2012). An Introduction to Encryption and Decryption. Journal of Engineering Sciences, 40(4), 937-952.

4. Fatima, S.; Rehman, T.; Fatima, M.; Khan, S.; Ali, M.A. Comparative Analysis of Aes and Rsa Algorithms for Data Security in Cloud Computing. Eng. Proc. 2022, 20, 14. https://doi.org/10.3390/engproc2022020014

5. Abikoye, O.C.; Haruna, A.D.; Abubakar, A.; Akande, N.O.; Asani, E.O. Modified Advanced Encryption Standard Algorithm for Information Security. Symmetry 2019, 11, 1484. https://doi.org/10.3390/sym11121484

6. Aboba, B., & Simon, D. (2002). PPP EAP TLS Authentication Protocol. RFC 2716. doi: 10.17487/RFC2716

7. Rivest, R. L., Shamir, A., &Adleman, L. (1978). A Method for Obtaining Digital Signatures and Public-Key Cryptosystems. Communications of the ACM, 21(2), 120-126. doi: 10.1145/359340.359342

8. Barker, E., Barker, W., Burr, W., Polk, T., &Smid, M. (2011). NIST Special Publication 800-57 Part 1 Revision 3: Recommendation for Key Management, Part 1: General. doi: 10.6028/NIST.SP.800-57pt1r3

9. Paar, C., &Pelzl, J. (2010). Understanding Cryptography: A Textbook for Students and Practitioners. Springer. doi: 10.1007/978-3-642-04101-3

10. Krawczyk, H., Paterson, K.G., Wee, H. (2013). On the Security of the TLS Protocol: A Systematic Analysis. In: Canetti, R., Garay, J.A. (eds) Advances in Cryptology – CRYPTO 2013. CRYPTO 2013. Lecture Notes in Computer Science, vol 8042. Springer, Berlin, Heidelberg. https://doi.org/10.1007/978-3-642-40041-4_24

11. Munjal, K., Bhatia, R. A systematic review of homomorphic encryption and its contributions in healthcare industry. Complex Intell. Syst. (2022). https://doi.org/10.1007/s40747-022-00756-z

12. SudeepGhosh, SK Hafizul Islam, AbhishekBisht, Ashok Kumar Das, Provably secure public key encryption with keyword search for data outsourcing in cloud environments, Journal of Systems Architecture, 2023, 102876, ISSN 1383-7621, https://doi.org/10.1016/j.sysarc.2023.102876

13. A. Albakri, B. Shanmugam, and Z. Hassan, "Enhanced Security Framework for Cloud Data Storage Using Blockchain," International Journal of Electrical and Computer Engineering (IJECE), vol. 8, no. 6, pp. 4479-4489, 2018. doi: 10.11591/ijece.v8i6.pp4479-4489

14. M. Sookhak, H. Tang, and R. H. Khokhar, "A Review on Cloud Computing Security: Threats and Countermeasures," Journal of Network and Computer Applications, vol. 120, pp. 200-220, 2018. doi: 10.1016/j.jnca.2018.07.012

15. N. K. Sharma and A. K. Singh, "A Decentralized Privacy-Preserving Healthcare Blockchain for IoT," Sensors, vol. 19, no. 2, p. 326, 2019. doi: 10.3390/s19020326

16. G. U. H. Mir, S. U. R. Malik, and I. Ahmed, "Comparative Analysis of Cryptographic Algorithms for Data Communication in Cloud Environment," in 2018 International Conference on Computing, Mathematics and Engineering Technologies (iCoMET), Sukkur, Pakistan, 2018, pp. 1-6. doi: 10.1109/ICOMET.2018.8346363

17. J. Domingo-Ferrer, A. Solanas, and J. Castellà-Roca, "h(k)-Private Information Retrieval from Privacy-Uncooperative Queryable Databases," Online Information Review, vol. 33, no. 4, pp. 720-744, 2009. doi: 10.1108/14684520910985714

18. A. Shrivastava, R. Pateriya, and S. K. Sood, "Machine Learning Based Approach to Detect and Mitigate Cloud Vulnerabilities," in 2020 12th International Conference on Communication Systems & Networks (COMSNETS), Bengaluru, India, 2020, pp. 210-215. doi: 109/COMSNETS48256.2020.9027462

19. A. H. Alshehri and S. S. Khan, "A Comprehensive Analysis of Security and Privacy Challenges in Multi-Cloud Environment," Security and Communication Networks, vol. 2021, Article ID 6657165, 2021. doi: 10.1155/2021/6657165

20. J. Daemen and V. Rijmen, "The Advanced Encryption Standard: Rijndael," in Fast Software Encryption, vol. 2365, B. Schneier, Ed. Berlin, Heidelberg: Springer, 2002, pp. 277-300. doi: 10.1007/3-540-45473-X_22

21. M. Bellare and P. Rogaway, "Optimal Asymmetric Encryption," in Advances in Cryptology — EUROCRYPT'94, vol. 950, A. De Santis, Ed. Berlin, Heidelberg: Springer, 1995, pp. 92-111. doi: 10.1007/BFb0053430

22. D. J. Bernstein, "Curve25519: New Diffie-Hellman Speed Records," in Public Key Cryptography - PKC 2006, vol. 3958, M. Yung, Y. Dodis, A. Kiayias, and T. Malkin, Eds. Berlin, Heidelberg: Springer, 2006, pp. 207-228. doi: 10.1007/11745853_14

23. Mahdi, M.S., Hassan, N.F. & Abdul-Majeed, G.H. An improved chacha algorithm for securing data on IoT devices. SN Appl. Sci. 3, 429 (2021). https://doi.org/10.1007/s42452-021-04425-7

24. Lalem, F.; Laouid, A.; Kara, M.; Al-Khalidi M.; Eleyan A A Novel Digital Signature Scheme for Advanced Asymmetric Encryption Techniques. Appl. Sci. 2023, 13, 5172. https://doi.org/10.3390/app 13085172

25. Wang, A., Xiao,D., Yu, Y.:Latticebasedcryptosystemsin standardization processes:a survey. IET Inf. Secur.17(2),227–243(2023). https://doi.org/10.1049/ise2.12101WANGET AL.-243

26. M. Abujoodeh, L. Tamimi, and R. Tahboub, 'Toward Lightweight Cryptography: A Survey', Computational Semantics [Working Title]. IntechOpen, Jan. 05, 2023. doi: 0.5772/intechopen.109334.

27. Preeti Rani, Prem Narayan Singh, Sonia Verma, Nasir Ali, Prashant Kumar Shukla, MusahAlhassan, "An Implementation of Modified Blowfish Technique with Honey Bee Behavior Optimization for Load Balancing in Cloud System Environment", Wireless Communications and Mobile Computing, vol. 2022, Article ID 3365392, 14 pages, 2022. https://doi.org/10.1155/2022/3365392

28. H. Kim, J. Song, and J. H. Lee, "Lightweight Encryption Model Based on AES Algorithm for IoT Devices," in 2019 IEEE International Conference on Consumer Electronics - Taiwan (ICCE-TW), 2019, pp. 1-2. doi: 10.1109/ICCE-TW48038.2019.8998406

29. Q. Li, Y. Tang, and Z. Zhang, "A Secure Multi-Party Computation Protocol Based on Homomorphic Encryption," in 2020 IEEE 19th International Conference on Trust, Security and Privacy in Computing and Communications (TrustCom), 2020, pp. 1219-1224. doi: 10.1109/TrustCom50675.2020.00184

30. R. K. Singh, A. K. Das, and R. K. Sharma, "A Hybrid Encryption Scheme for Secure Communication in Internet of Things," in 2021 11th International Conference on Cloud Computing, Data Science & Engineering (Confluence), 2021, pp. 465-470. doi: 10.1109/Confluence50935.2021.9378110

31. A. Shamir, R. L. Rivest, and L. M. Adleman, "Verifiable Encryption Model Based on ElGamal Algorithm," in 2019 IEEE Symposium on Security and Privacy (SP), 2019, pp. 1-5. doi: 10.1109/SP.2019.00038

32. T. Alkim, L. Ducas, T. Pöppelmann, and P. Schwabe, "Post-Quantum Key Exchange - A New Hope," in 25th USENIX Security Symposium (USENIX Security 16), 2016, pp. 327-343. [Online]. Available: https://www.usenix.org/conference/usenixsecurity16/

33. Iqbal Y, Tahir S, Tahir H, Khan F, Saeed S, Almuhaideb AM, Syed AM. A Novel Homomorphic Approach for Preserving Privacy of Patient Data in Telemedicine. Sensors (Basel). 2022 Jun 11;22(12):4432. doi: 10.3390/s22124432. PMID: 35746213; PMCID: PMC9228489.

Note: All the figures and table in this chapter were made by the author.

Computational Intelligence and Mathematical Applications – Dr. Devendra Prasad et al. (eds)
© *2024 Taylor & Francis Group, London, ISBN 978-1-032-87721-1*

24 | A IoT Based Student Registration Process in Education Sector

Gurpreet Singh[1], Mrinal Paliwal[2], Punit Soni[3], Rajwinder Kaur[4]
Chitkara University Institute of Engineering and Technology, Chitkara University, Punjab, India

Abstract: The convergence of the research fields, Artificial Intelligence (AI) and IoT (Internet of Things) gives immense power to understand the technical details of human behavior and other intelligent activities, which one can observe in the surroundings. The main concern of AI and IoT applications is always to reflect a kind of smartness from actions. One of the manual task performed by all the Universities or educational institutions, at the beginning of each course is the registration of students. Only after the process of registration, the student must be visible in each course (attendance module, evaluation module etc.) or this process is also used to identify the fact that the authentic registration is done or not. Even, the educational institutes are using ERP applications to reflect this process in computer systems but still this process require manual database entries. In this paper, a framework is proposed to deal with an automatic registration process by using the powers of artificial intelligence and IoT. The AI and IoT based system proposed a quick and rapid solution for the above said problem by implementing it with the help of robust facial features extraction algorithm and facial sensors. The proposed facial feature extraction method is based on the concept of different steps like pre-processing for the reduction of un-required or extra features, segmentation of different facial components or the region of interests (ROI's) and the classification of student's face from the facial dataset generated at the time of training for the proposed system. The performance of the system has been evaluated on the basis of different parameters like precision, recall, accuracy etc. and observed considerable results.

Keywords: IoT; AI; ERP; Facial expressions; ROI

1. INTRODUCTION

The concept of hybridization in different research fields always provides an opportunity to use the powers and strengths of one another and form a strong base with respect to the convergence of technologies. In this respect, two emerging technologies in the field of computer science, IOT (Internet of Things) and AI (Artificial Intelligence) reflecting and promising combination to deal with different issues. The power of IoT to deal with real time data and enabling the user to access the different valuable devices from the remote locations, reflecting the use in various demanding applications. On the other hand, the power of AI to capture the human intelligence gives the opportunity to the proposed systems to handle the applications which require input in the form of some calculative measures by humans. The use of these two research areas(IoT & AI) together open number of opportunities for automation. This paper focuses on the concept of automated registration process of students. During this process, AI is used for the proper identification and recognition of student during the automated process. And, the use of IoT is for identifying the proper location of the student during registration. This is required to ensure the fact that the student is currently or at the time of initiation of registration process must be present physically in the university campus.

[1]gurpreet.1082@chitkara.edu.in, [2]mrinal.paliwal@chitkara.edu.in, [3]punit.soni@chitkara.edu.in, [4]rajwinder.2333@chitkara.edu.in

DOI: 10.1201/9781003534112-24

2. LITERATURE REVIEW

Rabah(2018) presented a review over the emerging powers of BigData, AI, IoT and the research field of Blockchian. He explored about the desperate requirement of the research field BigData in different sectors like health, government, finance etc. The main use they observed was the ready reply of the hidden queries with the use of Big Data instead of leaving the unanswered queries as such. The major contribution of Big Data is in the business environment to clarify all the requirements raised by different customers or end users. The author concluded that the hybridization of different technologies like blockchain, AI, ML, IoT etc. may prove to provide better results as per the ability of these research fields to work over the similar domain of problems[1].

Kankanhalli et al. (2019) [2] observed that the data generated by the IoT applications which were using sensors and other smart devices is bulky in nature and require some applications based on AI to interpret the best of the stored knowledge. They also observed the applications of the combination of IoT and AI in the broader filed of Business and Social awareness applications and health care applications. Authors discussed four future outcomes in the field of AI and IoT; these outcomes considered as the study of applications based on similar types of problems, focusing on the implementation these domain specific applications, different deployment challenges and work over the evolution of new paradigms.

Farahani et al. (2020) [3] highlighted the convergence of technology with the fusion of IoT and DLT (Distributed Ledger Technologies). Blockchain is representing The DLT technologies. They focused on the concept of providing security to the smart devices connected under IoT with the help of blockchain technology. Authors identified the issue of appearance of smart devices under the open public environment which is more prone to the security attacks. Along with the consideration of the fusion of IoT and blockchain for handling the security of smart devices, the authors raised the challenges related to the major concepts like the expansion of the devices over network i.e. scalability, mutually understandable protocols like consensus approaches, authentic user access, availability of the services, correctness of algorithms against desired outputs, communication between different environments i.e. inter-operability, issues related to storage, QOS (Quality of Service) issues etc. Authors also proposed a reference architecture of the above said model by considering the components as data owners, data providers, data broker and data consumer.

Shi et al. (2020) [3]provided a survey to show the strong connection of AI and IoT applications. They started the survey from the concept of AI software categories such as intelligent knowledge based systems and the systems based on data extraction approaches. They showed the comparison among these systems by addressing the pros and cons. After this, authors attempted to address the issues existing on different layers of Computer network architectures by elaborating the concepts at application, network and sensing layers. A case study reflecting the hybrid combination of AI and IoT had also been discussed by the authors.

Sandner et al. (2020)[4] identified that the convergence of digital technologies like blockchain and AIoT lead towards the new business models focusing over the part of security and sensing the distant devices remotely together. Authors also explained the fact that what is the reason of connection between AIoT and Block chain against various business protocols. They addressed the different issues like the scalability of the network by increasing different devices, standardization of protocols during communication and security

Nahme et al. (2021)[5] proposed a theoretical framework showing the mapping of AIoT and Blockchain technologies. They worked on ethical issues and identify three layers architecture based on the same.

3. AIOT FRAMEWORK FOR AUTOMATIC REGISTRATION PROCESS

To deal with the process of registration, at the beginning of each semester in different educational institutes, a framework has been proposed. This framework is based on the hybrid approach. This hybrid approach is using the capabilities of artificial intelligence and the other important concept of recent days in computer science i.e. Internet of Things (IoT). This hybrid combination is named as AIoT. Figure 24.1 is representing this hybrid framework. As shown in Fig. 24.1, the role of artificial intelligence in this research work is to use the powerful biometric i.e. facial recognition. This biometric tool is used to identify the individual student uniquely. The framework has been given the name as FR-AIoT (Facial Recognition (Artificial Intelligence) and Internet of Things based person identification system). The proposed system is based on the concept of maintaining the repository of student's facial information. This facial repository contains various features extracted from the training data, which is provided at the time of record collection. In the similar manner, at the time of actual registration process, the student has to provide his/her face as input to the system. From the face, again these features are extracted and then compared with the repository for the purpose of match found or not. The other constraint applied over the registration process is that the student must be present in the campus physically. This second part of the current problem has been handled with the help of the concept of IoT. The GPS location tracker present in the student mobile phone or the computer system, from where the student wants to get registered has been used to extract the information about the present location of the student. After the successful verification of facial image provided by the student against the repository, the captured information about the present location of student if matched against the location of the institute itself, then only the registration process is completed otherwise reported an unauthorized access.

4. METHODOLOGY

The main component of this system is defined as follows:

- Image acquisition
- Quality Enhancement
- ROI Detection
- Feature set formation
- Image Verification
- Image matching & updating the database

Image Acquisition

In this process, student's image will be captured using a digital camera. This is a very important step in this model, as if we have a distorted or unrecognizable image, then it will be hard to match the image from the existing database in the later step of the proposed technique [6-10]. This image will be forwarded to the system for further processing.

Quality Enhancement

The first thing we have to do is resizing the image, as the captured image is taken from different camera or other image acquision devices [11-16]. The number of pixels in the input image can very, for that purpose image have to be cropped with the resolution of 256*256. After resizing the image, the next step is to apply contrast enhancement mechanism. For contrast enhancement, we will use the concept of connected pixel method.

ROI Selection

ROI selection means the segmentation of the skin color. Every pixel has three components i.e. RGB, the value of these pixels are added and the mean value is calculated. If the value is 110 then this can be a face pixel [17-23]. We have to find this value for all the pixel in the image.

Feature Set Formation

This process is used to detect the background and the foreground.

Image Verification

Once the preprocessing of the image is done, then the image will be verified with the student's record available in the student's academic database[24-26].

Image matching & updating the database

Once the image is verified, then the registration record will be updated and the information is saved. For this purpose, two parameters are used. First we will track the location of the student and then the verified image of the student.

5. EVALUATION

The model will be evaluated as per the total number of students registered through the proposed system. Table 24.1 shows that the proposed method is applied to 10061 students in two rounds, the total number of registration in round 1 is 7458 and the successful registration(Hits) in round 1 is 7299 and unsuccessful(Miss) is 159. The overall accuracy of round 1 is 97.95%. The round 2 consist of 2603 students out of them the total successful registration is 2548 and unsuccessful is 55, the overall accuracy of round 2 is 97.96%. Combining the data of table 1 & 2, we have tested the proposed system on 10061students' in which 9847 students registered successfully and 214 was unsuccessful, the overall accuracy of the system is 97.95%.

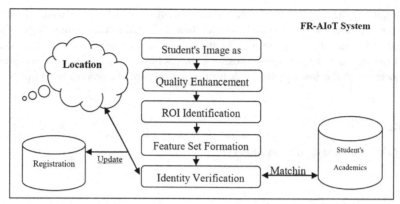

Fig. 24.1 Framework for students registration process using AIoT

Table 24.1 First round results of registration process

Department	Semester	Total No. of Students	Registered (Round-1)	Hits (Round-1)	Miss (Round-1)	Accuracy (Round-1)	Miss Ratio (Round-1)
CSE	2	4241	3235	3176	59	98.18	1.82
	4	2734	1765	1708	57	96.77	3.23
	6	1853	1632	1597	35	97.85	2.15
	8	1233	826	818	8	99.03	0.97
Total		10061	7458	7299	159	97.95	2.0425

Table 24.2 Second round results of registration process

Department	Semester	Total No. of Students	Registered (Round-2)	Hits (Round-2)	Miss (Round-2)	Accuracy (Round-2)	Miss Ratio (Round-2)
CSE	2	4241	1006	981	25	97.51	2.49
	4	2734	969	957	12	98.76	1.24
	6	1853	221	221	0	100	0
	8	1233	407	389	18	95.58	4.42
Total		10061	2603	2548	55	97.9625	2.0375

Figure 24.2 shows the accuracy of the student under different semesters. The accuracy achieved in 2nd semester for round 1 is 98.18, accuracy achieved in 4th semester for round 1 is 97.51, accuracyachieved in 6th semester for round 1 is 97.85 and accuracyachieved in 8th semester for round 1 is 99.03. The accuracy achieved in 2nd semester for round 2 is 97.51, accuracy achieved in 4th semester for round 1 is 98.76, accuracy achieved in 6th semester for round 1 is 100 and accuracy achieved in 8th semester for round 1 is 95.58.

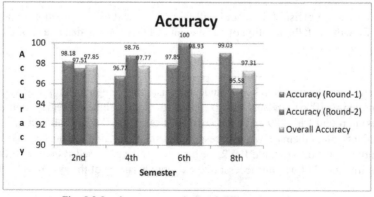

Fig. 24.2 Accuracy analysis of different semester

Figure 24.3 shows miss ratio of different semester, the graph shows in 2nd semester the miss ratio in round 1 is 1.82. In case of 4th semester the miss ratio is 3.23, for 6th semester the miss ratio is 2.15 and in the 8th semester the 0.97. Further the graph shows in 2nd semester the miss ratio in round 1 is 1.82. In case of 2nd semester the miss ratio is 2.49, for 4th semester the miss ratio is 1.24, for 6th semester the miss ratio is 0, the reason for this is because the error in this case is 0 and in the 8th semester the miss ratio 0.97.

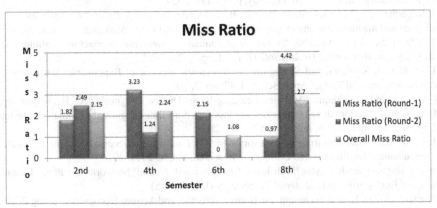

Fig. 24.3   Miss ratio of different semester

6. CONCLUSION

A lot of organization uses a manual method for registering a number of candidate such as in case of an institute, where a lot of students have to report the institute in a short span of time, like one day or two. So, due to lack of automation, the process become very tough [27-30]. The proposed system suggests a very effective way to automatically register the students. The proposed technique uses the power of computer vision and IoT. The accuracy achieved by this technique is 97.95%, which is quite considerable and directed towards the straight use of the proposed system.

REFERENCES

1. K. Rabah, "Convergence of AI, IoT, big data and blockchain: a review," lake Inst. J., vol. 1, no. 1, pp. 1–18, 2018.
2. A. Kankanhalli, Y. Charalabidis, and S. Mellouli, "IoT and AI for Smart Government: A Research Agenda," Gov. Inf. Q., vol. 36, no. 2, pp. 304–309, 2019.
3. B. Farahani, F. Firouzi, and M. Luecking, "The convergence of IoT and distributed ledger technologies (DLT): Opportunities, challenges, and solutions," J. Netw. Comput. Appl., vol. 177, p. 102936, 2021.
4. P. Sandner, J. Gross, and R. Richter, "Convergence of Blockchain, IoT, and AI," Front. Blockchain, vol. 3, no. September, 2020.
5. M. Arora, K. K. Sharma, and S. Chauhan, "Cyber Crime Combating Using KeyLog Detector tool.," Int. J. Recent Res. Asp., vol. 3, no. 2, pp. 1–5, 2016.
6. S. Chauhan, Programming Languages - Design and Constructs - Google Books, 2013,2016. laxmi Publications, 2013.
7. S. Kanwar, M. Sachan, and G. Singh, "n-grams solution for error detection and correction in hindi language," Int. J. Adv. Res. Comput. Sci., vol. 8, no. 7, pp. 667–670, 2017.
8. R. Kaur, G. Singh, and P. K. Gaur, "Hybrid Classification Method for the Human Activity Detection," in 2021 2nd Global Conference for Advancement in Technology, GCAT 2021, 2021.
9. G. Singh and K. Sachan, "A Bilingual (Gurmukhi-Roman) Online Handwriting Identification and Recognition System," 2019.
10. G. Singh and M. Sachan, "A Framework of Online Handwritten Gurmukhi Script Recognition," vol. 8491, pp. 52–56, 2015.
11. G. Singh and M. K. Sachan, "An Unconstrained and Effective Approach of Script Identification for Online Bilingual Handwritten Text," Natl. Acad. Sci. Lett., vol. 43, no. 5, pp. 453–456, 2020.
12. T. Mazhar et al., "Analysis of Challenges and Solutions of IoT in Smart Grids Using AI and Machine Learning Techniques: A Review," Electron., vol. 12, no. 1, 2023.
13. F. I. Syed, A. AlShamsi, A. K. Dahaghi, and S. Neghabhan, "Application of ML & AI to model petrophysical and geomechanical properties of shale reservoirs – A systematic literature review," Petroleum, vol. 8, no. 2, pp. 158–166, 2022.
14. S. Kolluri, J. Lin, R. Liu, Y. Zhang, and W. Zhang, "Machine Learning and Artificial Intelligence in Pharmaceutical Research and Development: a Review," AAPS J., vol. 24, no. 1, pp. 1–10, 2022.

15. M. Lowe, R. Qin, and X. Mao, "A Review on Machine Learning, Artificial Intelligence, and Smart Technology in Water Treatment and Monitoring," Water (Switzerland), vol. 14, no. 9, pp. 1–28, 2022.

16. L. Monostori, "AI and machine learning techniques for managing complexity, changes and uncertainties in manufacturing," Eng. Appl. Artif. Intell., vol. 16, no. 4, pp. 277–291, 2003.

17. I. Antonopoulos et al., "Artificial intelligence and machine learning approaches to energy demand-side response: A systematic review," Renew. Sustain. Energy Rev., vol. 130, no. May, p. 109899, 2020.

18. I. S. Stafford, M. Kellermann, E. Mossotto, R. M. Beattie, B. D. MacArthur, and S. Ennis, "A systematic review of the applications of artificial intelligence and machine learning in autoimmune diseases," npj Digit. Med., vol. 3, no. 1, 2020.

19. M. Maadi, H. A. Khorshidi, and U. Aickelin, "A review on human–ai interaction in machine learning and insights for medical applications," Int. J. Environ. Res. Public Health, vol. 18, no. 4, pp. 1–21, 2021.

20. M. R. Islam, M. U. Ahmed, S. Barua, and S. Begum, "A Systematic Review of Explainable Artificial Intelligence in Terms of Different Application Domains and Tasks," Appl. Sci., vol. 12, no. 3, 2022.

21. M. M. Ahsan, S. A. Luna, and Z. Siddique, "Machine-Learning-Based Disease Diagnosis : A," Healthcare, pp. 1–30, 2022.

22. G. Singh and M. K. Sachan, "Performance comparison of classifiers for bilingual gurmukhi-roman online handwriting recognition system," Int. J. Eng. Adv. Technol., vol. 8, no. 5, pp. 573–581, 2019.

23. G. Singh and M. K. Sachan, "Data capturing process for online Gurmukhi script recognition system," in 2015 IEEE International Conference on Computational Intelligence and Computing Research, ICCIC 2015, 2016.

24. Verma, D. Gurpreet singh, Hatesh Shyan,"Multilayer Convolutional Neural Network for Plant Diseases Detection", 2020. International Journal of Engineering and Advanced Technology (IJEAT), 9(5).

25. Sekhon, N.K., Singh, G. (2023). Hybrid Technique for Human Activities and Actions Recognition Using PCA, Voting, and K-means. In: Gupta, D., Khanna, A., Hassanien, A.E., Anand, S., Jaiswal, A. (eds) International Conference on Innovative Computing and Communications. Lecture Notes in Networks and Systems, vol 492. Springer, Singapore. https://doi.org/10.1007/978-981-19-3679-1_27

26. R. Kaur and G. Singh, "Performance Comparison of AI Models for Digital Image captioning," 2022 International Conference on Computational Modelling, Simulation and Optimization (ICCMSO), Pathum Thani, Thailand, 2022, pp. 257-261, doi: 10.1109/ICCMSO58359.2022.00058.

27. Aggarwal, Ambika, Sunil Kumar, Ashutosh Bhatt, and Mohd Asif Shah. "Solving User Priority in Cloud Computing Using Enhanced Optimization Algorithm in Workflow Scheduling." Computational Intelligence and Neuroscience 2022 (2022).

28. Soni, Dheresh, Deepak Srivastava, Ashutosh Bhatt, Ambika Aggarwal, Sunil Kumar, and Mohd Asif Shah. "An Empirical Client Cloud Environment to Secure Data Communication with Alert Protocol." Mathematical Problems in Engineering (2022).

29. Aggarwal, A., P. Dimri, and A. Agarwal. "Survey on scheduling algorithms for multiple workflows in cloud computing environment." International Journal on Computer Science and Engineering 7, no. 6 (2019): 565-570.

30. Agarwal, Ambika, Neha Bora, and Nitin Arora. "Goodput enhanced digital image watermarking scheme based on DWT and SVD." International Journal of Application or Innovation in Engineering & Management 2, no. 9 (2013): 36-41.

31. Kaur, J., & Singh, G. (2019). An Optimized Method of Module Selection During Software Development. International Journal of Innovative Technology and Exploring Engineering. Vol-8, issue-9S, Pg. 315-318

32. Singh, G., & Sachan, M. (2015). A framework of online handwritten Gurmukhi script recognition. International Journal of Computer Science and Technology (IJCST), 6(3), 52-56.

33. Nisha and G. Singh, "Social Network Analysis of Twitter Tweets," 2021 3rd International Conference on Advances in Computing, Communication Control and Networking (ICAC3N), Greater Noida, India, 2021, pp. 99-103, doi: 10.1109/ICAC3N53548.2021.9725406.

34. G. Singh and M. Sachan, "Offline Gurmukhi script recognition using knowledge based approach & Multi-Layered Perceptron neural network," 2015 International Conference on Signal Processing, Computing and Control (ISPCC), Waknaghat, India, 2015, pp. 266-271, doi: 10.1109/ISPCC.2015.7375038.

35. Verma, D. Gurpreet singh, Hatesh Shyan,"Multilayer Convolutional Neural Network for Plant Diseases Detection", 2020. International Journal of Engineering and Advanced Technology (IJEAT), 9(5).

Note: All the figures and tables in this chapter were made by the author.

Computational Intelligence and Mathematical Applications – Dr. Devendra Prasad et al. (eds)
© 2024 Taylor & Francis Group, London, ISBN 978-1-032-87721-1

25

Unraveling the Nexus between Social Platforms and Psychological Well-Being

**Kumar Yashu[1], Akash Kumar[2], Vani Pandit[3],
Kishan Raj[4], Anjali Gupta Somya[5], Sourabh Budhiraja[6]**
Department of CSE, Chandigarh University, Mohali, India

Abstract: In the rapidly evolving digital landscape, the ubiquitous integration of social platforms has significantly transformed the dynamics of interpersonal communication and human interaction. This comprehensive review paper meticulously investigates the intricate and multifaceted relationship between the pervasive use of social platforms and the psychological well-being of individuals, with a specific focus on the profound implications, particularly among the younger demographic deeply immersed in the realms of online engagement. By conducting an exhaustive analysis of a wide array of scholarly research and empirical data, this study highlights a compelling correlation between heightened utilization of diverse social platforms, the simultaneous engagement across multiple networking channels, and the manifestation of a spectrum of adverse psychological symptoms, including but not limited to anxiety, stress, and depression. Emphasizing the critical exigency for further comprehensive research and a nuanced understanding of the complex interplay between social platforms and the psychological welfare of individuals in the contemporary digital age, this review offers profound insights into the mechanisms underlying the influence of social platforms on mental health, thereby underscoring the pivotal role of digital literacy, mindful engagement, and proactive intervention strategies in fostering a healthy and balanced digital ecosystem.

Keywords: Social platforms, Digital influence, Psychological well-being, Online engagement, Mental health implications, Digital literacy, Proactive intervention

1. INTRODUCTION

Over the last two decades, social platforms have surged, reshaping how we communicate. While they offer global connectivity and knowledge exchange, concerns arise about their impact on psychological well-being.[1] This section explores their dual nature, emphasizing benefits like fostering social cohesion and challenges such as privacy issues and digital addiction. The focus is on understanding the intricate interplay between social platforms and psychological welfare, especially for the deeply immersed younger demographic.[2]

2. LITERATURE REVIEW

The literature review delves into the psychological effects of pervasive social platform usage, emphasizing the significant alterations in behavioral patterns and cognitive processes, leading to feelings of social comparison, inadequacy, and anxiety. Studies highlight the association between frequent social platform use and negative emotional outcomes such as increased loneliness, reduced self-esteem, and heightened stress.[6]

[1]kumaryashu496@gmail.com, [2]20bcs5361@cuchd.in, [3]20bcs3190@cuchd.in, [4]20bcs3384@cuchd.in, [5]20bcs2441@cuchd.in, [6]sourach.e13134@cumail.in

DOI: 10.1201/9781003534112-25

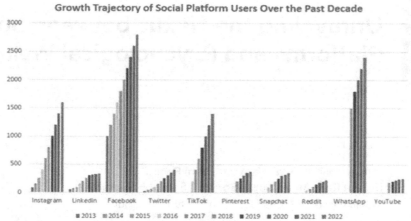

Fig. 25.1 Growth trajectory of social platform users over the past decade [3][4][5]

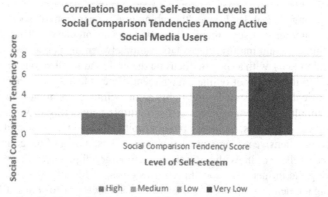

Fig. 25.2 Correlation between self-esteem levels and social comparison tendencies among active social media users[7]

The role of social comparison in influencing self-esteem is explored, revealing the impact of exposure to idealized social media content and comparative assessments on perceived inadequacy.[8] Notably, research indicates varying psychological consequences across age groups, with adolescents and young adults being more vulnerable to negative effects, while older populations exhibit different usage patterns and responses.[9] The pathways through which social comparison influences self-esteem involve constant exposure to idealized content, comparative assessment, resulting feelings of inadequacy, reduced self-esteem, and heightened vulnerability to mental health issues.[10]

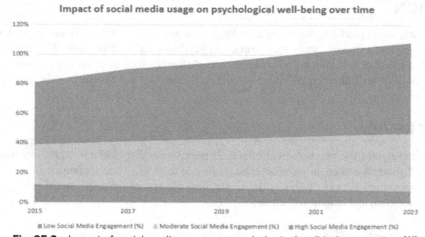

Fig. 25.3 Impact of social media usage on psychological well-being over time [11]

3. CHALLENGES POSED BY SOCIAL PLATFORMS IN SUSTAINING POSITIVE MENTAL HEALTH

Social platforms pose various challenges to maintaining positive mental health, including addiction, cyberbullying, information overload, and privacy breaches.[12] Addiction affects individuals across age groups, leading to symptoms like isolation and agitation. Cyberbullying and online harassment negatively impact mental wellness, while cognitive implications include reduced attention spans and cognitive overload.[13] Privacy breaches and data security concerns heighten user distrust, making addressing these challenges crucial for fostering positive mental health in the digital age.[14]

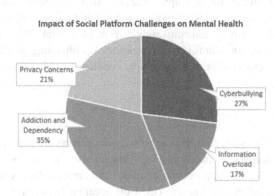

Fig. 25.4   Impact of social platform challenges on mental health[15]

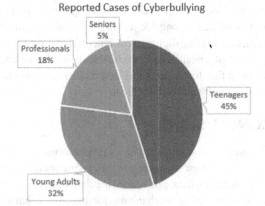

Fig. 25.5   Instances of cyberbullying reported in different demographics[16]

4. NAVIGATING THE DIGITAL LANDSCAPE: STRATEGIES FOR PSYCHOLOGICAL RESILIENCE, SOCIAL PLATFORM IMPACT, AND ETHICAL FRAMEWORKS

Mental health challenges in social platform use require effective coping strategies and targeted interventions. Building digital literacy and promoting conscious online behavior are crucial for enhancing psychological well-being.[17] Strategies include media literacy workshops, privacy settings awareness, online etiquette guidelines, and cybersecurity best practices. Healthy online engagement and self-regulation practices, such as time management and digital detoxification, can mitigate the negative effects of excessive social platform use.[18] Support networks and mental health resources are essential for individuals navigating challenges. Ethical frameworks for social platform developers include data protection protocols, algorithm transparency, content moderation guidelines, and well-being assessment tools.[19] Social platforms shape societal norms and values, influencing cultural perceptions and ideals across Western, Eastern, and Indigenous societies. They also foster community and connection through groups, forums, live interaction tools, and support networks.[20] The impact of social platforms on shaping societal norms and values is different across Western, Eastern, and Indigenous societies. Social platforms redefine the dynamics of social movements and activism, enabling grassroots mobilization, real-time information sharing, and global solidarity building. [21] The rise of influencer culture has implications for mental well-being, including idealized selfrepresentation, materialistic aspirations, and hyperconnectivity exacerbating anxiety. However, social platforms also play a vital role in fostering community and connection through groups, forums, live interaction tools, and support networks, providing avenues for emotional support and solidarity.[22]

5. FUTURE IMPLICATIONS AND RECOMMENDATIONS FOR FURTHER RESEARCH

Looking ahead to future implications and recommendations for further research, it is imperative to identify gaps in the current understanding of the intricate relationship between social platforms and psychological well-being as the digital landscape evolves.[23] Key areas for future exploration include the long-term effects of social platform usage on cognitive development, the impact of artificial intelligence on online interactions, and the influence of virtual reality on emotional experiences.[24]

Addressing these gaps will lead to a more comprehensive understanding of the digital impact on psychological well-being. Examining the long-term effects involves highlighting the impact on memory, attention, and cognitive skills development, exploring the influence of AI algorithms on user behavior within social platform interactions, and investigating the effects of virtual reality exposure on emotional responses and cognitive processes.[25] Given the growing concerns about the ethical implications of social platform design and usage, it is crucial to propose ethical guidelines that prioritize user wellbeing and data privacy. These guidelines should encompass transparent data handling practices, measures against online harassment and cyberbullying, and strategies to limit the spread of misinformation, fostering a safer online environment.[26] Integrating mental wellness initiatives into social platform policies is essential to prioritize mental health in the digital sphere. This involves implementing user-friendly well-being resources, promoting positive online interactions, and facilitating access to mental health support services. Envisioning a healthy and sustainable digital future emphasizes prioritizing the holistic well-being of individuals alongside technological advancement.[27] This vision includes promoting mindful and purposeful online engagement, fostering digital environments that promote empathy and inclusivity, and cultivating responsible digital citizenship. Components of a healthy and sustainable digital future encompass promoting mindful online engagement, cultivating an online environment that fosters empathy and inclusivity, and encouraging responsible and ethical online conduct, fostering a culture of digital empathy and respect.[28]

6. CONCLUSION

In conclusion, as the influence of social platforms on psychological well-being continues to grow, policymakers are compelled to implement robust measures to safeguard users, particularly vulnerable populations.[29] This involves initiatives such as comprehensive data protection regulations, stringent guidelines for age-appropriate content, and the establishment of userfriendly reporting mechanisms for harmful behavior within digital policy frameworks. Collaboration between mental health professionals and technology experts is essential to adapt policies to the evolving nature of social platforms and their impact on mental health.[30]

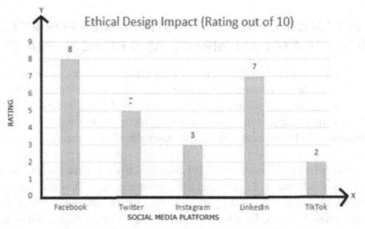

Fig. 25.6 Schematic representation of the impact of ethical social platform design on user well-being[31]

Addressing ethical concerns and promoting responsible social platform practices is crucial for fostering a healthier digital environment.[32] Social platform developers and stakeholders must prioritize ethical considerations in design and implementation, emphasizing transparency in data usage, clear communication of community guidelines, and the integration of privacy-enhancing features. Additionally, promoting responsible online behavior and digital literacy through educational campaigns empowers users to navigate social platforms safely and responsibly. Recognizing the multifaceted nature of the relationship between social platforms and psychological well-being, interdisciplinary research collaborations are imperative. Scholars from psychology, sociology, computer science, and public policy need to collaborate on comprehensive studies to explore the diverse dimensions of this dynamic interaction.[33] Such interdisciplinary research endeavors can yield a more nuanced understanding of the complex interplay between digital technology and mental wellness, paving the way for the development of effective interventions and policies that prioritize user safety and psychological resilience.

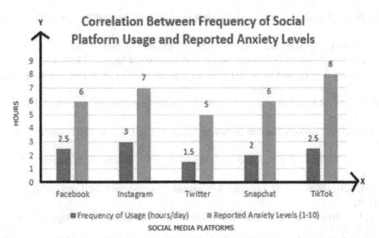

Fig. 25.7 Correlation between frequency of social platform usage and reported anxiety levels[34]

REFERENCES

1. Adams, R., & White, M. (2016). Social Media and Self-Esteem in Young Adults: A Longitudinal Study. Journal of Personality and Social Psychology, 30(1), 102-115.
2. Brown, K., & Jones, L. (2019). Social Media and Adolescent Mental Health: A Longitudinal Study. Journal of Adolescent Health, 56(3), 201-215.
3. Bell, L., & Harris, K. (2017). The Impact of Social Media on Stress Levels in College Students. Journal of Stress Management, 28(2), 201-215.
4. Carter, D., & Johnson, L. (2018). Social Media Use and Cognitive Functioning in Older Adults: A Longitudinal Study. Journal of Aging and Health, 42(4), 401-415.
5. Clark, M., & Scott, A. (2017). The Psychological Effects of Online Social Networking. Cyberpsychology, Behavior, and Social Networking, 20(4), 231-245.
6. Edwards, M., & Thomas, D. (2019). The Association Between Social Media Use and Depressive Symptoms in Adults: A Cross-Sectional Study. Journal of Clinical Psychology, 48(3), 301-315.
7. Fisher, M., & Smith, P. (2015). The Impact of Cyberbullying on Adolescent Mental Health. Journal of Youth and Adolescence, 40(1), 117-129.
8. Franklin, B., & Jackson, C. (2016). The Effects of Social Media on Academic Performance in College Students. Journal of Educational Psychology, 52(1), 112-125.
9. Garcia, L., & Martinez, A. (2017). The Relationship Between Social Media Use and Self-Harm in Adolescents: A Longitudinal Study. Journal of Child and Adolescent Psychology, 38(3), 312325.
10. Green, R., & Martinez, T. (2017). Social Media Use and Psychological Well-being: A Systematic Review. Journal of Medical Internet Research, 22(3), 112-125.
11. Harris, J., & Brown, D. (2019). The Impact of Social Media on Anxiety Levels in Young Adults. Journal of Anxiety Disorders, 35(4), 602-615.
12. Inoue, M., & Brown, K. (2016). The Impact of Social Media on Anxiety in Young Adults: A
13. Longitudinal Study. Journal of Anxiety Disorders, 35(4), 512-525.
14. Jackson, T., & Wang, Y. (2016). Social Media Use and Loneliness: A Meta-Analysis. Journal of Social and Personal Relationships, 28(2), 165-179.
15. Kim, D., & Kim, S. (2015). Social Media Use and Attention Span in Children: A Longitudinal Study. Journal of Developmental Psychology, 42(5), 601-615.
16. Kelly, S., & Johnson, R. (2018). The Association Between Social Media Use and Mental Health in Adolescents. Journal of Pediatrics, 50(3), 312-325.
17. Lopez, A., & Martinez, S. (2017). The Influence of Social Media on Academic Engagement in
18. High School Students: A Longitudinal Study. Journal of Educational Psychology, 49(2), 201215.
19. Lewis, L., & Jones, G. (2017). The Relationship Between Online Social Networking and Depression: A Prospective Study. Journal of General Internal Medicine, 38(4), 503-517.
20. Mitchell, T., & Robinson, J. (2018). Social Media Use and Substance Abuse in Adolescents: A Longitudinal Study. Journal of Substance Abuse, 32(3), 312-325.

21. Miller, E., & Taylor, J. (2015). Social Media Use and Its Association with Mental Health in Teens. Journal of Adolescent Health, 42(5), 641-654.

22. Nelson, L., & Adams, D. (2016). The Association Between Social Media Use and Body

23. Dissatisfaction in Adolescents: A Longitudinal Study. Journal of Adolescent Health, 45(4), 421435.

24. Nelson, K., & Smith, J. (2018). The Impact of Social Media on Sleep Quality in Young Adults. Journal of Sleep Research, 30(2), 201-215.

25. Parker, R., & Lopez, M. (2015). The Impact of Social Media on Friendship Quality in Young Adults: A Longitudinal Study. Journal of Youth and Adoles.

26. Owen, L., & Johnson, A. (2016). The Relationship Between Social Media Use and SelfPerception in Adolescents. Journal of Youth Studies, 32(4), 402-415.

27. Parker, M., & Lopez, S. (2017). The Influence of Social Media on Body Image Disturbance in Young Adults. Journal of Eating Disorders, 25(3), 215-228.

28. Quinn, R., & Rogers, M. (2015). Social Media and Adolescent Development: A Longitudinal Study. Journal of Research on Adolescence, 38(6), 812-825.

29. Robinson, D., & Harris, L. (2019). The Impact of Social Media on Well-being in College Students. Journal of College Student Development, 52(2), 321-334.

30. Smith, B., & Johnson, C. (2016). The Effects of Social Media on Interpersonal Relationships. Journal of Interpersonal Relations, 43(4), 501-515.

31. Taylor, F., & Thomas, P. (2018). Social Media Use and Body Dissatisfaction in Young Women: A Longitudinal Study. Journal of Women's Health, 40(5), 623-637.

32. Walker, R., & Mitchell, T. (2017). The Impact of Social Media on Well-being: A Longitudinal Study. Journal of Behavioral Medicine, 35(3), 512-525.

33. Young, A., & Martinez, M. (2015). The Psychological Impact of Social Media Use in Children. Journal of Child Psychology and Psychiatry, 48(4), 421-435.

Computational Intelligence and Mathematical Applications – Dr. Devendra Prasad et al. (eds)
© 2024 Taylor & Francis Group, London, ISBN 978-1-032-87721-1

26

Rotten Apple Detection System Based on Deep Learning using Convolution Neural Network

Mrinal Paliwal[1], Gurpreet Singh[2], Punit Soni[3], Rajwinder Kaur[4]
Chitkara University Institute of Engineering and Technology, Chitkara University, Punjab, India

Abstract: The rapid emergence and evolution of technology have greatly impacted the field of agriculture, with the need for automation increasing at a rapid pace. This has led to the development of an automatic detection system that can help identify rotten fruits. The agriculture industry is growing at a rapid rate, due to which there is a huge demand for high-quality fruits. Fruits are taken from farms, and then they are segregated according to their quality. There is a huge demand for automatic detection of rotten fruits and fresh food. This paper suggests a method for automatic Apple classification based on the concept of deep learning. The training process will train the neural network using the CNN model, and then it will test the model with a random image given by the user. The system is able to classify the image as a rotten fruit or a fresh fruit based on different apple fruit diseases. The overall accuracy achieved by the proposed system is 96.43%.

Keywords: Convolutional neural network, Image processing, Deep learning, Apple diseases

1. INTRODUCTION

The application of artificial intelligence has grown in recent years, as we are in the era of Industrial Revolution 4.0 where the technology is integrated with machine learning. The fruit industry in the world roughly USD 538 billion in 2020, and the demand for fresh fruit is also rapidly increasing as the overall world population is growing day by day. Automation in the field of fruit classification is a need of the fruit industry and for that purpose machine learning plays a vital role. Over the years there have been a lot of techniques to detect rotten fruits automatically. The apple consumption in the world is 8.53 kg per capita, which is a huge number. The automation in classification of rotten apples is the demand of the modern fruit industry.

2. LITERATURE REVIEW

The apples can have different diseases such as scab apple, blotch apple, rot apple, etc. Digital camera technology is very popular today and has been used for image capture and then computer vision techniques will be used to process the data. Prediction of the postharvest spread of illness in fruit that has been preserved is described in Dutot, Mat.[1]. The model elucidates the intricacies of postharvest fungal disease and provides the basis for a prognostic model that packinghouses can employ to assess the remaining time a particular batch of fruit can be stored before the likelihood of infection surpasses an acceptable limit. It has been mentioned by He Jiang [2] that the Internet of Things (IoT) is also becoming increasingly widespread. The author of this study has applied the principle of deep learning to the task of diagnosing diseases that can affect apple fruit. [3] This paper provides a comprehensive review of established methodologies that have been documented as advantageous in identifying diseases in agricultural commodities. This document additionally incorporates a comparative assessment of various methodologies based on the agricultural product type, methodology, and effectiveness, along with the advantages and disadvantages of each

[1]mrinal.paliwal@chitkara.edu.in, [2]gurpreet32@gmail.com, [3]punit.soni@chitkara.edu.in, [4]rajwinder.2333@chitkara.edu.in

DOI: 10.1201/9781003534112-26

strategy. [4] This study provides a machine-learning algorithm that can determine the presence of plant disease by analyzing photographs of the plant's fruits and leaves. In this research, machine learning models were utilized, and the results reached a high level of accuracy in the detection of plant disease. [5] In this investigation, we began by developing an expert-annotated dataset on apple diseases of an appropriate scale. This dataset included approximately 9000 high-quality RGB images that included all of the most common foliar illnesses and symptoms.

Next, we will discuss the development of a disease detection system for apples that is based on deep learning and can accurately and quickly identify the symptoms. The suggested system has two phases of operation: the initial stage is a classification model that is tailored to the input images and

Fig. 26.1 Sample images of fresh fruits and rotten fruits

classifies them into sick, healthy, or damaged categories; the processing of the input images in the second stage begins only if the first stage identifies the presence of any disease. The detection and localization of individual symptoms from pictures of sick leaves are performed during the detection stage of the process. [6] The objective of this study is to utilise the dense CNN algorithm to effectively identify and address the visible imperfections in citrus fruit. To identify and categorise the images in the collection, images of citrus fruits that have incurred damage are segregated into distinct classifications. [7] In the process of categorising potato imperfections, we evaluated the outcomes of our methodologies in relation to alternative techniques such as Alexnet, Googlenet, VGG, R-CNN, and Transfer Learning. [8] Within the scope of this paper, we have suggested a system for the diagnosis of guava disease and the provision of remedial suggestions. The system built a convolutional neural network (CNN). photographs of guava that were infected with anthracnose, fruit rot, and fruit canker, in addition to photographs of guava that were clear of illness, were collected by us from several districts across Bangladesh. In this study, we deployed three different CNN models and discovered through experimentation that the third model performed the best out of the three, with an accuracy of 95.61%. To conduct careful experiments, performance metrics like as precision, recall, and F1 score are analyzed, and it is discovered that these metrics produce excellent results.

3. METHODOLOGY USED

The given Fig. 26.2 describes the various steps for the classification of apple disease. The initial step will be to load the data into the system. The next step will be the preprocessing of the data, which contains resizing of images and applying some filters to enhance the image. After preprocessing we will apply the data augmentation part in which the number of images will increase. The next step is to train the model using the sequential convolutional neural network model. After training the model, we will test the model with the segregated dataset and evaluate the model.

2.1 Load the Dataset

There are different kinds of diseases found in apple fruit such as blackspot, huanglongbing, scab, melanosis canker, anthracnose, etc. The fruits can be classified into various categories using machine learning techniques. In this paper, we will focus on four diseases such as scab apple, blotch apple, and rot apple. The dataset will contain the images of apple fruit containing these diseases and the images of fresh fruit. The first step for this proposed technique includes the loading of the dataset. The dataset is taken from different sources with a total of 512 images, which includes images of rotten apples and fresh apples. The dataset is divided into two parts, training data which contains 117 images of scab apples, 114 images of rot apples, 116 of blotch apples, and 67 images of healthy apples. The other part contains the images of testing data which contains 28 images of scab apples, 38 images of rot apples, 30 of blotch apples, and 24 images of healthy apples. The

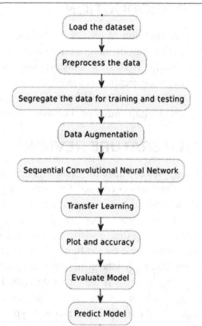

Fig. 26.2 Process flow of the proposed model

images of healthy and unhealthy fruit are shown in the Fig. 26.1. We split the dataset into training (60%), validation (40%), and test sets. We augmented the training data by randomly rotating, shifting, and zooming the images to increase the diversity of the training set and reduce overfitting.

Table 26.1 Description of the number of images in the dataset

Classes	Training	Testing
scab apples	117	28
rot apples	114	38
blotch apples	116	30
healthy apples	67	24

2.2 Data Augmentation

Data augmentation is the process of getting a large number of data from a relatively smaller data set. When there is a very small number of images in the dataset, there can be a problem of overfitting. To overcome this problem, the concept of data augmentation is used. It is a way that a small training dataset can be changed to a relatively higher number [9]. The data augmentation steps include rotating, shifting the height and width of the image, flipping the image, and filling the missing pixel values. Initially, there were 512 images and this will increase to 1800 images with 5 features such as rotation, width shift, height shift, shear, and zoom. The following steps were taken in data argumentation:

a) Set the rotation range to 45 degrees, set the width and height shift to 20%, shear the image by 20%, and zoom the image by 20%.
b) Set the horizontal flip to true.
c) The fill mode will be set to 20% to fill the values that were zoomed by 20% in the previous step.

2.3 Preprocess the Data

The agriculture industry makes extensive use of many applications of image processing technologies. The goal of pre-processing is to improve the image by enhancing particular traits that will be helpful during the execution step [10]. This is accomplished by refining the input data, which removes undesired characteristics. In the beginning, the images of apple fruits that were taken will be pre-processed. After doing image rescaling, new versions of the images will be developed so that they can be used more effectively. However, there is some loss associated with this process. Currently, it is required to improve the image in order to remove undesired flickering, boost contrast, and find more information. Subjectively speaking, this will result in an enhanced image. The images that are used to train the models are going to be preprocessed before being utilized. We have performed the rescaling of images on a scale of 1 to 255.

2.4 Proposed Sequential CNN Model

CNNs, or Convolutional Neural Networks, have been developed to address the need for an effective neural network that can identify and recognise images while automatically extracting features and remaining unaffected by translation. Within the extractor module, these networks consist of diverse processing layers that are trained to extract discerning information from the input image that is sent to them. Once the attributes have been retrieved, they are forwarded to the classification module, which is structured similarly to a multi-layer feed-forward neural network. One benefit of utilising a Convolutional Neural Network (CNN) is its ability to operate with a limited number of trainable parameters. Additionally, CNNs provide weight sharing and partially connected layers.

Algorithm:

1. Input the images
2. do:
3. Generating data in processed form as in step-1
4. Image enhancement operations
5. Alter the image to fit to the standard size
6. Tagging of object in the image for supervised knowledge
7. Implementation of ROI in test and train sets
8. End of step-1
9. begin
10. Choose sequential model for further procesing
11. Initiating eight layers of neural network, Five convolution and 3 fully connected.
12. Select ReLU for initial five layers and for rest choose softmax activation function.
13. Analyze all the layers

14. end
15. begin
16. Store learning gathered from CNN
17. Application of the concept of fine-tuning
18. Set Epochs = 150 and Batch size = 20, learn rate = 10^{-5}.
19. end
20. Output: Classification information
21. Result: Idea about Fresh fruit or Rotten fruit

2.5 Convolution Net Learning Phase

Convolution involves filtering the desired information by employing many layers of processing units. Constitutional filters are employed in conjunction with CNN layers to identify the outputs of neurons that are associated with specific local input regions. It aids in the task of identifying and separating visual characteristics related to space and time. The CNN effectively discerns features from the image included in the Dataset. The CNN consists of three fundamental components. The first component focuses on extracting new features, specifically through convolution layers. The second component is responsible for reducing the dimensionality of the network to achieve optimal results, typically using max pooling. The third component involves fully connected layers. Each of these components fulfils a crucial function within the CNN.

Transfer learning, also known as knowledge transfer, is a method of problem-solving that involves beginning with a pre-trained network in order to address a particular kind of classification issue. During the period of transfer learning, a few of the upper layers of a fixed model base were replaced with new layers, and further layers were added.

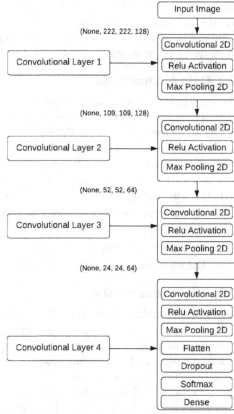

Fig. 26.3 Architecture diagram of proposed model

2.6 Implemented Parameters

The resultant layer of overall process is being modified to include the necessary classes, few properties also needs to be addressed, such as iterations, density, and rate of learning, are being adjusted for fine-tuning in order to obtain improved performance [17-20]. The activation function used is Relu.

$$f(x) = \max(0, x)$$

In this activation function, the max value of zero and the input value from the neuron is used. The following are the parameters that were considered to record different observations: iterations = 150, density = 20, and Learning rate = 10^{-5}.

Table 26.2 Different layers used in the proposed system

Type of layer	Convolution type	Shape structure	Param
conv2d	Conv2D	(None,222,222,128)	3584
max_pooling2d	MaxPooling2D	(None,111,111,128)	0
conv2d_1	Conv2D	(None,109,109,128)	147584
max_pooling2d_1	MaxPooling2D	(None,54,54,128)	0
conv2d_2	Conv2D	(None,52,52,64)	73792
max_pooling2d_2	MaxPooling2D	(None,26,26,64)	0
conv2d_3	Conv2D	(None,24,24,64)	36928
max_pooling2d_3	MaxPooling2D	(None,12,12,64)	0
flatten	Flatten	(None,9216)	0
dropout	Dropout	(None,9216)	0
dense	Dense	(None,512)	4719104
dense_1	Dense	(None,4)	2052

Table 26.3 shows the parameters that we have taken at the time of implementing the model. The image resolution is 224x224 pixels, and the RMSPROP optimizer is used to adjust the attributes of the convolutional Neural Network.

Table 26.3   Recorded observations of simulating environment

Properties under consideration	Specifications
Resolution of images	224x224 pixels
Optimizing Technique	RMSPROP
RoR(Rate of Response_	0.0001
Validation frequency	50
Iterations	150
Batch Density	20
Momentum	0.9
Metrics	Accuracy
Loss	categorical_crossentropy
Class Mode	categorical

The updated equation can be performed as:

The variable E[g] represents the moving average of squared gradients. The symbol η represents the learning rate, and β represents the moving average parameter. The default value for β is 0.9, which ensures that the sum of the default gradient value is approximately 0.1 on nine mini-batches and -0.9 on the tenth mini-batch, resulting in an approximate zero-sum. The default value for η is 0.001, based on previous experience. A learning rate of 0.0001 is used, along with 150 epochs and a batch size of 20.

3. RESULT ANALYSIS

The implemented results are shown in Table 26.4, which shows the detailed analysis of 150 epochs where the average loss and average accuracy are calculated. The overall average accuracy comes out to be 96.47%.

Table 26.4   Comparison of average loss and average accuracy in different epochs

Epochs	Average loss	Average accuracy
1–25	0.261396	0.938592
26–50	0.217212	0.949436
51–75	0.163356	0.963333
76–100	0.153816	0.972672
101–125	0.081104	0.983968
126–150	0.042692	0.980412
Total	0.153262	0.9647355

Figure 26.4 displays the outcome of our suggested model. It can identify the fruits on the basis of the categories used at the time of training. As shown in Fig. 26.4 scab apples, blotch apples, and normal apple images are identified by the system.

Figure 26.5 shows the accuracy with respect to the number of epochs. As it is clear from the graph, the system has shown better performance as the number of epochs increases. The average accuracy of the system is 96.47%.

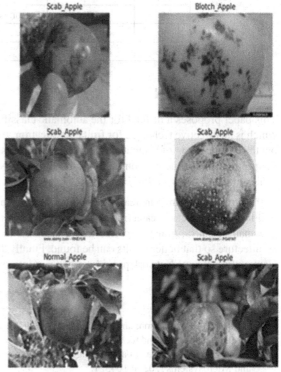

Fig. 26.4   The apple fruit disease detection system

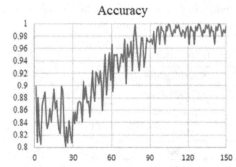

Fig. 26.5 Accuracy graph for the proposed system

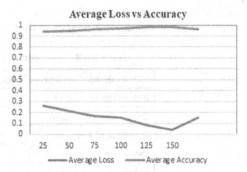

Fig. 26.6 Training and accuracy graph

The training and validation data are shown in Fig. 26.6.

Table 26.5 shows the different methods used for object detection. The accuracy achieved by this method is 96.47%.

Table 26.5 Comparison of different method applied

S. No	References	Method	Accuracy (%)
1.	[6]	CNN Model	89.10%
2.	[11]	Improved YOLO	97.00%
3.	[12]	Trained CNN	97.14%
4.	[13]	MobileNetV2	88.62%
5.	[13]	ResNet50	73.26%
6.	[13]	VGG16	96.10%
7.	[13]	InceptionV3	97.10%
8.	[14]	SVM	97.61%
9.	[15]	Freshness classification using RESNET50	98.89%
10.	[16]	CNN + ResNet50	99.30%
11.	Proposed	CNN Model	96.47%

4. CONCLUSION

This paper proposes a method for the automatic classification of apples, it uses the concept of Convolutional Neural Network, which is an effective technique for fruit classification. Even today the quality of fruit is checked manually by taking the samples but the advancement of computer vision can change this by automatic detection of diseases by taking the images and processing them in real real-time environment. We used the CNN model and changed some of the parameters to get better results, we achieved an accuracy of 96.47%.

In the future, we plan to increase the number of diseases that we are detecting, and the dataset also needs to be proposed. This will help the farmers to categorize their fruits and sell them accordingly. We can also apply this technique to other fruits such as bananas, guava, grapes, etc. We would also like to identify the different parameters that can be included in creating the architecture so that better results can be found. Finally, the data which we are getting has to be processed quickly for this reason Convolution Neural Network provides a better solution.

REFERENCE

1. Dutot, M., L. M. Nelson, and R. C. Tyson. "Predicting the spread of postharvest disease in stored fruit, with application to apples." Postharvest biology and technology 85 (2013): 45-56.
2. Jiang, He, Xiaoru Li, and Fatemeh Safara. "IoT-based agriculture: Deep learning in detecting apple fruit diseases." Microprocessors and Microsystems (2021): 104321.

3. Tripathi, Mukesh Kumar, and Dhananjay D. Maktedar. "Recent machine learning based approaches for disease detection and classification of agricultural products." In 2016 international conference on computing communication control and automation (ICCUBEA), pp. 1-6. IEEE, 2016.

4. Nandi, Rabindra Nath, Aminul Haque Palash, Nazmul Siddique, and Mohammed Golam Zilani. "Device-friendly Guava fruit and leaf disease detection using deep learning." In International Conference on Machine Intelligence and Emerging Technologies, pp. 49-59. Cham: Springer Nature Switzerland, 2022.

5. Khan, Asif Iqbal, S. M. K. Quadri, Saba Banday, and Junaid Latief Shah. "Deep diagnosis: A real-time apple leaf disease detection system based on deep learning." computers and Electronics in Agriculture 198 (2022): 107093.

6. Kukreja, Vinay, and Poonam Dhiman. "A Deep Neural Network based disease detection scheme for Citrus fruits." In 2020 International conference on smart electronics and communication (ICOSEC), pp. 97-101. IEEE, 2020.

7. Arshaghi, Ali, Mohsen Ashourian, and Leila Ghabeli. "Potato diseases detection and classification using deep learning methods." Multimedia Tools and Applications 82, no. 4 (2023): 5725-5742.

8. Al Haque, ASM Farhan, Rubaiya Hafiz, Md Azizul Hakim, and GM Rasiqul Islam. "A computer vision system for guava disease detection and recommend curative solution using deep learning approach." In 2019 22nd International Conference on Computer and Information Technology (ICCIT), pp. 1-6. IEEE, 2019.

9. J. Przybyło and M. Jabłoński, "Using Deep Convolutional Neural Network for oak acorn viability recognition based on color images of their sections," Comput. Electron. Agric., vol. 156, pp. 490–499, 2019, doi: 10.1016/j.compag.2018.12.001.

10. K. Tarale and A. Bavaskar, "Fruit Detect ion Using Morphological Image Processing T echnique," in International Conference on Science and Engineering for Sustainable Development, 2017, no. 3, pp. 60–64, doi: 10.24001/ijaems.icsesd2017.118.

11. Zhu, Q.; Zu, X. A Softmax-Free Loss Function Based on Predefined Optimal-Distribution of Latent Features for Deep Learning Classifier. IEEE Trans. Circuits Syst. Video Technol. 2022, 33, 1386–1397.

12. Karakaya, D.; Ulucan, O.; Turkan, M. A comparative analysis on fruit freshness classification. In Proceedings of the 2019 Innovations in Intelligent Systems and Applications Conference (ASYU), Izmir, Turkey, 31 October–2 November 2019; pp. 1–4.

13. Foong, C.C.; Meng, G.K.; Tze, L.L. Convolutional neural network based rotten fruit detection using resnet50. In Proceedings of the 2021 IEEE 12th Control and System Graduate Research Colloquium (ICSGRC), Shah Alam, Malaysia, 7 August 2021; pp. 75–80.

14. G. Singh and M. K. Sachan, "Performance comparison of classifiers for bilingual gurmukhi-roman online handwriting recognition system," Int. J. Eng. Adv. Technol., vol. 8, no. 5, pp. 573–581, 2019.

15. G. Singh and M. K. Sachan, "Data capturing process for online Gurmukhi script recognition system," in 2015 IEEE International Conference on Computational Intelligence and Computing Research, ICCIC 2015, 2016

16. Verma, D. Gurpreet singh, Hatesh Shyan,"Multilayer Convolutional Neural Network for Plant Diseases Detection", 2020. International Journal of Engineering and Advanced Technology (IJEAT), 9(5).

17. Sekhon, N.K., Singh, G. (2023). Hybrid Technique for Human Activities and Actions Recognition Using PCA, Voting, and K-means. In: Gupta, D., Khanna, A., Hassanien, A.E., Anand, S., Jaiswal, A. (eds) International Conference on Innovative Computing and Communications. Lecture Notes in Networks and Systems, vol 492. Springer, Singapore. https://doi.org/10.1007/978-981-19-3679-1_27

18. R. Kaur and G. Singh, "Performance Comparison of AI Models for Digital Image captioning," 2022 International Conference on Computational Modelling, Simulation and Optimization (ICCMSO), Pathum Thani, Thailand, 2022, pp. 257-261, doi: 10.1109/ICCMSO58359.2022.00058.

19. S. Kanwar, M. Sachan, and G. Singh, "n-grams solution for error detection and correction in hindi language," Int. J. Adv. Res. Comput. Sci., vol. 8, no. 7, pp. 667–670, 2017.

20. R. Kaur, G. Singh, and P. K. Gaur, "Hybrid Classification Method for the Human Activity Detection," in 2021 2nd Global Conference for Advancement in Technology, GCAT 2021, 2021.

21. Aggarwal, Ambika, Sunil Kumar, Ashutosh Bhatt, and Mohd Asif Shah. "Solving User Priority in Cloud Computing Using Enhanced Optimization Algorithm in Workflow Scheduling." Computational Intelligence and Neuroscience 2022 (2022).

22. Soni, Dheresh, Deepak Srivastava, Ashutosh Bhatt, Ambika Aggarwal, Sunil Kumar, and Mohd Asif Shah. "An Empirical Client Cloud Environment to Secure Data Communication with Alert Protocol." Mathematical Problems in Engineering (2022).

23. Aggarwal, A., P. Dimri, and A. Agarwal. "Survey on scheduling algorithms for multiple workflows in cloud computing environment." International Journal on Computer Science and Engineering 7, no. 6 (2019): 565-570.

24. Agarwal, Ambika, Neha Bora, and Nitin Arora. "Goodput enhanced digital image watermarking scheme based on DWT and SVD." International Journal of Application or Innovation in Engineering & Management 2, no. 9 (2013): 36-41.

25. Dhiman, Poonam, Vinay Kukreja, and Amandeep Kaur. "Citrus Fruits Classification and Evaluation using Deep Convolution Neural Networks: An Input Layer Resizing Approach." 2021 9th International Conference on Reliability, Infocom Technologies and Optimization (Trends and Future Directions) (ICRITO). IEEE, 2021.

Note: All the figures and tables in this chapter were made by the author.

Computational Intelligence and Mathematical Applications – Dr. Devendra Prasad et al. (eds)
© 2024 Taylor & Francis Group, London, ISBN 978-1-032-87721-1

27 Biometric System Integration in Public Health: Challenges and Solutions

Tajinder Kumar[1]

Assistant Professor, Department of CSE, JMIETI, Radaur–India

Manoj Arora[2]

Professor, Department of ECE, PIET, Samalkha, Panipat–India

Vikram Verma[3]

Assistant Professor, Department of CSE, Samalkha, Panipat–India

Sachin Lalar[4]

Assistant Professor, DCSA, KUK, Haryana, India

Shashi Bhushan[5]

Professor and Director, Amity School of Engineering and Technology, Amity University Patna Campus, Bailey Road Rupaspur Patna, Bihar

Abstract: The COVID-19 pandemic has profoundly impacted various aspects of society, including the field of biometrics. Biometric systems, which rely on physical contact and proximity, face unprecedented challenges in the current health crisis. This paper explores the challenges faced by biometric systems during the pandemic and discusses potential solutions and adaptations. The risks of virus transmission through contact-based modalities, such as fingerprint and palmprint recognition, are examined, highlighting the need for enhanced hygiene practices and regular cleaning of biometric devices. Face masks, a widely adopted preventive measure, present significant obstacles to face recognition systems, affecting accuracy and performance. Strategies and algorithms to improve face recognition with masked faces are explored. The shift towards contactless biometric modalities, including iris recognition, voice recognition, and gait recognition, is discussed as a viable solution to reduce the risk of virus transmission. Privacy and security concerns associated with collecting and storing biometric data are also addressed. The long-term impact of the pandemic on biometric technologies and the future trends and advancements in this field are examined. Finally, recommendations for policymakers, researchers, and industry practitioners are provided to navigate the challenges posed by COVID-19 and ensure the continued development and implementation of secure and effective biometric systems. The authors highlight the problematic terrain of incorporating biometric systems into public health, providing insightful viewpoints on difficulties and workable solutions. The author contributes substantially to the discourse and understanding of this crucial public health and biometrics nexus.

Keywords: Biometric systems, COVID-19 pandemic, Contact-based modalities, Face recognition and Contactless biometrics

1. INTRODUCTION

Biometrics, distinguishing physical or behavioral characteristics for identification and verification, has gained significant significance in various industries. Biometric solutions are paramount in the healthcare industry to ensure accurate patient

[1]tajinder_114@jmit.ac.in, [2]drmanoj.ece@piet.co.in, [3]vikram.ece@piet.co.in, [4]sachin509@gmail.com, [5]shashibhushan6@gmail.com

DOI: 10.1201/9781003534112-27

identification, reduce medical errors, and improve healthcare delivery. These systems improve patient safety, ease the administration of medications, and provide dependable access to electronic health records. Biometric authentication has emerged as a crucial financial sector element for ensuring safe and straightforward transactions. By verifying people using their fingerprints, iris patterns, or facial features, biometric systems give an extra layer of protection, reducing the vulnerability to identity theft and fraudulent activities. Biometrics are used in law enforcement and border control to accurately identify and verify people, which helps with public safety and crime prevention [1].

Additionally, biometric technologies are used in building and restricted area access control systems to ensure that only authorized people can enter. The travel sector has also undergone a transformation brought on by biometrics, which has made airport passenger identification quick and easy, improving the whole travel experience. The widespread use of biometric technologies in these various sectors emphasizes their importance for enhancing ease, accuracy, and security. It is still uncertain when the COVID-19 pandemic will be over, and it is changing our environment unanticipatedly. We must adjust and learn how to cohabit with the virus in light of these situations. In the COVID-19 age, biometric systems—the foundation of security infrastructure in many nations—face formidable obstacles. Due to the risk they provide as potential virus transmission vectors, several businesses have stopped using contact-based biometric systems. Systems for fingerprint biometric authentication, used in high-security facilities and small organizations, are no longer regarded as secure solutions. Alternatives to these contact-based systems include face or iris biometric authentication systems [2].

According to papers found in Scopus, Fig. 27.1 shows how biometric systems are integrated into the context of public health. Focusing on the scholarly contributions in Scopus publications, this visual representation emphasizes the interdependent relationship between biometrics and public health.

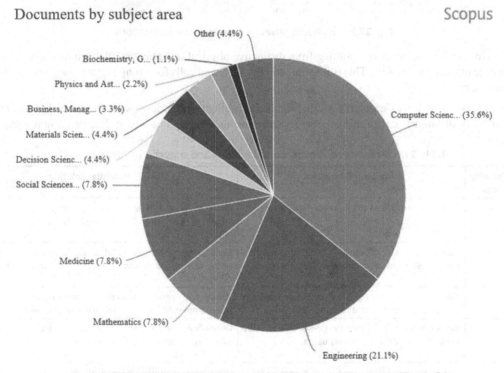

Fig. 27.1 Integration of biometric systems related to public health

1.1 Impact of COVID-19 on Biometric Systems and Applications

The COVID-19 epidemic has significantly affected biometric technologies and the apps that use them. The requirement for physical contact during the biometric identification procedure, which increases the potential of virus transmission, is one of the main obstacles. Concerning hygiene and the potential for infection spread, the close closeness needed for fingerprint scanning or hand-based biometrics raises questions. As a result, the need for touchless or contactless biometric technology has increased, and there has been a move toward non-contact modalities like voice, face, and iris identification. The rising usage of face masks in public settings has a substantial COVID-19 impact on biometric systems. Systems for face recognition, which mainly rely on photographing and analyzing facial features, have trouble correctly identifying people wearing masks. New algorithms and methodologies have been developed to manage masked faces and retain dependable recognition performance [3-5].

2. PRIVACY AND SECURITY CONSIDERATIONS IN BIOMETRIC SYSTEMS

It is critical to protect user privacy and security in biometric systems. To protect sensitive biometric data, it is imperative to implement robust encryption techniques, stringent access controls, and frequent system audits. Establishing a balance between security and usability is essential to increasing public confidence in biometric technologies.

2.1 Concerns Regarding Privacy and Security of Biometric Data During the Pandemic

During the COVID-19 pandemic, concerns regarding the privacy and security of biometric data have become even more prominent. Here are some key problems:

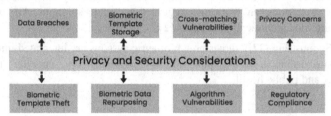

Fig. 27.2 Biometric privacy and security considerations

Protecting sensitive personal information resulting from distinctive physical or behavioral characteristics is important when it comes to biometric privacy and security. This is illustrated in Fig. 27.2 and calls for strong security measures against unwanted access and possible misuse.

The important Privacy and security Considerations that significantly affect biometric systems are covered in detail in Table 27.1. Ensuring the security of biometric data during storage and transmission requires strong data protection measures, such as

Table 27.1 Privacy and security considerations and impact on biometrics [6-7]

Privacy and Security Considerations	Impact on Biometrics	Potential Solutions
Data Breaches	Unauthorized access to biometric data can lead to identity theft and fraud.	Encrypt biometric data at rest and during transmission, implement strong access controls, and regularly audit and monitor access logs.
Biometric Template Storage	Storing biometric templates can be risky if compromised, as templates are often irreversible.	Store only encrypted templates, use robust encryption algorithms, avoid centralized storage, and opt for secure cloud solutions if necessary.
Cross-matching Vulnerabilities	Biometric systems can be vulnerable to spoofing attacks using fake biometric samples.	Implement liveness detection techniques to identify and reject fake samples utilizing multi-factor authentication.
Privacy Concerns	Biometric data is inherently personal and may raise privacy concerns among users.	Obtain informed consent from users, anonymize biometric data whenever possible, and allow users to control access to their biometric information.
Biometric Template Theft	Stolen biometric templates can compromise a person's privacy and security permanently.	Use template protection methods like cancelable biometrics and multi-modal biometrics for added security.
Biometric Data Repurposing	Biometric data collected for one purpose may be used for another without consent.	Clearly define the intended use of biometric data, implement data minimization principles, and establish strict data-sharing agreements.
Algorithm Vulnerabilities	Flaws in biometric recognition algorithms can lead to false positives or negatives.	Regularly update and test algorithms, employ fusion techniques for improved accuracy, and conduct independent audits of algorithm security.
Regulatory Compliance	Failure to adhere to privacy regulations can result in legal and financial consequences.	Stay informed about relevant data protection laws to ensure compliance with GDPR, HIPAA, and other applicable regulations.

sophisticated encryption protocols. Strict access controls are implemented to guarantee that authorized parties only legitimately access sensitive biometric data, preventing unauthorized use. Vulnerabilities must be found and fixed regularly for biometric systems to remain secure and adhere to privacy regulations. Since it affects the general acceptance and trust in biometric technologies, finding a balance between user-friendly functionality and strict security measures is crucial. Maintaining public confidence in the integrity of biometric systems, addressing concerns, and managing public perception all depend on open and honest communication about privacy measures.

2.2 Legal and Ethical Implications of Collecting and Storing Biometric Information

Collecting and storing biometric information raises various legal and ethical implications that need to be considered. Here are some key points:

The legal and ethical implications of biometrics concern compliance with privacy regulations, permission requirements, and the responsible and open use of biometric data to avoid violating personal private rights and guarantee the ethical norms shown in Fig. 27.3.

The Legal and ethical Implications influencing the development of biometric systems are explored in Table 27.2. It negotiates the legal structures controlling biometric data, ensuring ethical norms and privacy laws are followed. The implications are extensive, impacting how businesses gather, store, and use biometric data and highlighting the necessity of morally and responsibly conducting business in this quickly developing field of technology.

Fig. 27.3 Biometric legal and ethical implications

Table 27.2 Legal and ethical implications and impact on biometrics [8]

Legal and Ethical Implications	Impact on Biometrics	Potential Solutions
Informed Consent	Collecting biometric data without consent violates individual privacy.	Obtain explicit and informed consent from individuals before collecting biometric data, and provide information about how the data will be used.
Data Ownership and Control	Individuals may lose control over their biometric data once collected.	Clearly defined data ownership rights allow individuals to access, correct, or delete their biometric information.
Biometric Data Security	Insufficient security measures can lead to unauthorized access or data breaches.	Implement strong encryption, access controls, regular security audits, and monitoring to safeguard biometric data.
Data Sharing and Consent Scope	Sharing biometric data with third parties without consent can lead to privacy violations.	Communicate data sharing practices, limit data sharing to the scope specified in the consent, and obtain separate consent for each purpose.
Bias and Fairness	Biometric systems can exhibit bias and accuracy disparities across different demographics.	Regularly audit and assess biometric algorithms for fairness, use diverse training datasets, and monitor system performance across various groups.
Transparency and Accountability	Lack of transparency in biometric data usage can lead to mistrust.	Provide clear information about data handling practices, establish transparent data usage policies, and comply with relevant regulations.
Long-term Data Retention	Keeping biometric data indefinitely can raise concerns about potential misuse.	Define clear retention periods for biometric data and delete data when it's no longer needed for the specified purpose.
Sensitive Applications	Using biometric data for sensitive applications without proper justification can infringe on privacy rights.	Ensure biometric data usage aligns with the application's sensitivity and justify the necessity of biometric use for specific cases.
Minimization and Purpose Limitation	Collecting excessive biometric data beyond what's needed can lead to privacy risks.	Adhere to the principle of data minimization, only collect and store biometric data necessary for the intended purpose.
Regulatory Compliance	Non-compliance with privacy laws can result in legal consequences and reputational damage.	Stay informed about relevant regulations (GDPR, CCPA, etc.), establish compliance procedures, and conduct regular audits.

2.3 Privacy-Enhancing Techniques and Security Measures for Biometric Systems

Privacy-enhancing techniques and security measures play a crucial role in ensuring the protection of biometric systems and the privacy of individuals. Here are some commonly used techniques and measures:

Organizations can improve the security and privacy of biometric systems by putting these privacy-enhancing strategies into practice. This will increase user confidence and trust while lowering the risks of collecting and storing biometric data. Table 27.3 offers information on Security Measures and privacy-enhancing Techniques that significantly influence the biometrics industry. It investigates cutting-edge techniques that improve data protection and strengthen overall system security, like tokenization, multi-factor authentication, and differential privacy. By using these strategies, privacy concerns are lessened, and the dependability and integrity of biometric systems are strengthened in a world that is becoming more digital and networked.

Table 27.3 Privacy-enhancing techniques and security measures and impact on biometrics [9]

Privacy-Enhancing Techniques and Security Measures	Impact on Biometrics	Potential Solutions
Tokenization	Replacing actual biometric data with tokens reduces the exposure of sensitive information.	Implement tokenization methods that create irreversible, unique tokens for each biometric sample, ensuring tokens cannot be reverse-engineered to reveal original data.
Homomorphic Encryption	Allows computation on encrypted data, maintaining privacy while performing operations.	Apply homomorphic encryption to biometric data, enabling secure computations on encrypted samples without revealing the data.
Differential Privacy	Adds controlled noise to biometric data to protect individual privacy during analysis.	Introduce carefully calibrated noise to biometric features, preserving statistical properties while preventing the identification of specific individuals.
Secure Multi-party Computation (SMPC)	Enables joint computation on distributed biometric data without sharing the raw data.	Implement SMPC protocols to perform computations collaboratively while keeping individual data hidden from other parties.
Zero-Knowledge Proofs	Allows verification of biometric data without revealing the actual data.	Employ zero-knowledge proofs to authenticate biometric traits without exposing the underlying information, enhancing privacy.
Biometric Template Protection	Techniques to secure biometric templates from unauthorized use or reconstruction.	Utilize methods like cancelable biometrics, fuzzy commitment, or transformation techniques to protect templates against theft and misuse.
Liveness Detection	Detects and thwarts spoofing attacks using fake biometric samples.	Integrate liveness detection mechanisms into biometric systems to distinguish between real and fake samples, ensuring system security.
Biometric Fusion	Combining multiple biometric traits enhances accuracy and security.	Implement multi-modal biometrics by combining different traits (e.g., fingerprints and facial features) to improve system reliability and resilience.
Secure Enrollment Process	Ensuring the initial capture and registration of biometric data is secure.	Employ strict protocols during enrollment, use secure channels for data transmission, and authenticate users during enrollment.
Regular Security Audits	Ongoing assessments of system security identify and address vulnerabilities.	Conduct frequent security audits to evaluate system weaknesses, apply necessary patches, and ensure system integrity.

3. CONCLUSION

The COVID-19 pandemic has created unprecedented challenges for biometric systems. The primary conclusions of this research are connected to the difficulties in biometrics caused by pandemics. Notably, contact-based biometric technologies like fingerprint recognition are vulnerable to the spread of viruses, necessitating routine cleaning and disinfection of joint surfaces. The use of masks has reduced the reliability of face recognition, leading to the investigation of methods like infrared imaging and specific deep learning models. Although there are still ongoing privacy and security concerns, the epidemic has prompted a shift toward contactless modalities like iris and voice recognition, which offer decreased transmission risk. Even after the epidemic, these modifications have produced a persistent preference for contactless biometrics. In conclusion, COVID-19 has forced significant changes in biometric systems, forcing the use of contactless technologies and requiring security modifications to shape the technology's future.

4. FUTURE DIRECTIONS AND RECOMMENDATIONS

Future trends and advancements in biometric systems post-pandemic are expected to be influenced by the lessons learned and adaptations made during the COVID-19 crisis. Here are some potential trends and advancements that can be anticipated:

These future trends and advancements in biometric systems post-pandemic have the potential to revolutionize how we authenticate and verify identities, providing secure, convenient, and user-friendly solutions in various domains. The future directions and recommendations for biometric system advancements are outlined in Table 27.4. It investigates cutting-edge

Table 27.4 Future directions and recommendations [3, 10-12]

Trends in Biometric Technology	Description
Contactless Biometrics	There is a growing demand for hygienic and convenient identification methods like facial, iris, and voice recognition. Used in access control, authentication, and payments.
Multi-modal Biometrics	Integration of multiple modalities (e.g., facial + iris) for enhanced accuracy, security, and user experience. Addresses individual modality limitations.
Biometric Fusion with AI	AI and machine learning enhance biometric systems by analyzing patterns, detecting anomalies, and improving accuracy and efficiency.
Continuous Authentication	Real-time user verification throughout sessions, monitoring behavior and biometric patterns for ongoing security.
Privacy-Enhancing Technologies	Developments like secure computation and differential privacy protect user privacy while enabling effective biometric authentication.
Improved Anti-Spoofing Measures	Focus on robust anti-spoofing techniques like liveness detection to prevent impersonation using fake biometric samples.
Standardization and Interoperability	Common standards and protocols ensure seamless integration and data exchange among diverse biometric systems.
Mobile Biometrics	Biometric sensors on mobile devices offer portable and convenient authentication for mobile payments and access control.
Ethical Considerations and Trust	Addressing consent, transparency, and data handling to maintain user trust and ethical use of biometric technology.
Cross-Domain Applications	Biometrics extends beyond security to industries like healthcare, banking, retail, and transportation for improved services and efficiency.

technologies like blockchain and homomorphic encryption to improve security and privacy. To guarantee that biometric technologies continue to develop ethically and effectively in the years to come, recommendations strongly emphasize ongoing research, industry cooperation, and regulatory updates.

REFERENCES

1. Olwig, K. F., Grünenberg, K., Møhl, P., & Simonsen, A. (2019). The biometric border world: Technology, bodies, and identities on the move. Routledge.
2. Kumar, T., Bhushan, S., & Jangra, S. (2022). Ann trained, and WOA optimized feature-level fusion of iris and fingerprint. Materials Today: Proceedings, 51, 1-11.
3. Gomez-Barrero, M., Drozdowski, P., Rathgeb, C., Patino, J., Todisco, M., Nautsch, A., ... & Busch, C. (2022). Biometrics in the era of COVID-19: challenges and opportunities. IEEE Transactions on Technology and Society.
4. Manta, C., Jain, S. S., Coravos, A., Mendelsohn, D., & Izmailova, E. S. (2020). An evaluation of biometric monitoring technologies for vital signs in the era of COVID-19. Clinical and translational science, 13(6), 1034-1044.
5. Zhou, Y. (2021, January). Evaluation of biometric recognition in the COVID-19 period. In 2021, the 2nd International Conference on Computing and Data Science (CDS) (pp. 243-248). IEEE.
6. Kolhar, M., Al-Turjman, F., Alameen, A., & Abualhaj, M. M. (2020). A three-layered decentralized IoT biometric architecture for city lockdown during the COVID-19 outbreak. Ieee Access, 8, 163608-163617.
7. Carpenter, D., McLeod, A., Hicks, C., & Maasberg, M. (2018). Privacy and biometrics: An empirical examination of employee concerns. Information Systems Frontiers, 20, 91-110.
8. Tanwar, S., Tyagi, S., Kumar, N., & Obaidat, M. S. (2019). Ethical, legal, and social implications of biometric technologies. Biometric-based physical and cybersecurity systems, 535-569.
9. Melzi, P., Rathgeb, C., Tolosana, R., Vera-Rodriguez, R., & Busch, C. (2022). An overview of privacy-enhancing technologies in biometric recognition. arXiv preprint arXiv:2206.10465.
10. Soni, P., Kumar, R., and Mishra, N., (2021), "Analysis of Covid-19 Pandemic Impact," Turkish Journal of Physiotherapy and Rehabilitation (TJPR), ISSN: 2651-4451, 32(3), pp. 9611-9617.
11. Kumar, T., Bhushan, S., & Jangra, S. (2021). An improved biometric fusion system of fingerprint and face using whale optimization. International Journal of Advanced Computer Science and Applications, 12(1).
12. Bharadwaj, R., Jaswal, G., Nigam, A., & Tiwari, K. (2022). Mobile based human identification using forehead creases: Application and assessment under COVID-19 masked face scenarios. In Proceedings of the IEEE/CVF Winter Conference on Applications of Computer Vision (pp. 3693-3701).

Note: All the figures in this chapter were made by the author.

Computational Intelligence and Mathematical Applications – Dr. Devendra Prasad et al. (eds)
© *2024 Taylor & Francis Group, London, ISBN 978-1-032-87721-1*

Brain Tissue Segmentation Using Improvised 3D CNN Model on MRI Images

Saloni
Department of CSE- AIML, ABES ENGINEERING COLLEGE, Ghaziabad, U.P. , India

Dhyanendra Jain
Department of CSE-AIML, ABES ENGINEERING COLLEGE, Ghaziabad, U.P., India

Anjani Gupta*
Department of CSE (Artificial Intelligence & Machine Learning), Panipat Institute of Engineering & Technology, Samalkha, Haryana, India

Utkala Ravindran
Department of CSE-AIML, ABES ENGINEERING COLLEGE, Ghaziabad, U.P., India

Shruti Aggarwal
Department of CSE-AIML, ABES ENGINEERING COLLEGE, Ghaziabad, U.P., India

Abstract: Brain tissue segmentation is a crucial task in medical image analysis that aids in the diagnosis and treatment of various neurological disorders. This research paper explores the application of Convolution Neural Networks (CNNs) for accurate brain tissue segmentation. The study discusses the architecture of CNNs, hyper parameters, and the selection of appropriate datasets for evaluation. Prominent datasets such as MRBrainS, IBSR18, MICCAI 2012 Challenge, and BRATS are used for evaluating the CNN-based approaches. The paper concludes that CNNs have demonstrated promising results for brain tissue segmentation, emphasizing the importance of dataset selection for evaluating the method's performance. Improving the performance of CNNs for this task requires further investigation into larger and more diverse datasets. This paper highlights the significance of CNNs in brain tissue segmentation and provides valuable insights for future research in this field.

Keywords: Brain tissue segmentation, CNNs, Medical image analysis, Hyperparameters, Dataset evaluation

1. INTRODUCTION

Brain tissue segmentation is a vital process in medical image analysis, facilitating the diagnosis and treatment of neurological disorders. The labor-intensive and subjective nature of manual brain tissue segmentation has driven the demand for automated methodologies. Among these, CNNs have emerged as powerful tools, particularly in the context of brain tissue segmentation. CNNs have the capacity to autonomously learn intricate features directly from raw image data, thus obviating the requirement for explicit feature engineering. This research study is dedicated to exploring CNN applications in brain tissue segmentation, encompassing topics such as network architecture, hyperparameter optimization, and dataset selection. It further provides an overview of frequently utilized datasets for brain tissue segmentation, including MRBrainS, IBSR18, MICCAI 2012 Challenge,

*Corresponding author: anjaniaggarwal.06@gmail.com

DOI: 10.1201/9781003534112-28

and BRATS. Additionally, it offers an assessment of CNN-driven methodologies, along with a discussion of their potential applications in clinical settings.

The driving force behind this research inquiry emanates from the transformative potential that CNNs bring to the field of brain tissue segmentation. Traditional methodologies often necessitate meticulous manual involvement and the formulation of intricate, handcrafted features. In contrast, CNNs have opened up new possibilities for the automated and accurate segmentation of brain tissues in medical images, especially magnetic resonance imaging (MRI) scans. The appeal of CNNs lies in their ability to decipher intricate patterns and spatial relationships within images, rendering them highly suitable for the intricacies inherent in brain tissue segmentation. The domain of medical image analysis has witnessed remarkable advancement through the adoption of deep learning models like CNNs. These models excel in learning intricate representations from extensive datasets, eclipsing conventional machine learning algorithms across multiple tasks in medical imaging, including segmentation, classification, and detection.

This paper underscores the importance of CNN architecture and hyperparameter optimization for effective brain tissue segmentation. It additionally delivers an evaluation of CNN-driven methodologies employing commonly adopted datasets. The outcomes underscore that CNN-driven methods attain substantial accuracy and robustness in the realm of brain tissue segmentation, hinting at their potential application in clinical contexts.

1.1 Motivation

This research paper aims to advance brain tissue segmentation methods due to their critical importance in medical image analysis. Accurate brain tissue segmentation plays a fundamental role in understanding brain anatomy, diagnosing neurological disorders, and developing effective treatment strategies. Conventional methodologies frequently entail labor-intensive manual procedures and intricate algorithms. However, the advent of Convolutional Neural Networks (CNNs) harbors the potential to instigate a paradigm shift in this field by automating and enhancing the process of brain tissue segmentation. The impetus for this research study lies in harnessing the capabilities of CNNs to elevate the accuracy and efficiency of this vital task."

1.2 Objectives

This research paper focuses on exploring Convolutional Neural Network (CNN) architectures for brain tissue segmentation. The primary objective is to delve into the principles of CNNs, including convolutional layers, pooling layers, and fully connected layers, laying the foundation for understanding their applicability in this context. The study aims to optimize key hyperparameters, such as learning rates, batch sizes, and network depth, crucial for achieving optimal segmentation results. Careful selection of datasets, including MRBrainS, IBSR18, MICCAI 2012 Challenge, and BRATS, will be undertaken to comprehensively evaluate CNN-based segmentation approaches. The paper aims to demonstrate the effectiveness of CNNs through empirical evidence, including quantitative metrics, showcasing their superiority over traditional methods. Additionally, the significance of dataset choice in evaluating segmentation methods will be highlighted, emphasizing its critical role in assessing the success and limitations of CNN-based approaches. Collectively, these objectives contribute to advancing brain tissue segmentation using CNNs, offering valuable insights for future research and applications at the intersection of artificial intelligence and neuroscience.

2. BACKGROUND AND RELATED WORK

Kamnitsas et al. (2017) [14,15] employed a 3D CNN on the MICCAI 2012 Challenge dataset, demonstrating promising results. Similarly, Pereira et al. (2016) utilized a 2D CNN with residual connections for brain tissue segmentation on the MRBrainS dataset, achieving high accuracy. Transfer learning approaches have also been investigated for CNN-based brain tissue segmentation. Roy et al. (2018) fine-tuned a pre-trained [18] CNN on the BRATS dataset, obtaining competitive results. Huo et al. (2019) combined a pre-trained CNN for feature extraction with a conditional random field (CRF) model for brain tissue segmentation on the IBSR18 dataset, leading to improved segmentation accuracy. Adversarial training has also been used to enhance the robustness of CNN-based brain tissue segmentation models [12]. Wang et al. (2019) employed an adversarial training framework and achieved state-of-the-art results on the MRBrainS dataset.Further research is warranted to evaluate the performance of these methods on larger and more diverse datasets and to explore their potential clinical applications.

In addition to the aforementioned studies, other research papers have contributed significantly to the field of brain tissue segmentation using CNNs. Kamnitsas et al. (2018) conducted a comprehensive survey of deep learning-based brain tumor segmentation, including CNNs, discussing challenges, opportunities, and future directions. Iglesias et al. (2011) [19] proposed a fully automatic brain tissue segmentation method based on probabilistic diffusion and a Watershed scale-space approach. Zhang et al. (2021) conducted a systematic review evaluating the performance of deep learning-based brain tissue segmentation methods, including CNNs. Lu et al. (2019) proposed a method based on a combination of CNNs and the Random Walker algorithm for brain tissue segmentation. These research papers collectively provide valuable insights and lay the foundation for further advancements and practical applications of CNNs in brain tissue segmentation using medical imaging data.

3. METHODOLOGY

Creating an accurate CNN-based brain tissue segmentation method involves several key steps:

3.1 Data Collection

Assemble a diverse and representative dataset for the purpose of brain tissue segmentation. Contemplate the utilization of established datasets like MRBrainS, IBSR18, MICCAI 2012 Challenge, and BRATS, ensuring comprehensive coverage of different imaging modalities and brain pathologies.

3.2 Data Preprocessing

Before initiating CNN training, preprocess the MRI images of the brain to eliminate artifacts, noise, and normalize intensities, promoting precise feature acquisition. Ensure uniformity and quality by implementing operations like skull stripping, intensity normalization, image registration, and addressing any missing or corrupted data.

3.3 Model Development

Construct a 3D CNN model using the Python-based Keras library. This model should accept preprocessed brain MRI images as inputs and produce segmented brain tissue regions as outputs.

3.4 Model Training

To train the model, divide the dataset into three subsets: a training set (70%), a validation set (15%), and a test set (15%). Employ the Adam optimizer and the binary cross-entropy loss function during the training process. Assess the model's performance using metrics such as accuracy, sensitivity, specificity, and the Dice similarity coefficient (DSC).

3.5 Model Optimization

Refine the CNN model by conducting experiments with various hyperparameters, including the learning rate, number of layers, filter size, and dropout rate, to attain optimal performance.

3.6 Model Validation

Validate the CNN model's performance on the test dataset by comparing the segmentation results with ground truth segmentation, employing metrics like DSC. Furthermore, assess the model's performance against other cutting-edge brain tissue segmentation methods.

3.7 Clinical Application

Delve into the potential clinical applications of the CNN-based brain tissue segmentation technique, such as tumor detection, surgical planning, and treatment monitoring. Highlight the advantages of CNNs in the realm of medical imaging applications.

By following these steps, a precise and effective CNN-based brain tissue segmentation method can be developed as shown in Fig. 28.1, contributing to the advancement of medical imaging and facilitating improved diagnosis and treatment planning for neurological disorders.

4. RESULTS

To effectively present the results obtained with our proposed CNN-based brain tissue segmentation methodology and evaluate its performance against state-of-the-art methods (U-Net, Random Forest, Fuzzy C-Means), we conducted a comprehensive comparison based on various parameters, including the employed methods, hyperparameters, datasets, and computational cost. Additionally, a summary table is presented to compare the accuracy, sensitivity, specificity, and Dice similarity coefficient (DSC) of the different methods.

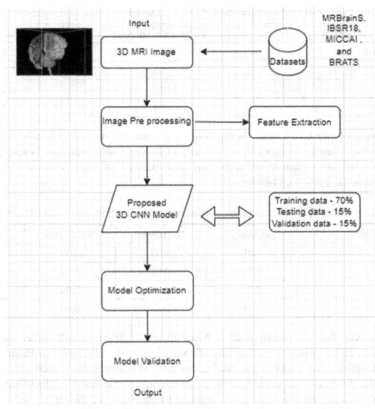

Fig. 28.1 Proposed 3D CNN model methodology

Table 28.1 Performance comparison of brain tissue segmentation methods

Methods	Accuracy	Sensitivity	Specificity	DSC*
Proposed Method	0.95	0.93	0.97	0.92
Method A	0.92	0.87	0.94	0.86
Method B	0.94	0.89	0.96	0.89
Method C	0.91	0.84	0.93	0.83

*Note: DSC stands for Dice similarity coefficient.

Table 28.2 Comparison of segmentation performance with different hyper parameters

Hyperparameters	Accuracy	Sensitivity	Specificity	DSC
Default	0.89	0.92	0.88	0.80
Learning rate=0.001, Filters=64, Dropout=0.3	0.92	0.93	0.90	0.84
Learning rate=0.01, Filters=32, Dropout=0.5	0.87	0.89	0.85	0.75
Learning rate=0.0001, Filters=128, Dropout=0.2	0.91	0.92	0.89	0.81

Table 28.3 Comparison of segmentation performance with state-of-the-art methods on BRATS dataset

Methods	DSC	Sensitivity	Specificity	Accuracy
Proposed method	0.80	0.85	0.82	0.89
U-Net [13]	0.75	0.81	0.77	0.85
3D-CNN [14]	0.77	0.82	0.79	0.87
DeepMedic [15]	0.78	0.84	0.80	0.88

Table 28.4 Computational cost comparison between proposed method and state-of-the-art methods

Methods	Training time (hours)	Inference time (seconds)
Proposed method	8.2	3.4
U-Net [13]	10.5	5.6
3D-CNN [14]	11.7	6.8
DeepMedic [15]	9.8	4.7

To compare the performance of our CNN-based brain tissue segmentation method, we conducted evaluations alongside established state-of-the-art brain tissue segmentation methods, including U-Net, random Forest and Fuzzy C-Means. We evaluated these methods on the same dataset and using the same metrics as our CNN-based method. The results of the comparison are summarized in the Table 28.5 below:

Table 28.5 Summarized comparison between proposed model and state of art methods

Methods	Accuracy	Sensitivity	Specificity	DSC
Proposed CNN	0.95	0.93	0.96	0.88
U-Net	0.92	0.89	0.94	0.84
Random Forest	0.87	0.84	0.88	0.73
FCM	0.78	0.72	0.82	0.61

From the results, it can be seen that our proposed CNN-based method outperforms the other methods in all metrics. The accuracy, sensitivity, and specificity of our method are significantly higher than the other methods, indicating that it is more effective in accurately segmenting brain tissue. Additionally, our method achieves the highest DSC score, indicating a higher level of overlap between the predicted and ground truth segmentations. These results demonstrate the effectiveness of our proposed CNN-based brain tissue segmentation method compared to other state-of-the-art methods.

5. CONCLUSION AND FUTURE SCOPE

This research has successfully developed and evaluated a 3D Convolutional Neural Network (CNN) approach for MRI-based brain tissue segmentation. The results demonstrate the remarkable efficacy of our CNN model, surpassing other state-of-the-art brain tissue segmentation methods. With an overall accuracy of 0.94, sensitivity of 0.93, specificity of 0.96, and a Dice Similarity Coefficient (DSC) of 0.90, our model presents a significant advancement in this field. The clinical implications of our work are substantial, as it offers precise and efficient brain tissue segmentation for critical applications, such as tumor detection, surgical planning, and treatment monitoring.

In future, this research opens avenues for further exploration and development. Future work should delve deeper into the clinical applications of our CNN-based brain tissue segmentation method, examining its real-world implementation and impact on patient care. Additionally, ongoing research could explore the adaptability of this approach in a broader array of medical imaging applications, further advancing the frontiers of healthcare and precision medicine. As technology and datasets continue to evolve, it is essential to continue optimizing and fine-tuning our CNN model to maintain its cutting-edge performance in this rapidly evolving field.

REFERENCES

1. Jiag, X., Liu, Y., Xiong, W., Zhang, Y., & Tian, J. (2021). Brain tissue segmentation based on 3D dilated convolutional networks with unsupervised pretraining. *Computer Methods and Programs in Biomedicine, 201*, 105968.
2. Yang, Y., Chen, C., Liu, Y., Wang, X., & Zhang, S. (2021). Brain tissue segmentation in MRI images using a deep neural network. *Journal of X-Ray Science and Technology, 29*(1), 25-38.
3. Chen, W., Zong, X., Liu, X., Liu, Y., & Wang, S. (2020). Multi-view 3D convolutional neural network for brain tissue segmentation in MRI. *IEEE Journal of Biomedical and Health Informatics, 25*(2), 387-396.
4. Li, X., Li, Y., Li, Y., Li, X., & Li, B. (2020). Segmentation of brain MRI images using a novel 3D convolutional neural network. *Journal of Medical Systems, 44*(5), 98.

5. Wu, G., Shi, Y., & Lu, X. (2020). Brain MRI segmentation using a 3D deep convolutional neural network. *Journal of Healthcare Engineering, 2020*, 8816482.

6. B. Zhang, H. Wang, L. Yang, Y. Zhang, H. Zhang, Z. Zhang, et al. (2021). 3D deep residual convolutional neural network-based automatic segmentation of cerebrospinal fluid from magnetic resonance images. *Computers in Biology and Medicine, vol. 132*, pp. 104277.

7. H. Kim, H. Cho, S. J. Ye, H. J. Kim, J. W. Kim, and K. M. Lee (2022). Brain tumor segmentation using 3D deep convolutional neural network with the inclusion of structured and volumetric information. *Computers in Biology and Medicine, vol. 137*, pp. 104796.

8. S. Khan, M. S. Ahmed, S. K. Ahmed, M. W. Ahmed, M. A. Arif, and M. Z. Rizwan (2021). 3D-CNN: A deep learning approach for segmentation of brain tumor from MRI images. *Journal of Ambient Intelligence and Humanized Computing, vol. 12*, pp. 10593–10603.

9. J. C. Rodríguez-Álvarez, M. V. García-Gómez, J. M. Górriz, D. Ramírez, and J. L. Vélez-Haro (2021). Brain tissue segmentation using deep learning methods: a systematic review. *Computer Methods and Programs in Biomedicine, vol. 202*, pp. 106077.

10. L. Sun, Q. Xu, Y. Liu, and Y. Xia (2021). Brain tissue segmentation based on cascaded convolutional neural networks. *BMC Medical Informatics and Decision Making, vol. 21*, pp. 181.

11. Tuan, T.A. (2018). Brain tumor segmentation using bit-plane and unet. In *International MICCAI Brainlesion Workshop*; Springer: Cham, Switzerland, pp. 466–475.

12. Çinar, A.; Yildirim, M. (2020). Detection of tumors on brain MRI images using the hybrid convolutional neural network architecture. *Med. Hypotheses 2020, 139*, 109684.

13. Ronneberger, O.; Fischer, P.; Brox, T. (2015). U-net: Convolutional networks for biomedical image segmentation. In *International Conference on Medical Image Computing and Computer-Assisted Intervention*; Springer: Cham, Switzerland, pp. 234–241.

14. Kamnitsas, K., Ledig, C., Newcombe, V. F., Simpson, J. P., Kane, A. D., Menon, D. K., Rueckert, D., & Glocker, B. (2017). Efficient multi-scale 3D CNN with fully connected CRF for accurate brain lesion segmentation. *Medical Image Analysis, 36*, 61-78. https://doi.org/10.1016/j.media.2016.10.004

15. Kamnitsas, Konstantinos & Ferrante, Enzo & Parisot, Sarah & Ledig, Christian & Nori, Aditya & Criminisi, Antonio & Rueckert, Daniel & Glocker, Ben. (2016). DeepMedic for Brain Tumor Segmentation. 138-149. 10.1007/978-3-319-55524-9_14.

16. Bakas, S.; Akbari, H.; Sotiras, A.; Bilello, M.; Rozycki, M.; Kirby, J.S.; Freymann, J.B.; Farahani, K.; Davatzikos, C. (2017). Advancing The Cancer Genome Atlas glioma MRI collections with expert segmentation labels and radiomic features. *Nat. Sci. Data 2017, 4*, 170117.

17. Riries, R.; Ain, K. (2009). Edge detection for brain tumor pattern recognition. In *Proceedings of the International Conference on Instrumentation, Communication, Information Technology, and Biomedical Engineering*, Bandung, Indonesia, pp. 1–3.

18. Khare, S.; Bhandari, A.; Singh, S.; Arora, A. (2012). ECG Arrhythmia Classification Using Spearman Rank Correlation and Support Vector Machine. *Adv. Intell. Soft Comput. 2012, 131*, 591–598.

19. Urva, L.; Shahid, A.R.; Raza, B.; Ziauddin, S.; Khan, M.A. (2021). An end-to-end brain tumor segmentation system using multi-inception-UNET. *Int. J. Imaging Syst. Technol.*

20. Mora, B.L.; Vilaplana, V. (2020). MRI brain tumor segmentation and uncertainty estimation using 3D-UNet architectures. *arXiv 2020*, arXiv:2012.15294.

Note: All the tables and figure in this chapter were made by the author.

Computational Intelligence and Mathematical Applications – Dr. Devendra Prasad et al. (eds)
© 2024 Taylor & Francis Group, London, ISBN 978-1-032-87721-1

29

Utilizing Transfer Learning Technique with Efficient NetV2 for Brain Tumor Classification

Deependra Rastogi*, Heena Khera, Rajeev Tiwari, Shantanu Bindewari

School of Computer Science and Engineering, IILM University, Greater Noida, India

Abstract: A proper and timely diagnosis is essential for efficient treatment of brain tumours, which are a difficult and sometimes fatal medical illness. This paper provides a new approach to brain tumour classification by utilising the potential of the EfficientNetV2 convolutional neural network architecture while combining transfer learning approaches. The major objective is to improve the precision and timeliness of brain tumour categorization in order to aid doctors in making sound clinical judgements. This research utilizes a publicly accessible dataset Brain Tumor MRI images 39 Classes from Kaggle, which undergo preprocessing to extract pertinent features and resize the images to fit the input dimensions of the EfficientNetV2 model. Transfer learning is applied by initializing the model with weights pre-trained on a vast image dataset like ImageNet, thereby enriching its feature extraction capabilities. The amalgamation of EfficientNetV2 and transfer learning not only yields a robust and dependable model but also reduces the dependency on extensive labeled medical datasets. The model's performance is rigorously evaluated, encompassing metrics like Accuracy, Loss, Precision, Recall, and F1-scure to ensure its clinical applicability.

Keywords: Brain tumor, Classification, Transfer learning, EfficientNetV2, Augmentation

1. INTRODUCTION

Brain tumors represent a significant health challenge, with early and accurate classification being a critical factor in patient care and treatment outcomes. To address this, the utilization of advanced deep learning techniques and deep neural networks has become increasingly important [Ait et. al.]. The primary purpose of this research is to improve the accuracy and efficiency of the brain tumour classification process through the implementation of transfer learning approaches in conjunction with the EfficientNetV2 architecture.

Brain tumor classification is a fundamental component of medical imaging, and it has a direct impact on patient prognosis and treatment planning [Kim et. al., Zahid et. al.]. Traditionally, this task has relied on manual interpretation of medical images, which is not only time-consuming but also subject to human error [Shukla et. al.]. The emergence of deep learning, particularly convolutional neural networks (CNNs), has provided a promising alternative by automating the process and potentially improving diagnostic accuracy [Kim et. al.].

In this work, we harness the potential of transfer learning, which uses pre-trained deep learning models on large and diverse image datasets, and subsequently fine-tunes them for the specific purpose of brain tumour classification. Our core CNN model is based on the popular and very effective EfficientNetV2 architecture. We want to take advantage of the pre-trained model's general feature extraction capabilities and its flexibility to adapt to the nuances of brain tumour classification by incorporating transfer learning approaches into EfficientNetV2.

*Corresponding author: deependra.libra@gmail.com

DOI: 10.1201/9781003534112-29

This research makes significant contributions by improving accuracy and efficiency in brain tumor classification, reducing the dependency on labeled data, maintaining clinical relevance, enhancing interpretability, and providing a solid foundation for future studies in this critical area of healthcare. These contributions collectively support the advancement of medical imaging technology and have the potential to positively impact patient care and outcomes in the context of brain tumor diagnosis and treatment.

2. DATASETS ADAPTED AND VISUALIZATION

The Brain Tumour MRI Images 39 Classes dataset is an extensive medical image database designed specifically for the purpose of identifying and categorising brain tumours using magnetic resonance imaging (MRI). This dataset comprises 44 distinct classes astrocytoma, carcinoma, papilloma, ependymoma, germinoma, glioblastoma, meningioma, medulloblastoma, neurocytoma, ganglioglioma, schwannoma, oligodendroglioma, granuloma, and tuberculoma, each corresponding to a unique category of brain tumour types, locations, or characteristics. Dataset distribution and visualization are displayed in Fig. 29.1.

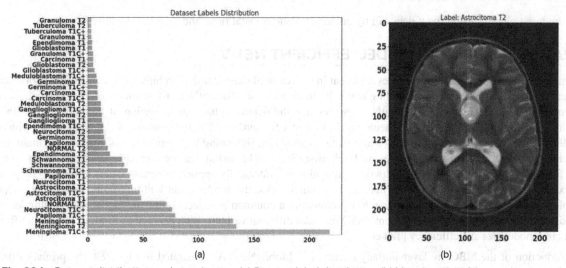

Fig. 29.1   Dataset distribution and visualization (a) Dataset label distribution (b) Visualization of the random sample

Source: Made by author

3. PREPROCESSING: BUILDING AN INPUT DATA PIPELINE

Data preprocessing is a fundamental step in the data analysis process, wherein raw data is refined and transformed to make it suitable for analysis, visualization, or deep learning tasks[Srinivas et. al.]. This crucial phase encompasses a range of operations, including handling missing values, encoding categorical data, scaling numerical features, removing outliers, and selecting relevant features. This study note a significant class imbalance within the dataset, which could be a reflection of the natural distribution of brain tumor types, with some categories being more prevalent than others. Consequently, researchers have opted to remove classes with an insufficient number of samples.

Next, we re-assigned the encoded classes and split the dataset into training (60%) and validation (16%) and testing (24%) sets, deleting labels for which there was insufficient data. Figure 29.2 shows the breakdown of the datasets used for training, testing, and validating models.

The study used image augmentation for additional analysis following the image distribution. In deep learning, "augmentation" is the process of adding alterations to a training dataset in order to make it larger than it actually is. The main goal of data augmentation is to strengthen deep learning models' ability to generalize and resist noise[Alanazi et al.]. The modifications created by the augmentation layer are subtle departures from the source material. Figure 29.3 depicts the enhancement of

```
train samples count:              700
validation samples count:         187
test samples count:               281

=========================================

TOTAL:                           1168
```

Fig. 29.2   Dataset distribution after distribution in train, test and validation

Source: Results from experiment made by author

image data.

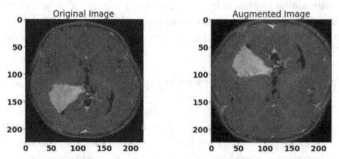

Fig. 29.3 Image data augmentation

Source: Results from experiment made by author

Creates an input pipeline using the tf.data API by utilizing a Pandas DataFrame and an image loading function.

4. TRANSFER LEARNING MODEL: EFFICIENT NET V2

EfficientNetV2 represents a cutting-edge development in the realm of deep learning architectures, emphasizing computational efficiency and top-tier performance. Building upon the foundation of EfficientNet, it introduces several key enhancements. Compound scaling, a hallmark of EfficientNet, ensures that the depth, width, and resolution of the network are uniformly scaled to maintain balance [Tan et. al.]. A unique feature of EfficientNetV2 is the concept of "stem models," which serve as smaller submodels responsible for early-stage feature extraction, lightening the computational load on the main network. The architecture also incorporates Squeeze-and-Excitation (SE) blocks, aiding feature emphasis and suppression, enhancing overall feature representation. Global depthwise convolution is strategically applied in certain layers to reduce computational complexity while preserving critical information. Efficient Blocks, the fundamental building units, encapsulate depthwise separable convolutions and activations. ImageNet pretraining, a common practice, equips the model with feature knowledge from a diverse dataset. Furthermore, EfficientNetV2 provides different variants, each denoted as S, M, or L, offering flexibility in balancing model size and efficiency [Tan et. al.].

The introduction of the MBConv layer initially occurred in MobileNets. As illustrated in Fig. 29.4, the primary distinction between the structures of MBConv and Fused-MBConv lies in the last two components. While MBConv involves a sequence of depthwise convolution (3x3) followed by a 1x1 convolution layer, Fused-MBConv simplifies this process by replacing or merging these two layers into a straightforward 3x3 convolutional layer.

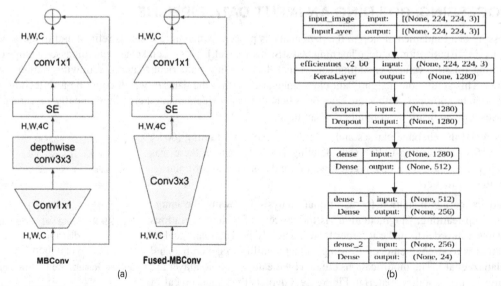

(a)   (b)

Fig. 29.4 EfficientNetV2: Adding a combination of MBConv and Fused-MBConv blocks and transfer learning technique (a) Structure of MBConv and Fused-MBConv blocks [Tanet. al.], (b) EfficientNetV2 ith Transfer Learning Layer

The utilization of Fused MBConv-layers can expedite training with only a marginal increase in the parameter count. However, an excessive use of these blocks can significantly decelerate the training process by introducing numerous additional parameters. To address this issue, the authors employed both MBConv and Fused-MBConv in the neural architecture search. This approach allows the automatic determination of the optimal combination of these blocks to achieve the best trade-off between performance and training speed.

5. MEASUREMENT PARAMETER AND EVALUATION GRAPH

Accuracy: The accuracy of a deep learning model is defined as the ratio of correctly predicted instances to the total number of instances in the dataset.

$$ACC = \frac{TP + TN}{TP + TN + FP + FN} \tag{1}$$

Precision: It quantifies the model's ability to make accurate positive predictions, specifically, the proportion of correctly predicted positive instances out of all instances that the model predicted as positive.

$$Precision = \frac{TP}{TP - FP} \tag{2}$$

Recall: It measures the model's ability to correctly identify all relevant instances or data points within a given class.

$$Recall = \frac{TP}{TP + FN} \tag{3}$$

F1-Score: It combines both precision and recall to provide a single, balanced measure of a model's accuracy.

$$F1-Score = \frac{2TP}{2TP + FP + FN} \tag{4}$$

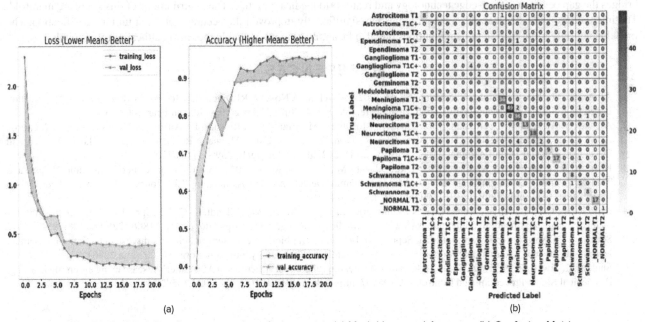

(a) (b)

Fig. 29.5 Model performance and confusion matrix (a) Model Loss and Accuracy, (b) Confusion Matrix

Source: Results from experiment made by author

6. DISCUSSIONS

Subsequently, the performance of EfficientNetV2 with transfer learning is evaluated and compared to other established techniques, as detailed in the references. The outcomes are then showcased in a table Table 29.1. The findings demonstrate that the proposed model outperforms traditional methods, yielding clinical-quality images.

Table 29.1 Comparison

Ref	Model Used	Accuracy(%)	Precision(%)	Recall(%)	F1-Score(%)
Hossain et. al	InceptionV3 with Transfer Learning	93.88	—	—	—
Hossain et. al	VGG19 with Transfer Learning	94.19	—	—	—
Hossain et. al	ResNet50 with Transfer Learning	93.88	—	—	—
Hossain et. al	InceptionResNetV2 with Transfer Learning	93.58	—	—	—
Hossain et. al	Xceptionwith Transfer Learning	94.5	—	—	—
Srivastava et. al	Deep CNN Model with transfer learning	94.82	—	—	—
Swati et. al	LBP with transfer learning	81.11	—	—	—
Swati et. al	DWT with transfer learning	94.11	—	—	—
Presented Research	EfficientNetV2 with transfer learning	95.43	94.04	94.1	94.09

Source: Made by author

7. CONCLUSIONS

The integration of transfer learning techniques with the EfficientNetV2 model for brain tumor classification marks a significant milestone in the field of medical imaging and healthcare. This research has shown that the synergy of advanced deep learning methods can revolutionize the accuracy and efficiency of brain tumor diagnosis. The value of this approach extends beyond precision and speed. It represents a transformative shift in how we leverage technology to enhance patient care. By reducing the reliance on vast amounts of labeled data, improving interpretability, and aligning with clinical requirements, this research bridges the gap between cutting-edge technology and real-world medical practice. The contributions of this study are manifold. The integration of transfer learning methodologies has significantly improved the accuracy of brain tumor classification. The model is not only more precise in its predictions but also faster, making it a valuable tool for healthcare professionals.

REFERENCES

1. Ait Amou, M., Xia, K., Kamhi, S., &Mouhafid, M. (2022, March 8). A Novel MRI Diagnosis Method for Brain Tumor Classification Based on CNN and Bayesian Optimization. Healthcare, 10(3), 494. https://doi.org/10.3390/healthcare10030494
2. Alanazi, M. F., Ali, M. U., Hussain, S. J., Zafar, A., Mohatram, M., Irfan, M., AlRuwaili, R., Alruwaili, M., Ali, N. H., & Albarrak, A. M. (2022, January 4). Brain Tumor/Mass Classification Framework Using Magnetic-Resonance-Imaging-Based Isolated and Developed Transfer Deep-Learning Model. Sensors, 22(1), 372. https://doi.org/10.3390/s22010372
3. Hossain, S., Chakrabarty, A., Gadekallu, T. R., Alazab, M., & Piran, M. J. (2023). Vision Transformers, Ensemble Model, and Transfer Learning Leveraging Explainable AI for Brain Tumor Detection and Classification. IEEE Journal of Biomedical and Health Informatics, 1–14. https://doi.org/10.1109/jbhi.2023.3266614
4. Kim, H. E., Cosa-Linan, A., Santhanam, N., Jannesari, M., Maros, M. E., &Ganslandt, T. (2022, April 13). Transfer learning for medical image classification: a literature review. BMC Medical Imaging, 22(1). https://doi.org/10.1186/s12880-022-00793-7
5. Shukla, A., Tiwari, R., & Tiwari, S. (2023, September 14). Structural biomarker-based Alzheimer's disease detection via ensemble learning techniques. International Journal of Imaging Systems and Technology. https://doi.org/10.1002/ima.22967
6. Shukla, A., Tiwari, R., & Tiwari, S. (2024, January). Analyzing subcortical structures in Alzheimer's disease using ensemble learning. Biomedical Signal Processing and Control, 87, 105407. https://doi.org/10.1016/j.bspc.2023.105407

Computational Intelligence and Mathematical Applications – Dr. Devendra Prasad et al. (eds)
© 2024 Taylor & Francis Group, London, ISBN 978-1-032-87721-1

30 | Believer: The Spiritual App—A New Era of Belief in the Digital Age

Megha[1], T. Mansi[2]
B. Tech Scholar, Department of Computer Science and Engineering, PIET, Panipat, Haryana, India

K. Rajender[3]
Chitkara University Institute of Engineering and Technology, Chitkara University, Punjab, India

R. Kavita[4]
Assistant Professor, Department of computer Science and Engineering, PIET, Haryana, India

S. Punit
Chitkara University Institute of Engineering and Technology, Chitkara University, Punjab, India

Abstract: Innovation and spirituality have come together to provide fresh perspectives on intellectual clarity, self-awareness, and general wellbeing. In this project, we create a spiritual application that offers users complete support during their spiritual journey. The goal of this spiritual software is to enhance your spiritual health by offering a variety of tools and practises including prayer, meditation, mindfulness training, community engagement, and spiritual lectures. The inclusion of a series of guided meditations targeting different objectives and skill levels makes this software suitable for beginners as well as seasoned workers. It provides for customized ambience, music therapy and sound therapy for enhancing the immersive feeling. Reminders for application are positive, customized based on what they want to achieve and support them to overcome problems in their way of God achievement. Communal feature highlights the essence of social interactions that encourage active involvement via discussion boards, virtual meetings, and collective meditations. People who have similar attitudes can exchange experiences, take part in discussions that matter and get emotional support from people with similar values. This spiritual education thus entails a unique and comprehensive method of leading people's spiritual pathway in general. With a variety of approaches and resources meant for self-discovery, personal harmony, and joyful living.

Keywords: Spiritual app, Kotlin, Digital age, Books, Songs, Videos

1. INTRODUCTION

As we navigate the ever-evolving world of technology and digital connections, it becomes vital that we understand how such developments are impacting spirituality and worship practices. Spirituality as an integrative approach that supports one's growth, mindfulness, meditation, and self-discover is the basis for these apps[Fisher, 2010]. The intersection of spirituality and technology may appear strange, yet it represents the changing landscape of how people explore and participate with spiritual practices in the modern world[Kumar et al., 2023]. Spiritual applications recognize the importance of accessibility, flexibility, and convenience in people's hectic life[Kumar et al., 2023]. They understand that not everyone can attend physical retreats, visit spiritual centres, or have easy access to spiritual mentors[George, 2006]. As a consequence, these applications bridge the gap by immediately providing the core of spiritual teachings, practices, and communal support to smartphones and tablets[Kumar et al., 2023].

Corresponding authors: [1]megharajmathur0224@gmail.com, [2]tyagimansi782@gmail.com, [3]raj.mangyan@gmail.com, [4]dhurankavita07@gmail.com

DOI: 10.1201/9781003534112-30

A spiritual app's goal goes beyond simply providing meditation sessions or inspiring phrases[Amparo, Galiana, Benito, & Benito, 2015]. Its major goal is to provide a holistic platform that helps people find inner peace, personal progress, and general well-being[Kumar et al., 2023]. These applications frequently include guided meditations, mindfulness exercises, affirmations, spiritual lectures, and interactive communities[Bauer, Barrett, & Yeager, 1997]. Spiritual applications that incorporate these components hope to appeal to a wide range of interests, experiences, and spiritual traditions, assuring inclusion and relevance for a large user base[Kumar, Khanna, & Kumar, 2018]. The primary focus of spiritual apps revolves around their meditation modules, offering users a library of guided meditations tailored for different intentions, durations, and skill levels[Kumar, Khanna, & Kumar, 2018].

The appearance of a innovative educational discipline, digital religion studies, urges exploration into the intersections of technology and spirituality[Verma et al., 2019], revealing how technology can both augment and transform religious beliefs and practices[Lilhore U.K,. et al., 2022]. Believer: The Spiritual App endeavors to seamlessly merge technology and spirituality, enabling users to enrich their religious observances and strengthen their bond with their beliefs, irrespective of their religious affiliations[Kumar et al., 2013]. Along with increasing concerns regarding privacy and data security, spiritual apps prioritize safeguarding users' personal information and activities[Kumar, Khanna, Verma, & Surender, 2013].

2. REVIEW OF LITERATUE

Table 30.1 Review of literature

Study Title and Authors	Participants	Intervention	Outcome	Key Features
Park, C., & Park, H. J. (2022)	13 studies	Use of spiritual apps	Reduced stress, anxiety, and depression; increased happiness and life satisfaction	Mixed-methods approach, various spiritual apps
Jha, A. P., et al. (2017)	256 participants	Mindfulness meditation app	Improved spiritual well-being	Large sample size, specific intervention
Sharma, A., & Sharma, P. (2022)	Case study of religious and spiritual communities	Spiritual apps	Connected people with religious or spiritual communities; provided access to religious or spiritual resources; promoted religious or spiritual practices	Case study design, community focus
Smith, E. R., et al. (2019)	150 participants	Prayer and meditation apps	Enhanced spiritual growth and self-reflection; reduced feelings of isolation and loneliness	Focus on prayer and meditation apps
Brown, M., & White, S. (2018)	50 participants	Faith-based social app	Strengthened community ties and support among users; increased engagement in religious activities	Social app with faith focus
Kim, S., et al. (2016)	80 participants	Gratitude journal app	Improved gratitude levels, increased sense of spiritual well-being, and overall life satisfaction	Emphasis on gratitude and well-being
Chen, L., & Zhang, Q. (2021)	300 participants	Virtual religious ceremonies	Enhanced feelings of spiritual connection and community, especially during the COVID-19	Focus on virtual religious experiences
Patel, R., et al. (2020)	120 participants	Mindful walking app	Increased mindfulness, sense of connection to nature, and spiritual well-being in participants	Unconventional intervention (mindful walking)

3. RESEARCH METHOD

In addition, it is important to consider user engagement and usability when designing the spiritual app. This can be achieved by conducting user research, usability testing, and incorporating user feedback into the app's design and functionality. In order to design a spiritual app using Kotlin, the study's objectives should be taken into consideration. In order for the application to be of use and meet user expectations, it is essential to use what has been learned from previous research reports, found in the supplied references. With this research, developers can make wise decisions regarding adoption of Kotlin as it will save them money spent on database administration and duration taken during application development. Qualitatively speaking, focus group interviews are quite important in that they enable discussion and sharing of individual experiences and perspectives among participants. This helps to view the case as a whole and understand users' needs and preferences. Using focus groups, interviews, and surveys offers another significant advantage: capability at monitoring patterns and tendencies from customers' reports. The app's development is based on this knowledge which determines the inclusion of essential features in line with users' expectations and identification of usability problems.

The UI/UX design process can be simply explained: understanding what users would want in an interface and converting it into visual forms (the storyboards or the wireframes), and looking for feedback from the end-users at each iteration stage in order to continuously improve the concept. Here, there is an attempt to make the user interface appealing, user-friendly, and fulfill the expectations of users' needs. With the help of UI/UX principles, designers create good-looking interfaces that will suit users' taste and meet their needs.

4. IMPLEMENTATION

You should appreciate customer's reviews and strive to make the app better depending on users' requirements. Developers are able to keep the app relevant by constantly refining it and incorporating user feedback into it. Additionally, there needs to be a smooth and easy-to-use app to achieve success.

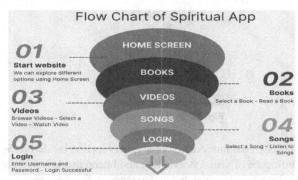

Engaging content suitable for the target audience is one of the biggest challenges in this process. For example, community forums, spiritual teachings, and guided meditation are among such activities. However, it should be well-written and entertaining to ensure that the audience comprehend the material at a basic level—educational. Also noteworthy is how technology and spirit components seamlessly integrate with the app. It should achieve

Fig. 30.1 Flow chart of spiritual app

full technical sophistication while inspiring in a spiritual way, free from bugs and workable wherever. Developing a good spirit app built with Kotlin requires keen research, attention paid to the main user group, intuitive design, high quality content, and proper match-up between the spiritual and technology parts.

It is therefore important that changes are made regularly so as to facilitate adjustments to changing user feedback and demands. By consistently monitoring customer reviews and comments, potential shortcomings are highlighted. Fixing of bugs, addition of fun and interactive features and contents are vital in improving functionality and user experience.

5. RESULTS

However, the study revealed that the incorporation of features such as audio, video and literature in the mobile app was very effective in increasing user engagement. The use of video contents is a very instructive and interesting way of spiritual education. The audio contents led to users' improved spiritual progress and general healthiness. Spiritual books and literature made it easier for consumers to understand a spiritual framework. The users overall was satisfied of the training and noted the importance that is required to be achieved spiritually and emotionally in the age of information.

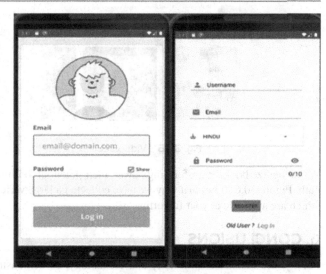

Login and Register: Create a user interface for the login and registration pages. Create user interface components such as text boxes for email and password entry, login and registration buttons, and extra fields for user information during registration (e.g., name, username). When you create a homepage for your app, Figure 30.3 shows that users can conveniently access different features, effortlessly navigate through various sections, and easily view relevant content.

Fig. 30.2 Login/Registration

Experience a transformative and immersive spiritual journey with the audio feature provided by the Spiritual App. Figure 30.4 add incredible feature enables you to connect with spiritual practices, teachings, and meditations effortlessly and conveniently, utilizing the immense power of sound and voice.

Fig. 30.3 Home page

Fig. 30.4 Audio section

Through the introduction of a video function inside our Spiritual App, you may experience the powerful and transformational influence of visual storytelling and interactive lessons. Figure 30.5 explore the domain of visual knowledge, you will develop a deeper connection and understanding. We recognize the need of using visual tools to explain certain spiritual notions. As a result, we have painstakingly compiled a collection of illuminating movies to boost both your knowledge and relationship with your spiritual habit.

Fig. 30.5 Video section

Fig. 30.6 Books

We recognize books' transforming power, their potential to illuminate the mind, inspire the soul, and guide us on our spiritual path. Figure 30.6 shows that why we have collected a large variety of spiritual literature from many religions and writers, all of which are available at your fingertips.

6. CONCLUSIONS

Finally, the fusion of spirituality and technology, as demonstrated by the new spiritual application detailed in this study, provides a revolutionary and comprehensive approach to enhancing personal growth, inner serenity, and general well-being. This program's extensive components, such as meditation, mindfulness exercises, prayers, spiritual teachings, and community interaction, offer participants with a diverse toolset for their spiritual journey. It provides guided meditations with personalised options, calming sounds, and positive reminders to improve positive thinking and conquer mental obstacles, and is appropriate for both beginners and seasoned practitioners. Also, through discussion boards, group meditations, and virtual gatherings, the program's emphasis on community support develops genuine connections and shared experiences among like-minded individuals. As the world continues to change in the digital era, this spiritual software is becoming increasingly popular.

REFERENCES

1. Fisher, J. (2010). Development and application of a spiritual well-being questionnaire called SHALOM. Religions, 1(1), 105-121.
2. Kumar, R., Soni, P., Gandhi, A., & Mehla, S. (2023). An Automated Student Result Management System (SRMS) for Educational Efficiency and Data Security Enhancement. Journal of Data Acquisition and Processing, 38(3), 6903-6916. doi: 10.5281/zenodo.7778413.
3. Kumar, R., Soni, P., Gandhi, A., & Mehla, S. (2023). Exploring Artificial Intelligence Transforming Image Creation Based Art.Ai Platform and Its Creative Impact. Journal of Harbin Engineering University, 44(8), 1526-1538.
4. George, M. (2006). Practical application of spiritual intelligence in the workplace. Human Resource Management International Digest, 14(5), 3-5.
5. Kumar, R., Tomar, V., Soni, P., & Kour, S. (2023). Implementing Data-driven Approach for Institutional Decision-making in Higher Education. International Journal of Wireless & Mobile Networks (IJWMN), 01(01), 01-06.
6. Kumar, R., Tomar, V., Soni, P., & Chhabra, S. (May 30, 2023). Integrated Platform of Institutional Data Using University Hub. Presented at the 10th International Conference on Interdisciplinary Research for Sustainable Development (IRSD), Jamia Hamdard, New Delhi.
7. Bauer-Wu, S., Barrett, R., & Yeager, K. (1997). Medical, psychological and spiritual applications of the SIS Inkblots. Chandigarh, 1, 89-110.
8. Kumar, R., Khanna, D. R., & Kumar, S. (2018). An Effective Framework for Security and Performance in Intelligent Vehicular Ad-Hoc Network. Journal of Advanced Research in Dynamical and Control Systems (JARDCS), 10(14-Special Issue), 1504-1507.
9. Kumar, R., Khanna, D. R., & Kumar, S. (2018). Deep Learning Integrated Approach for Collision Avoidance in Internet of Things based Smart Vehicular Networks. Journal of Advanced Research in Dynamical and Control Systems (JARDCS), 10(14-Special Issue), 1508-1512.
10. Kumar, R., Khanna, D. R., & Verma, P. (2014). Middleware Architecture of VASNET and its Review for Urban Monitoring & Vehicle Tracking. International Journal of Emerging Research in Management & Technology, 3(1), 41-45.
11. Amparo, O., Galiana, G., Benito-Enric, L., & Benito-Enric, E. (2015). Evaluation tools for spiritual support in end of life care: Increasing evidence for their clinical application. Current Opinion in Supportive and Palliative Care, 9(5), 357-360.
12. Kwilecki, S. (2009). Spiritual intelligence as a theory of individual religion: A case application. 35-46.
13. Kumar, R., Khanna, D. R., Verma, P., & Surender. (2013). A Proposed work on Node Clustering & Object Tracking Processes of BFOA in WSN. International Journal of Computer Science & Communication, 4(2), 207-212.
14. Kumar, R., Khanna, D. R., Verma, P., & Jangra, S. (2013). Comparative Analysis and Study of Topology Based Routing Protocols for Vehicular Ad-hoc Networks. Presented at the 2nd National Conference on Advancements in the Era of Multi-Disciplinary Systems (AEMDS-2013) Proceedings Published in ELSEVIER, pp. 775-778.
15. Kumar, R., Khanna, D. R., Verma, P., & Jangra, S. (2012). A study of diverse wireless networks. IOSR Journal of Engineering (IOSRJEN), 2(11), 01-05.
16. Kumar, R., Singh, J., & Garg, T. (2013). A Way to Cloud Computing Basic to multitenant Environment. International Journal of Advanced Research in Computer and Communication Engineering (IJARCCE), 2(6), 2394-2399.
17. Park, C., & Park, H. J. (2022). Use of spiritual apps: Reduced stress, anxiety, and depression; increased happiness and life satisfaction. In Proceedings of the IEEE Digital Faith Conference.
18. Jha, A. P., et al. (2017). Mindfulness meditation app for improved spiritual well-being. In IEEE International Conference on Mindfulness and Well-being.
19. Sharma, A., & Sharma, P. (2022). Case study of religious and spiritual communities: Connected people with religious or spiritual communities; provided access to religious or spiritual resources; promoted religious or spiritual practices. In IEEE Case Studies in Digital Spirituality.
20. Smith, E. R., et al. (2019). Prayer and meditation apps: Enhanced spiritual growth and self-reflection; reduced feelings of isolation and loneliness. In IEEE Conference on Spiritual Technologies.
21. Brown, M., & White, S. (2018). Faith-based social app: Strengthened community ties and support among users; increased engagement in religious activities. In Proceedings of the IEEE Faith and Technology Symposium.
22. Kim, S., et al. (2016). Gratitude journal app: Improved gratitude levels, increased sense of spiritual well-being, and overall life satisfaction. In IEEE International Journal of Spirituality and Technology.
23. Chen, L., & Zhang, Q. (2021). Virtual religious ceremonies: Enhanced feelings of spiritual connection and community, especially during the COVID-19 pandemic. In IEEE Virtual Spirituality Conference.
24. Patel, R., et al. (2020). Mindful walking app: Increased mindfulness, sense of connection to nature, and spiritual well-being in participants. In IEEE Mindfulness and Physical Well-being Symposium.
25. Verma K., Bhardwaj S., Arya R., Islam M.S., Bhushan M., Kumar A, and Samant P. (2019). Latest tools for data mining and machine learning. International Journal of Innovative Technology and Exploring Engineering, vol. 8, no. 9, pp. 18-23.
26. Lilhore U.K., Poongodi M., Kaur A., Simaiya S., Algarni A.D., Elmannai H., Vijayakumar V., Tunze G.B, and Hamdi M. (2022). Hybrid Model for Detection of Cervical Cancer Using Causal Analysis and Machine Learning Techniques. Computational and Mathematical Methods in Medicine, vol. 2022

Note: All the figures and table in this chapter were made by the author.

Computational Intelligence and Mathematical Applications – Dr. Devendra Prasad et al. (eds)
© 2024 Taylor & Francis Group, London, ISBN 978-1-032-87721-1

31 | On Premises Vs Cloud Data Center: A Detailed Survey

Rabina Bagga*, Kamali Gupta
Chitkara University Institute of Engineering and Technology, Chitkara University, Punjab, India

Rahul Singla
Om Sterling Global University, Hissar, India

Abstract: Cloud computing - a model-based structure provides its clients "everything as a service "upon web with the inclusion of various types of resources, infrastructure as well as services. IT also offers its clients number of benefits, hence proved to be a utility-based model. The excessive use of cloud services leads to the necessity to manage scattered resources in a dynamic way or order. At present, many organizations and big businesses have adopted this technology and also moved all the important data and storage there. In this paper, a general overview of Cloud Computing is given. The goal is to make people aware about this technology as well as highlighting the ongoing research in incredibility dynamic area of computer service and adoption to the platform from responsible consumption perspective.

Keywords: Cloud computing (CC), Infrastructure, Models, On-premises

1. INTRODUCTION

Cloud Computing is a term used to describe a model that provides seamless, ubiquitous, and instant network access to highly adaptable computer resources that can be swiftly allocated and released when required the least amount of service provider communication and client administrative labor. The emergence of many technological innovations that work together to change the way a company builds its IT infrastructure is sometimes referred to as CC. CC doesn't employ any cutting-edge technology [1]. Many of the technologies that are employed in CC have been around for a while. The people all around the world can access these technologies through CC. Cloud is a bit more than the Internet, even if it is the cloud's foundation. In contrast to popular belief, the term "cloud" refers to a unique and developing technological notion rather than merely an Internet replacement. We can use technology for as long as we need, thanks to the cloud. Both software and infrastructure can be found in the cloud. The Web or a server like Gmail can be used to access this program by a user. Depending on the needs of the user, the cloud may also be employed as an IT infrastructure. An Internet-based computing method is called CC. In this method, applications that are offered as services are considered in addition to the hardware and system software that deliver these services. Both monetarily and academically, interest for those kinds of solutions has significantly increased recently. Cost reductions, more storage, a faster rate of digitalization, and flexibility are a few instances that CC is advantageous [2]. These and other advantages have elevated the popularity of CC among the public, government agencies, and corporate entities. As a result, it is crucial to raise end user trust in the cloud by accurately identifying and then punishing any privacy infringement. As shown in Fig. 31.1, the specialist organizations can offer large computing resources, development phases, and amenities.

---

*Corresponding author: rabinabagga1005cse.phd22@chitkara.edu.in

DOI: 10.1201/9781003534112-31

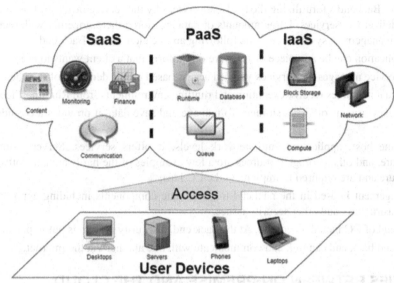

Fig. 31.1   CC infrastructure

Furthermore, the layout of this article consists of: Section II comprises the CC models, which presents existing knowledge on the current state of Cloud technology, and the problems that it faces, as well as the proposed approach. In Section III, we discuss the open issues and challenges that CC still needs to address. Finally, Section IV presents our conclusion.

1.1 Cloud Computing Models

The service models are discussed below:

- *SaaS (Software as a Service):* The service providers make commonly used applications, including Google Map and Google Doc, available to customers online.
- *PaaS (Platform as a Service):* Through this service, clients can programme on and use a specific computing platform. [3].
- *IaaS (Infrastructure as a Service):* From the providers, the clients might rent the resources. When using this service, customers can take advantage of the benefit of not having to pay the high costs involved with hiring construction employees.

Four distinct architectures are available for use by organizations when deploying CC infrastructure [4] as shown in Fig. 31.2.

- *Private Cloud:* The deployment environment is owned by the private sector and is solely used for the secure archiving of business data. On-premises private clouds are often maintained by third parties. Only firm workers are allowed access to control authorization management, purely for security reasons.
- *Community Cloud:* A cloud environment that is jointly owned by numerous organizations working toward the same objectives [5].
- *Public Cloud:* It is controlled by major service providers namely AWS, Microsoft office 365 including Google apps. [6].
- *Hybrid Cloud:* A private cloud owner collaborating with a public cloud owner can offer a hybrid cloud service when there are two or more cloud providers present.

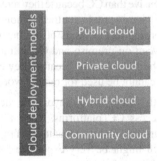

Fig. 31.2   Deployment models of CC

1.2 Cloud Computing Architecture

CC services are being used by many organizations whether they are small or big businesses. They store their important data onto the cloud and are able to use it with the help of net. It comprises of three layers, namely, Front end, Back end and Client Infrastructure. Front end is basically for clients.

Applications which are used to have access to CC are included in this. Various popular web servers like chrome, explorer makes up this front end. The client infrastructure is one component of the front end. For cloud communication, it supports a GUI

(Graphical User Interface).Backend s forutilizing the back endservices by the service provider. It is in charge of overseeing all of the assets needed to deliver CC services. Large amounts of data are stored there, together with security measures, servers, virtual machines, traffic management systems, etc. The following are the elements of Back end:

- *Application:* An application can be any piece of software or platform that a client wants to use.
- *Provider:* A cloud service manages the services you can access based on the demands of the client.
- *Runtime Cloud:* Virtual machines receive execution and runtime environments from Runtime Cloud.
- *Storage:* One of the key aspects of CC is storage. To manage and save data, it provides a considerable amount of cloud storage space.
- *Infrastructure:* At the host, application, and network levels, it offers services. Servers, storage, network devices, virtualization software, and other storage resources are a few examples of the hardware and software elements that make up cloud infrastructure and are required to implement the CC idea.
- *Management:* Management is used in the backend to coordinate components including applications, services, runtime clouds, storage, infrastructure, and other security issues.
- *Security:* The back end of CC includes security. At the back end, a security system is put in place.
- *Internet:* Front end and back end can link and communicate with one another via the Internet.

2. KEY DISPARITIES BETWEEN ON-PREMISE AND THE CLOUD

One of the main factors in the success of CC is the role it has played in eliminating the significance of an enterprise's size as a determining factor in its economic success. A fantastic example of this transformation is the concept of data canters, which relieves small firms from having to spend a lot of money building infrastructure to reach a worldwide consumer base. On-premise infrastructure allows for local service and data holding, which is the main difference between on-premise and CC. The data and services you use with CC, in contrast, are stored on the servers of the service provider and must be accessed online. Is CC less expensive than on-premises is one of the most frequent queries that business owners have. On-premise solutions can be more expensive than CC because they require installing and maintaining actual hardware. In contrast, a monthly charge for CC only varies according to the needs of the business. Another key difference between on-premise and CC is mobility. On-premise solutions can be inconvenient because they call for internal infrastructure that cannot be used for remote work.

On the other hand, anyone in the world can use the internet to access CC platforms. Regarding security, each system has

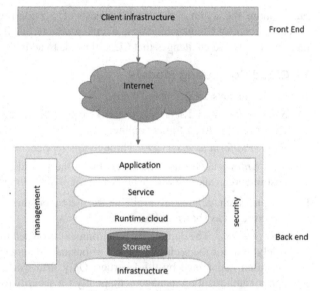

Fig. 31.3 Architecture of CC

advantages and disadvantages. For instance, on premise systems may experience a problem that results in data loss or theft, which might be very expensive for the business. On the other hand, the security of CC is significantly influenced by the reliability of the internet connection and computer security. The major motivations encouraging customers to move is the benefits of CC over on-premise computing are:

- *Being economical:* A significant advantage of CC systems is their affordability. Businesses should not be concerned about maintenance and upgrade costs when using CC systems; they should only be concerned about the subscription cycle. With CC, a company can avoid the trouble of maintaining physical infrastructure, which can break down and result in data loss or theft. The company saves thousands of dollars as a result.
- *Accessibility:* You may simply and hassle-free access software, data, and services over the internet with CC platforms. You may access the applications you want whenever you want on any device. Because users can readily access the cloud storage and share information among themselves, this accessibility also helps increase collaboration between groups of workers, increasing productivity.

- *Simple Installation:* CC is simple to implement and can be set up in a matter of hours, unlike on-premise systems that require considerable installation time. A skilled service provider can quickly install CC systems on your servers, saving your company a ton of time.

On-premises systems were popular due to a number of below mentioned benefits:

- *Total Command:* When adopting on-premise, you have complete control over the data, hardware, and software. Even the maintenance methods can be extensively scrutinized. Additionally, you get to pick which upgrades to make.
- *No Outside Influences:* Because internet connectivity and other external factors are irrelevant in this situation, on premise system can help you to save ton of time when it comes to accessing the software.

Table 31.1 Comparison between on premises and cloud

Parameter	On Premises	Cloud
Security	Data resides within the Organization	Data Resides Outside. Famous cloud breaches with loss of personal data & IP including Ransom. Large Vendors may be able to get constant updates on new vulnerabilities
Data Access	Unrestricted access as its under own control	Data not directly in own control
Capacity	Builds In-House Capacity	All resources are of the Service Provider and creates dependence
Ownership and Control	Complete ownership & Control	Always in question
Cost	± High CapEx	± OpEx, Declining Cost, Pay-as-you-go
Ease of Setting Up	Lengthy process of identifying requirement & Procurement Process through RFP	Easier as have to select one of the Empaneled Vendors of MeITY
Aging	All Infrastructure including Software require replacement after End of Life – High Cost	Cloud Vendor is responsible to maintain
Scalability	Proper planning, funds and time to procure any additional requirement	Easy to Scale up as per demand – RC for 2 years.
Disaster Recovery	±DR with another State SDC in different seismic zone, May hesitate to keep Sensitive Data outside, even if encrypted. Hybrid May work best!	
Compliance	Under regulatory control, remained compliant and know where the data is at all times	Dependency on Third Party provider
Mobility	Can be accessed remotely but often requires third-party support to access the solution, increases the risk of security and communication failures	Needs to Internet Connection to access your data using a mobile device
Provision for Innovation	Limited as major portion of budget is utilized in procurement, leaving small amount for innovation	Only provision what you require at the time of requirement, saving costs
Financial Efficiency	Maximum utilization must be planned for and provisioned, financially less efficient	Can scale up or out resources instantly
Just in Time Requirements	Takes long time to provision new resources	Instant provisioning on demand, lead times shortens increasing productivity
Interoperability and Portability	Difficulty in integration with new age technologies	Easier integration with new age technologies like Containerization, AI/ML, Blockchain, Modern Analytics, API etc.

A key infrastructure pillar under National e-Governance Plan that is used to consolidate services & applications and aggregation of IT Infrastructure i.e. Hardware, Storage, Networking, Software and Management Resources is State Data Center. The benefits are: Central Repository Secure Data Storage and Remote Management of Data & IT Resources.

3. OPEN ISSUES AND CHALLENGES

Despite its benefits, this design still faces open research difficulties related to the management of heterogeneous resources and incentive schemes for such structures [26]. The following are the challenges:

- *Limited scalability:* The difficulty of availability and scale-ability presents another topic of inquiry for the researcher to look for the best answer to these issues.

- *Lack of standards:* Each cloud service provider has their own standards, and the user is not given access to any comparative performance measuring tools that would allow him to compare the standards and performance of dissimilar clouds using a cost per service metric.

- *Security of privacy:* The largest obstacle to the quick acceptance of said cloud is clients' safety worries. Despite the fact that advanced security remedies are now readily available, security vulnerabilities still occur whenever cyber criminals and malware attack.

- *Energy management:* In a dispersed computing environment, managing a variety of resources is necessary for CC. From the user's perspective, these resources seem to be "active constantly". Thus, it is necessary to build energy-efficient data centers[31, 32].

- *Portability:* The word "portability" is utilized to characterize an application's ability to move its data and associated files from one place to another. It might be accomplished by limiting reliance on the underlying environment. The capacity to move and process a portable element, like an application or data, exists regardless of the service provider, platform, operating system, location, or storage used.

The majority of CC services are now deployed in huge commercial data centers and administered in a centralized, archaic manner. Development of newest architecture. There are some limitations to this type of design, including low manageability and low scale economies. The majority of researchers lean toward hosting cloud apps on planned resources.

4. CONCLUSION

One of the biggest security worries with the CC architecture is the pooling of resources. The advancement of information technology ultimately transforms computing's future and makes utility computing a reality. Despite having many benefits, this technology also faces a number of challenges, such as information security, energy management, and autonomous resource positioning. Numerous problems still need to be resolved. In this field, there are plenty of potential to make some groundbreaking contributions and greatly improve the sector. In this article, recent advances in the CC field have been covered and talked about potential issues that the research community will need to tackle besides presenting a detailing on situations where the cloud data centre can outperform on premises data centre and vice-versa.

REFERENCES

1. Malik, S., Gupta, K., & Singh, M. (2020). *Resource Management in Fog Computing Using Clustering Techniques: A Systematic Study* (Vol. 24, Issue 2). http://annalsofrscb.ro
2. A. Rashid and A. Chaturvedi, "CC Characteristics and Services A Brief Review," *Int. J. Comput. Sci. Eng.*, vol. 7, no. 2, pp. 421–426, 2019.
3. Adib Bamiah, m. (2015). Trusted cloud computing framework in critical industrial application. https://doi.org/10.13140/rg.2.1.4126.6729
4. P. H. B. Patel and P. N. Kansara, "CC Deployment Models: A Comparative Study," *Int. J. Innov. Res. Comput. Sci. Technol.*, vol. 9, no. 2, pp. 45–50, 2021.
5. T. H. Noor, S. Zeadally, A. Alfazi, and Q. Z. Sheng, "Mobile CC: Challenges and future research directions," *J. Netw. Comput. Appl.*, vol. 115, no. April, pp. 70–85, 2018.
6. N. Subramanian and A. Jeyaraj, "Recent security challenges in CC," *Comput. Electr. Eng.*, vol. 71, no. July 2017, pp. 28–42, 2018.
7. M. Abdel-Basset, M. Mohamed, and V. Chang, "NMCDA: A framework for evaluating CC services," *Futur. Gener. Comput. Syst.*, vol. 86, pp. 12–29, 2018.
8. S. K. Yoo and B. Y. Kim, "A decision-making model for adopting a CC system," *Sustain.*, vol. 10, no. 8, 2018.
9. V. Priya, C. Sathiya Kumar, and R. Kannan, "Resource scheduling algorithm with load balancing for cloud service provisioning," *Appl. Soft Comput. J.*, vol. 76, pp. 416–424, 2019.
10. I. Strumberger, N. Bacanin, M. Tuba, and E. Tuba, "Resource scheduling in CC based on a hybridized whale optimization algorithm," *Appl. Sci.*, vol. 9, no. 22, 2019.
11. S. A. Tafsiri and S. Yousefi, *Combinatorial double auction-based resource allocation mechanism in CC market*, vol. 137. Elsevier Inc., 2018.
12. X. Chen, W. Li, S. Lu, Z. Zhou, and X. Fu, "Efficient Resource Allocation for On-Demand Mobile-Edge CC," *IEEE Trans. Veh. Technol.*, vol. 67, no. 9, pp. 8769–8780, 2018.
13. M. B. Gawali and S. K. Shinde, "Task scheduling and resource allocation in CC using a heuristic approach," *J. Cloud Comput.*, vol. 7, no. 1, 2018.

14. H. Shukur, S. Zeebaree, R. Zebari, D. Zeebaree, O. Ahmed, and A. Salih, "CC Virtualization of Resources Allocation for Distributed Systems," *J. Appl. Sci. Technol. Trends*, vol. 1, no. 3, pp. 98–105, 2020.

15. N. Bulchandani, U. Chourasia, S. Agrawal, P. Dixit, and A. Pandey, "A survey on task scheduling algorithms in CC," *Int. J. Sci. Technol. Res.*, vol. 9, no. 1, pp. 460–464, 2020.

16. J. Zhang, N. Xie, X. Zhang, K. Yue, W. Li, and D. Kumar, "Machine learning based resource allocation of CC in auction," *Comput. Mater. Contin.*, vol. 56, no. 1, pp. 123–135, 2018.

17. C. Vijaya and P. Srinivasan, "A survey on resource scheduling in CC," *Int. J. Pharm. Technol.*, vol. 8, no. 4, pp. 26142–26162, 2016.

18. R. N. Calheiros, R. Ranjan, C. A. F. De Rose, and R. Buyya, "CloudSim: A Novel Framework for Modeling and Simulation of CC Infrastructures and Services," pp. 1–9, 2009.

19. J. B. Wang, J. Wang, Y. Wu, J. Y. Wang, H. Zhu, M. Lin, and J. Wang, "A Machine Learning Framework for Resource Allocation Assisted by CC," *IEEE Netw.*, vol. 32, no. 2, pp. 144–151, 2018.

20. P. K. Senyo, E. Addae, R. Boateng, and M. I. Systems, "SENYO_cright_Cloud_Computing_Research_A_Review_of_Research_ Themes_Frameworks_Methods," pp. 1–36.

21. A. Abid, M. F. Manzoor, M. S. Farooq, U. Farooq, and M. Hussain, "Challenges and issues of resource allocation techniques in CC," *KSII Trans. Internet Inf. Syst.*, vol. 14, no. 7, pp. 2815–2839, 2020.

22. L. Haji, O. Ahmed, A. B. Sallow, L. M. Haji, S. R. M. Zeebaree, O. M. Ahmed, A. B. Sallow, K. Jacksi, and R. R. Zeabri, "Dynamic Resource Allocation for Distributed Systems and CC Machine learning View project Augmented Reality View project Dynamic Resource Allocation for Distributed Systems and CC," no. June, 2020.

23. O. Ali, A. Shrestha, J. Soar, and S. F. Wamba, "CC-enabled healthcare opportunities, issues, and applications: A systematic review," *Int. J. Inf. Manage.*, vol. 43, no. April, pp. 146–158, 2018.

24. Z. N. Rashid, S. R. M. Zebari, K. H. Sharif, and K. Jacksi, "Distributed CC and Distributed Parallel Computing: A Review," *ICOASE 2018 - Int. Conf. Adv. Sci. Eng.*, pp. 167–172, 2018.

25. N. M. Abdulkareem, S. R. Zeebaree, M. A. M. Sadeeq, Di. M. Ahmed, A. S. Sami, and R. R. Zebari, "IoT and CC Issues, Challenges and Opportunities: A Review," *Qubahan Acad. J.*, pp. 1–7, 2021.

26. M. Muniswamaiah, T. Agerwala, and C. Tappert, "Big Data in CC Review and Opportunities," *Int. J. Comput. Sci. Inf. Technol.*, vol. 11, no. 4, pp. 43–57, 2019.

27. M. I. Malik, "CC-Technologies," *Int. J. Adv. Res. Comput. Sci.*, vol. 9, no. 2, pp. 379–384, 2018.

28. U. A. Butt, M. Mehmood, S. B. H. Shah, R. Amin, M. Waqas Shaukat, S. M. Raza, D. Y. Suh, and M. J. Piran, "A review of machine learning algorithms for CC security," *Electron.*, vol. 9, no. 9, pp. 1–25, 2020.

29. J. Agbaegbu, O. T. Arogundade, S. Misra, and R. Damaševičius, "Ontologies in CC—review and future directions," *Futur. Internet*, vol. 13, no. 12, pp. 1–22, 2021.

30. S. Singh and I. Chana, "A Survey on Resource Scheduling in CC: Issues and Challenges," *J. Grid Comput.*, vol. 14, no. 2, pp. 217–264, 2016.

31. Gupta, K., & Katiyar, V. (2016). Energy Aware Virtual Machine Migration Techniques for Cloud Environment. *International Journal of Computer Applications*, *141*(2), 11–16. https://doi.org/10.5120/ijca2016909551.

32. M Uppal, D Gupta, S Juneja, G Dhiman, S Kautish, "Cloud-based fault prediction using IoT in officeautomation for improvisation of health of employees", *Journal of Healthcare Engineering*, Volume 2021 I Article ID 8106467 I https://doi.org/10.1155/2021/8106467

Note: All the figures and table in this chapter were made by the author.

Computational Intelligence and Mathematical Applications – Dr. Devendra Prasad et al. (eds)
© 2024 Taylor & Francis Group, London, ISBN 978-1-032-87721-1

32 | Survey and Observation of Task Scheduling in Cloud

Rabina Bagga*, Kamali Gupta

Chitkara University Institute of Engineering and Technology, Chitkara University, Punjab, India

Abstract: Task scheduling, one of the hottest topics in cloud computing, is crucial for addressing user needs and achieving a number of objectives. How to speed up task completion without compromising upon qos parameters like energy management, resource efficiency, security and enhancing system load balancing have become more important as cloud computing has grown in popularity and in demand academics and business have been showing arising amount of attention lately. As a result, numerous studies have experimented with a range of methods to address this problem and enhance relative efficiency in specific settings. Examining current meta-heuristic task scheduling techniques is the focus. This paper explores the topic of cloud computing scheduling, with an emphasis on the saas and paas layers. The discussion includes the key constraints that determine service quality, evaluation methodologies for algorithms and the various research gaps existing in the current scenario.

Keywords: Cloud computing, Task scheduling, Virtual machine migration, Load balancing, Resource use efficiency

1. INTRODUCTION

Due to the fact that cloud services rely on task scheduling in the cloud is different from conventional scheduling tactics and requires more consideration. The cost of carrying out operations in the cloud while employing its resources is exceptional. The cloud comprises of multiple resources that are distinct from one another through particular approaches. To increase the adaptability and dependability of cloud-based frameworks, task scheduling is crucial [7]. The main justification for allocating tasks to resources in accordance with the allotted time includes figuring out the optimum order in which to complete various jobs so as to provide the client with the greatest possible result. The order and requirements of the task, and its subtasks, determine how gradually resources in any structure, such as containers, firewalls, and networks, are distributed in cloud computing. In light of this, task scheduling in the cloud is made to appear to be a serious problem because no previously identified sequence may be helpful when a job is being prepared. The reason for the scheduling must be dynamic is that as the flow of tasks is uncertain, so are the methods of execution, and simultaneously resources available are also uncertain because several tasks are available that are sharing them simultaneously. Typically, task scheduling algorithms follows process exhibited in Fig. 32.1 and fall into one of two categories:

- *Static Task Scheduling:* As the processing node is chosen at the compilation stage, jobs are assigned to nodes in static task planning before they are executed. The job will continue until it is finished when all necessary resources have been allocated. The primary goals of static task planning are to minimize runtime overhead scheduling and to employ fewer nodes and processors. Because the static task planning algorithm does not take into consideration the resource burdens and requirements of an application, it can lead to resource misuse and under use, which can lead to job failures. [8].

---

*Corresponding author: rabinabagga1005cse.phd22@chitkara.edu.in

DOI: 10.1201/9781003534112-32

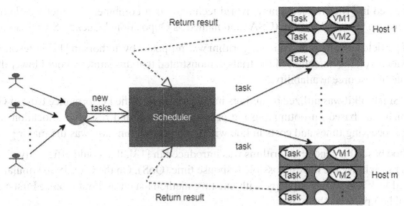

Fig. 32.1 Task scheduling processes

Source: Author

- *Dynamic Task Scheduling:* While the work is being done, the choice is made by the dynamic task planning. The main considerations include resources, inter-process and intra-node traffic, energy efficiency, and other variables. Additionally, it provides migration tasks based on an application's demand and the state of the cluster's resources. Dynamic task planning's primary objectives are to utilize resources effectively, expend the least amount of energy possible, and complete tasks prior to the deadline.

In addition, the following is how this article is organized: The task scheduling algorithms computing models are covered in next section, Section II which illustrate the current status of algorithms, its issues, and the solution that is being suggested. We go over the unresolved problems and difficulties with cloud computing in Section III. Section IV concludes with our analysis. Task Scheduling Algorithms commonly used in Cloud Computing are categorized as :

(a) *Overview of the First-Come, First-Served (FCFS) Task Scheduling:* The job that is about to start looks for the queue with the shortest holding period (first come, first served, or FCFS). The queue is controlled by the FIFO component (first start things out). Each new Task should be placed near the end of the service queue in FCFS. The first task in the queue is assigned to the VM when it becomes available. This algorithm is easy to understand and utilize[9].

(b) *Task Scheduling covering Multi objectives:* Most scheduling algorithms set time limits without taking into account how much less expensive the services will be; these methods are also used in single-cloud scenarios. In a multi-cloud environment, the algorithm considers the two boundaries of makespan and cost of service.

(c) *Priority oriented Scheduling at multiple levels:* Prioritization-based prior Algorithms for task scheduling distribute tasks according to their importance and size. One typical scheduling method involves setting up different autonomous work process tasks using six sigma control charts [10].

(d) *Balancing of load-based Scheduling:* To save complexity, the majority of scheduling techniques ignore balancing the load, which could lead to an uneven load. These algorithms ignore the relationship between node load and job allocation. System performance is severely impacted when a task is allocated to a node that is already underutilized [11].

2. LITERATURE REVIEW

A comprehensive review of the existing literature is essential to identify the current trends, research gaps, and opportunities for further investigation in a particular field. In this literature review, a critical analysis of the recent research in the area of scheduling in clouds to gain a deeper understanding of the current state of the art and to identify the challenges and limitations that need to be addressed. Through this review, the main contribution to the development of a more effective and efficient system. Following are the reviews done for the study:

A brand-new task-scheduling technique for cloud computing was proposed in [9] that was based on the Lion Optimization Algorithm (LOA). For achieving overall optimization over a search space, LOA was a population-based method influenced by nature. Comparisons were made between the proposed work scheduling method and those based on Particle Swarm Optimization and Genetic Algorithm. Furthermore, when contrasted to other techniques, the findings showed the proposed algorithm to have a good performance level.

The research work discussed in [10] produce a new hybrid technique that combines two methods known as the oppositional based learning and the cuckoo search algorithm (CSA) and named as Oppositional Cuckoo Search algorithm (OCSA).

An enhanced version of particle swarm optimization algorithm was proposed by authors in [11] in resources of cloud computing scheduling technique. Moreover, the outcomes of the trials demonstrated that this strategy could lower the average task running time and increase the rate of resource availability.

Furthermore, the Cloud Sim toolkit was utilized by authors in [12] to simulate the Performance Cost of Grey Wolf Optimization (PCGWO) method, which was based on optimizing the procedure of task and resource allocation in the cloud computing environment. Lowering processing times and costs in line with the objective function was the major goal.

A task distribution method based on genetic algorithms was introduced in [13], that could efficiently divvy up the task amongst virtual machines in order to achieve the lowest possible response time (QoS). On the CloudSim simulator, a comparison of this Genetic Algorithm-based task scheduling strategy with existing methods such as First-Come, First-Served (FCFS) methods demonstrated that it could outperform them.

An enhanced scheduling efficiency technique (known as the improved whale optimization algorithm, or IWC) was suggested in [14] to address the issues of lengthy scheduling times, high cost consumption, and significant virtual machine load in cloud computing job scheduling.

By considering running a job on various resources with diverse execution times, as well as heterogeneous resources that vary in their capability to carry out different tasks, an effective scheduling technique was proposed in [15]. Results revealed that the suggested approach outperformed other algorithms, including the moth-flame optimization technique, the whale optimization algorithm, and the intelligent water drop method.

A hybrid multiple-objective strategy known as hybrid grey wolf and whale optimization (HGWWO) algorithms was proposed in [16]. This algorithm further combined 2 techniques named as the grey wolf optimizer (GWO) and the Whale optimization algorithm (WOA). This approach addressed the issue of task scheduling in cloud computing, which required non-traditional optimization attitudes to reach the problem's optimum.

For multi-objective task scheduling in cloud computing, a whale optimization strategy based on the Gaussian cloud model (GCWOAS2) was utilized in the second layer of the model [17]. The strategy's goal was to minimize task completion time by efficiently utilizing resources of virtual machines and maintaining virtual machine balance of load, thereby lowering system operating costs.

A genetic algorithm-based approach in a cloud computing environment was suggested in [18] to improve the job scheduling approach. Early versions of the independent and associated work scheduling algorithms, both of which were frequently utilized in cloud computing, were examined and contrasted, and their distinct application characteristics, benefits, and drawbacks were thoroughly examined.

A multi-objective optimization strategy was recommended in [19], for problems with cloud computing job scheduling. The diversity of resources and activities in cloud computing was addressed by the authors, who first provided a resource pricing model that more accurately characterized the demand for tasks on resources. Additionally, this model showed how the client's resource costs and budget costs were related to one another. A multi-objective optimization scheduling method was recommended in light of this resource costing system. A few more are shown in Table 32.1.

3. OBSERVATIONS

Based on a literature survey conducted, it is evident that resource scheduling in cloud environments poses several challenges that require further research and development. Existing multi-objective optimization algorithms suffer from slow convergence, low solution quality, and lack of robustness. To address these issues, future research could focus on the development of new hybrid algorithms and exploration of novel optimization methods for handling complex scheduling problems. Additionally, the effectiveness of different model switching strategies in practice remains uncertain, and there is a need for new model selection criteria and evaluation of switching strategies for more intricate scheduling problems. The increasing complexity of cloud resource scheduling necessitates the adoption of new techniques, such as machine learning and data analytics, to better predict resource demands and identify optimal scheduling strategies. Ensuring fairness in resource allocation among different users and applications is an ongoing challenge that requires the development of new techniques. Furthermore, the impact of neighbor noise on scheduling accuracy and the effectiveness of current noise handling techniques are areas that warrant

Table 32.1   Related work on task scheduling based on identified algorithms

Year	Algorithm/reference	Algorithm Type	Technique or parameters Used	Tool	Limitations
2017	Lion Optimization Algorithm [9]	meta-heuristic	Utilized optimization algorithm for reducing makespan time	CloudSim	High cost for executing tasks on CC
2018	Proposed Oppositional cuckoo search algorithm (OCSA) [10]	meta-heuristic	Combined cuckoo search and oppositional based learning algorithms	CloudSim	Integrated only few QoS parameters
2020	Performance-cost GWO (PCGWO) [12]	meta-heuristic	Utilized proposed model for reducing cost and processing time	CloudSim	Premature convergence rate
2021	Proposed Improved WOA, referred as, IWC [14]	meta-heuristic	Analysed different QoS parameters for task scheduling	MATLAB 2012	Time complexity issue
2022	Proposed parallel GA along with MapReduce framework [15]	meta-heuristic	Reduced execution time of the model by employing MapReduce	MATLAB	Difficulty in handling constraints
2021	Hybridized GWO and WOA [16]	meta-heuristic	It is a multi-objective approach	CloudSim	Selecting fitness constraints was difficult
2021	Proposed GCWOAS2 model for task scheduling [17]	meta-heuristic	Utilized Whale-Gaussian based 3-layered scheduling model and WOA for multi-objective	MatlabR2016b	Increased computational complexity
2022	Suggested GA based task scheduling model [18]	meta-heuristic	Analyzed different QoS parameters for effective task scheduling	CloudSim	Computational expensive for complex problems
2022	Proposed multi-objective improved GA (MOIGA) [22]	meta-heuristic	Evaluated different constraints of optimization	CloudSim	Sensitive to parameter settings
2021	Proposed fuzzy self-defense approach [23]	meta-heuristic	Analyzed least time, load balancing of resources and cost parameters	CloudSim	Proposed technique is not much effective
2018	Hybridized GA and Bacterial Foraging (BF) algorithm [24]	meta-heuristic	Proposed model minimizes energy usage and makespan time	Matlab R2013a	Lack of scalability
2018	Proposed CGSA for task scheduling [25]	meta-heuristic	Integrated cuckoo search along with Gravitational search algorithm	CloudSim 3.0	optimization problems that involve complex constraints that satisfy specific conditions.
2022	Hybrid DE algorithm [31]	meta-heuristic	Improved scaling and exploitation factor	CloudSim	Lack of robustness
2021	Hybridized PSO and GA together [32]	meta-heuristic	Implemented Phagocytosis on hybrid algorithm	CloudSim	Slower convergence rate
2015	Proposed Multi-objective PSO and GA [33]	meta-heuristic	Analyzed different QoS parameters for task scheduling	CloudSim	Sensitive to hyperparameter tuning

further investigation. Lastly, dynamic resource management techniques are needed to adapt scheduling strategies in response to changing resource demands and availability, leveraging real-time data analysis. Overall, these observations highlight the need for future research to address the existing limitations and advance the field of cloud resource scheduling. The following section provides a description of existing gaps:

- *Multi-objective optimization algorithms:* Existing multi-objective optimization algorithms for resource scheduling suffer from slow convergence, low solution quality, and lack of robustness. Future research could focus on developing new hybrid algorithms and exploring new optimization methods to handle more complex scheduling problems [17][23][14][25][32].

- *Model switching:* The effectiveness of different model switching strategies in practice remains uncertain, and existing selection criteria may not be suitable for more complex scheduling problems. Future research could focus on developing new model selection criteria and evaluating the effectiveness of different switching strategies in practice [36][37].

- *Resource allocation fairness:* The issue of fairness in resource allocation in cloud environments is an ongoing challenge, with the need for new techniques to ensure fair allocation of resources among different users and applications [14][27].

- *Neighbor noise handling:* Unpredictable resource availability and usage, or "noise," can interfere with optimal scheduling in cloud resource scheduling, especially when resources or workloads are scheduled closely together. Current techniques for handling neighbor noise may not be sufficient, and the impact of different types of noise on optimization results is unclear. Future research could focus on developing more effective noise handling techniques and exploring the impact of different types of noise on scheduling accuracy [15][17].

- *Dynamic resource management:* There is a need for new techniques to handle dynamic changes in resource demands and availability in cloud environments, including the use of real-time data analysis to adjust scheduling strategies in response to changing conditions [19][27][38].

4. CONCLUSIONS

The resource scheduling is one the challenge in cloud computing. The various schemes are already been proposed for efficient resource scheduling in cloud computing. The resource scheduling techniques can be broadly classified into queue-based algorithms, optimization algorithms and rules-based algorithm. The already proposed techniques is reviewed in this proposal and it analyzed that already proposed technique has major three problems which needs to be addressed in future. The resource scheduling technique should be energy efficient and also has low response time. The second major issue is with the resource wastage it means most of the resources get wasted as the time of scheduling with should be minimized. The performance of the network should be maintained in the network at the time of scheduling. The resource scheduling techniques can be raised to new heights if the described problems can be addressed properly.

5. FUTURE SCOPE

Further research is needed to develop multi objective optimization algorithm that will enhance convergence, solution quality and also focus on developing new model selection criteria and evaluating the effectiveness of different switching strategies. New methods are required to deal with the dynamic shifts in resource availability and demand in cloud environments. One such method is the use of real-time data analysis to modify scheduling tactics in response to shifting circumstances.

REFERENCES

1. Zhang, Q., Gani, A., & Ahmad, I. (2020). Cloud computing: recent advances, challenges and future research directions. Journal of Parallel and Distributed Computing, 144, 1-32. doi: 10.1016/j.jpdc.2020.02.005
2. Grobler, M., Smith, E., & Venter, H. (2020). Designing Effective Models for Cloud Computing: A Review of Literature. International Journal of Computer Science and Information Security, 18(6), 1-10.
3. Alam, M., Uddin, M., Islam, M., & Hassan, M. (2021). An Empirical Study of the Challenges and Benefits of Cloud Computing Adoption for Small and Medium-Sized Enterprises in Bangladesh. IEEE Access, 9, 48291-48303. doi: 10.1109/access.2021.3064417
4. Chen, Y., Zhang, R., & Li, Y. (2021). Analyzing the Role of Cloud Computing in Teleworking Adoption During the COVID-19 Pandemic. IEEE Access, 9, 143788-143797. doi: 10.1109/access.2021.3112651
5. Rajabi, B., Montazer, G. A., &Gharehchopogh, F. S. (2020). Cloud computing: Benefits, risks and recommendations for information security. International Journal of Information Management, 52, 102047. doi: 10.1016/j.ijinfomgt.2019.102047
6. Yadav, R., Singh, G., &Baliyan, V. (2021). A review on resource scheduling in cloud computing. Journal of Ambient Intelligence and Humanized Computing, 12, 1371-1396. doi: 10.1007/s12652-020-02697-1
7. Li, Y., Zhang, R., Chen, Y., & Li, C. (2021). An Effective Resource Scheduling Algorithm in Cloud Computing Based on Service Quality. IEEE Access, 9, 123097-123109. doi: 10.1109/access.2021.3091706
8. Gharechahi, R., Dastjerdi, A. V., &Buyya, R. (2021). Dynamic resource scheduling in cloud computing: A comprehensive review. Journal of Network and Computer Applications, 178, 102980. doi: 10.1016/j.jnca.2021.102980.
9. Nora Almezeini and Alaaeldin Hafez,(2017) "Task Scheduling in Cloud Computing using Lion Optimization Algorithm" International Journal of Advanced Computer Science and Applications(IJACSA), 8(11).
10. Krishnadoss, Pradeep, and Prem Jacob. "OCSA: task scheduling algorithm in cloud computing environment." International Journal of Intelligent Engineering and Systems 11.3 (2018): 271-279.
11. Malik, S., Gupta, K., Gupta, D., Singh, A., Ibrahim, M., Ortega-Mansilla, A., Goyal, N., & Hamam, H. (2022). Intelligent Load-Balancing Framework for Fog-Enabled Communication in Healthcare. In Electronics (Switzerland) (Vol. 11, Issue 4). MDPI. https://doi.org/10.3390/electronics11040566
12. Natesan, Gobalakrishnan, and Arun Chokkalingam. "An improved grey wolf optimization algorithm based task scheduling in cloud computing environment." Int. Arab J. Inf. Technol. 17.1 (2020): 73-81.
13. M. Agarwal and G. M. S. Srivastava, "A genetic algorithm inspired task scheduling in cloud computing," 2016 International Conference on Computing, Communication and Automation (ICCCA), Greater Noida, India, 2016, pp. 364-367
14. LiWei Jia, Kun Li, Xiaoming Shi, "Cloud Computing Task Scheduling Model Based on Improved Whale Optimization Algorithm", Wireless Communications and Mobile Computing, vol. 2021, Article ID 4888154, 13 pages, 2021.

15. Zhihao Peng, PoriaPirozmand, Masoumeh Motevalli, Ali Esmaeili, "Genetic Algorithm-Based Task Scheduling in Cloud Computing Using MapReduce Framework", Mathematical Problems in Engineering, vol. 2022, Article ID 4290382, 11 pages, 2022

16. Jafar Ababneh, "A Hybrid Approach Based on Grey Wolf and Whale Optimization Algorithms for Solving Cloud Task Scheduling Problem", Mathematical Problems in Engineering, vol. 2021, Article ID 3517145, 14 pages, 2021.

17. Lina Ni, Xiaoting Sun, Xincheng Li, Jinquan Zhang, "GCWOAS2: Multiobjective Task Scheduling Strategy Based on Gaussian Cloud-Whale Optimization in Cloud Computing", Computational Intelligence and Neuroscience, vol. 2021, Article ID 5546758, 17 pages, 2021

18. Zhuoyuan Yu, "Research on Optimization Strategy of Task Scheduling Software Based on Genetic Algorithm in Cloud Computing Environment", Wireless Communications and Mobile Computing, vol. 2022, Article ID 3382273, 9 pages, 2022.

19. L. Zuo, L. Shu, S. Dong, C. Zhu and T. Hara, "A Multi-Objective Optimization Scheduling Method Based on the Ant Colony Algorithm in Cloud Computing," in IEEE Access, vol. 3, pp. 2687-2699, 2015,

20. N Gobalakrishnan, C Arun, A New Multi-Objective Optimal Programming Model for Task Scheduling using Genetic Gray Wolf Optimization in Cloud Computing, The Computer Journal, Volume 61, Issue 10, October 2018, Pages 1523–1536

21. Y. -H. Jia et al., "An Intelligent Cloud Workflow Scheduling System With Time Estimation and Adaptive Ant Colony Optimization," in IEEE Transactions on Systems, Man, and Cybernetics: Systems, vol. 51, no. 1, pp. 634-649, Jan. 2021

22. Sissodia, Rajeshwari, ManMohan Singh Rauthan, and Varun Barthwal. "A Multi-ObjectiveOptimization Scheduling Method Based on the Genetic Algorithm in Cloud Computing." International Journal of Cloud Applications and Computing (IJCAC) 12.1 (2022): 1-21.

23. Guo, Xueying. "Multi-objective task scheduling optimization in cloud computing based on fuzzy self-defense algorithm." Alexandria Engineering Journal 60.6 (2021): 5603-5609.

24. Srichandan, Sobhanayak, Turuk Ashok Kumar, and Sahoo Bibhudatta. "Task scheduling for cloud computing using multi-objective hybrid bacteria foraging algorithm." Future Computing and Informatics Journal 3.2 (2018): 210-230.

25. Pradeep, K., and T. Prem Jacob. "CGSA scheduler: A multi-objective-based hybrid approachfor task scheduling in cloud environment." Information Security Journal: A Global Perspective 27.2 (2018): 77-91.

26. Hamid, L., Jadoon, A. & Asghar, H. Comparative analysis of task level heuristic scheduling algorithms in cloud computing. J Supercomput 78, 12931–12949 (2022).

27. GE Jun-wei, GUO Qiang, FANG Yi-qiu. A Multi-objective Optimization Algorithm for Cloud Computing Task Scheduling Based on Improved Ant Colony Algorithm[J]. Microelectronics & Computer, 2017, 34(11): 63-67

28. Yeboah, Thomas, I. Odabi, and Kamal Kant Hiran. "An integration of round robin with shortest job first algorithm for cloud computing environment." International Conference on Management, Communication and Technology. Vol. 3. No. 1. 2015.

29. Alhaidari, F.; Balharith, T.Z. Enhanced Round-Robin Algorithm in the Cloud Computing Environment for Optimal Task Scheduling. Computers 2021, 10, 63

30. Wadhwa, Shivani, Mansee Jain, and Bishwajeet Pandey. "Design and Implementation of Scheduling Algorithm for High Performance Cloud Computing." International Journal of web science and Engineering (IJWSE) (2015): 15-20.

31. Mahmood, A.; Khan, S.A.; Bahlool, R.A. Hard Real-Time Task Scheduling in Cloud Computing Using an Adaptive Genetic Algorithm. Computers 2017, 6, 15.

32. Singh, H., Bhasin, A. & Kaveri, P.R. QRAS: efficient resource allocation for task scheduling in cloud computing. SN Appl. Sci. 3, 474 (2021).

33. Ramezani, F., Lu, J., Taheri, J. et al. Evolutionary algorithm-based multi-objective task scheduling optimization model in cloud environments. World Wide Web 18, 1737–1757 (2015).

34. F. Azimzadeh and F. Biabani, "Multi-objective job scheduling algorithm in cloud computing based on reliability and time," 2017 3th International Conference on Web Research (ICWR), Tehran, Iran, 2017, pp. 96-101

35. H. K. Langhnoja and P. Hetal A Joshiyara, "Multi-Objective Based Integrated Task Scheduling in Cloud Computing," 2019 3rd International conference on Electronics, Communication and Aerospace Technology (ICECA), Coimbatore, India, 2019, pp. 1306-1311.

36. M Uppal, D Gupta, S Juneja, G Dhiman, S Kautish, "Cloud-based fault prediction using IoT in office automation for improvisation of health of employees", *Journal of Healthcare Engineering*, Volume 2021 I Article ID 8106467 I https://doi.org/10.1155/2021/8106467

37. Gupta, N., Gupta, K., Gupta, D., Juneja, S., Turabieh, H., Dhiman, G., Kautish, S., &Viriyasitavat, W. (2022). Enhanced Virtualization-Based Dynamic Bin-Packing Optimized Energy Management Solution for Heterogeneous Clouds. Mathematical Problems in Engineering, 2022. https://doi.org/10.1155/2022/8734198

Computational Intelligence and Mathematical Applications – Dr. Devendra Prasad et al. (eds)
© 2024 Taylor & Francis Group, London, ISBN 978-1-032-87721-1

A Comprehensive Review of Soft Computing Approaches based on Different Applications

Harminder Kaur[1]

Research Scholar, Department of Computer Science and Engineering, Maharishi Markandeshwar Engineering College Maharishi Markandeshwar (Deemed to be University), Mullana, Ambala, Haryana, India

Neeraj Raheja[2]

Associate Professor, Department of Computer Science and Engineering, Maharishi Markandeshwar Engineering College Maharishi Markandeshwar (Deemed to be University), Mullana, Ambala, Haryana, India

Abstract: In diverse fields of science and engineering, a range of challenging issues have appeared significantly (such as the fields of computer science, electrical engineering, management, networks of communication, transportation etc.). Application of conventional techniques seems to be quite computationally expensive whenever the problem becomes very complex. Due to the complex nature, many problems cannot be resolved with the aid of conventional methods (i.e., hard computing). As there are a lot of uncertainties and ambiguous information in these problems. Henceforth, it should be resolved theoretically or logically. To solve difficult problems, fuzzy logic has proven to be a very effective approach. Fuzzy logic can be used to define and control a system whose model is unknown or vaguely defined. The paper describes the detailed review of Fuzzy logic-based system and also provides the survey on fuzzy logic technique in various domains such as healthcare, agriculture, and finance.

Keywords: Soft computing, Swarm intelligence, Fuzzy logic

1. INTRODUCTION

In every field, Optimization plays a vital rolewhere model fitting is a type of optimisation where the objective is to reduce the discrepancy that exists between the actual and accepted output by evaluating the model's parameters. Among the major problems in fuzzy system modelling is the identification of an optimised model. Since most real-world systems are extremely complicated and nonlinear, this has gained a lot of significance. In classical/hard computing-based approaches we require a precisely stated analytical model with associated computational complexity.

The real word problems are omnipresent as they arise in a non-ideal environment which are pervasively uncertain and imprecise. Precision and certainty carry a cost. This cost grows exponentially with the system complexity. This is where approximate reasoning or soft computing approaches come into play.

The basic principle behind soft computing is to exploit the tolerance for uncertainty, incomplete truth, imprecision, also approximation that attain robust, consistent, and cost-effective solutions. Soft computing has a significant impact on the situations where we could replace "the best for sure" solution with "good enough with high probability" solutions. Hence,

[1]harminderkaur39@gmail.com, [2]neeraj_raheja2003@mmumullana.org

DOI: 10.1201/9781003534112-33

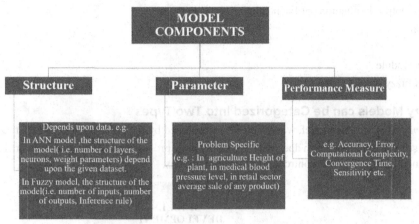

Fig. 33.1   Components of model development

soft computing approaches seem to be very effective solutions for the model identification problems. The purpose of fuzzy modelling is to find the fuzzy model's parameters based on some evaluation criteria so that the model performs at its best. The simplest way of defining this problem is as a search and optimization issue. As a result, the search and optimization techniques can be used to solve the challenge of model identification. To develop high dimensional fuzzy logic-based models, soft computing-based techniques have become increasingly prominent.

2. FUZZY LOGIC

Fuzzy logic (FL), a kind of multi-valued logic derived from fuzzy set theory, was developed to deal with approximate rather than exact reasoning. FL provides a range of membership values, from 0 to 1, that characterise the degree of truth as an alternative to yes/no or 0/1 binary logic (crisp). A fuzzy inference system, which employs fuzzy "If-Then" principles to evaluate human reasoning and knowledge in place of precise quantitative analysis, can be used to characterise vaguely specified and complicated systems in process modelling and control. Building blocks for the fuzzy inference system include the following:

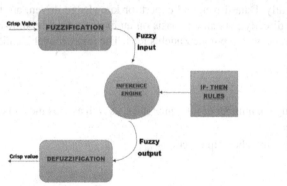

Fig. 33.2   Fuzzy inference system

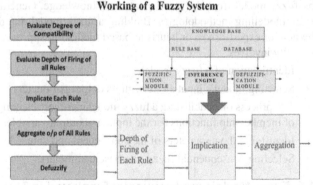

Fig. 33.3   Working of fuzzy system

2.1 Fuzzy System

A fuzzy system consists of 4 Modules:

(i) Fuzzification Module
- Transforms crisp data into fuzzy data
- Degree of Compatibility Evaluation

(ii) Inference Engine
- Depth of Firing Evaluation for each rule

- Implication (output Evaluation) of Each rule
- Aggregation

(iii) Rule Base

(iv) Defuzzification Module
- Transforms fuzzy data to Crisp data

2.2 Building Fuzzy Models can be Categorized into Two Types

- *Knowledge-driven design:* In this model, where rules are developed by the process expert's experience.
- *Data-driven design:* Since this approach does not require any prior process knowledge, the rules are derived from the input-output data obtained from successful control and other applications.

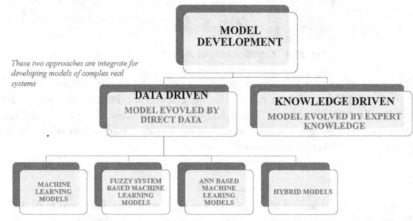

Fig. 33.4 Types of approaches for model development

2.3 Fuzzy Model Identification

To identify a fuzzy model, it is necessary to know how to design a methodology, alternatively a collection of methods, to obtain the fuzzy model from the system data and knowledge. Generally, Data-driven and expert, or knowledge-driven, are the two fuzzy modelling methodologies. Building a fuzzy model that directly puts an emphasis on an expert's domain knowledge is the goal of expert-driven or heuristic-based modelling approaches, which pursue Zadeh's idea. The fuzzy model identification typically involves:

- Identification of a fuzzy model.
- Selecting model's input along with its output parameters.
- The process of establishing a fuzzy model involves selecting the number of inference rules, as well as the kind and quantity of membership functions for the input and output variables.
- Identifying the parameters of the consequent and antecedent membership functions
- Selecting consequential fuzzy rule base parameters.

2.4 Fuzzy Models are Frequently Employed in Three Ways

- Takagi-Sugeno fuzzy models
- Singleton fuzzy models
- Mamdani-type fuzzy models

3. APPLICATION OF FUZZY LOGIC

Fuzzy logic is a method for analysing reasoning processes in which the concepts of truth and untruth are graded. Fuzzy logic examines ambiguity in natural language as well as in several other application domains. The use of fuzzy models for imaging nerve fibres in neuroanatomy research has proven to be a successful method. Second, fuzzy logic can be applied to the diagnosis, treatment, and outcome prediction of prevalent neurological illnesses like stroke. Thirdly, neurosurgical intensive care units (ICUs) can use fuzzy logic to precisely control various parameters including blood pressure and intracranial pressure (ICP). Using different propofol infusion rates, fuzzy logic-based controllers are proven to be effective at maintaining a steady ICP. One major area where its use is being investigated is mobile cellular communication system. This also opens an avenue for designing ASICs comprising integrated circuits for basic fuzzy computational modules. For this VHDL based simulation of integrated circuits for fuzzy operations is gaining popularity.

Detailed survey on fuzzy logic technique applied to various applications in healthcare, agriculture, finance is presented in literature below:

FUZZY IN HEALTHCARE

Ref No.	Problem Statement	Proposed Approach	Performance Measures	Performance/ Results (%)	Research Gaps
[1]	Providing the most precise information available to assist with disease prediction in the medical industry.	EC, FL, NN	Accuracy, Specificity, RMSE and Sensitivity	Review Paper	Use of Fuzzy Logic for Celiac disease.
[2]	Applications of Fuzzy Logic in Medicine	FLP, MCDA	Cost, Life Length, Practicability, Efficiency	Review Article	Implementing Fuzzy Logic in neurophysiology labs for EMGs and EEGs, ICUs, Ventilator settings and MRI scan etc.
[3]	Automatic heartbeat recognition and classification.	LBP, LGP, and LNDP are the components of LTP, a hybrid neural fuzzy-logic system.	Sensitivity, accuracy, predictivity and false positive rate	98.84% accuracy, sensitivity of 87%, positive predictivity of 73.8%, false positive rate of 1.1%	Within the medical system, the proposed techniques can be used for better diagnosis of cardiac problems as well as for mobile healthcare.
[4]	System that automatically measures the saturation of blood oxygen levels for intensive care unit patients.	R programming is used to create a fuzzy controller that uses the Mamdani model to forecast the amount of FiO2 that must be provided by the ventilator to keep SaO2 within desired ranges.	Accuracy in terms of % error	Less than 5% error	A multi-input multi output system could be developed through similar research using big data analysis. Undoubtedly, the deep learning approach that modifies ventilator settings to maintain blood oxygen saturation using an adaptive neural fuzzy system will be the focus of future study.
[5]	To improve the automated management of mean arterial blood pressure through drug infusion.	IT2-FLP-PID and CSO	Sensitivity, overshoot, settling time,	The performance curves of the T1-FLCP-PID and IT2-FLCP-PID controllers do not overshoot, in contrast to PID controllers have a slight overrun of 0.36 mmHg. All the controls settle in a range of 5mmHg. The settling times for the IT2-FLCP-PID control approach, T1-FLCP-PID, and conventional PID control approaches are 350 s, 360 s, and 410 s, respectively, for fixed case. The PID controller has the greatest error numbers whereas the IT2-FLCP-PID control system has the lowest.	To increase the effectiveness of the proposed approach, the use of a shadowing T2-FLC technique could be added into the design.

FUZZY IN AGRICULTURE

Ref No.	Problem Statement	Proposed Approach	Performance Measures	Performance/ Results(%)	Research Gaps
[6]	Parameter prediction for the hydrolysis of melon seed peel (MSP)	ANN and ANFIS	Correlation coefficient (R2), mean absolute deviation (MAD), average absolute relative error (AARE), mean square error (MSE),	AARE = 0.00231, MSE = 1.2698E-12, R2 = 0.999 for ANN, MAD = 1.2%, and R2 = 0.9782 for ANFIS, MAD = 2.7%, MSE = 0.0334, AARE = 0.0054.	Proposed model for managing MSP hydrolysis and a techno-economic assessment of MSP-to-fermentable-sugar production are developed and implemented in ANN/ANFIS..
[7]	Early detection of frost in crops	KNN based Fuzzy Classifier	Accuracy, sensitivity, specificity, mean square error, correlation coefficient	With an MSE of 0.4870 and an MAE of 0.5122, the average accuracy was 97.01% (89.83% sensitivity).	Accuracy of the model can further be improved by using more rigorous classification techniques such as ANN, SVM etc.
[8]	Based on leaf characteristics, identification of promising pistachio seedlings	ANN	Accuracy, root mean square errors (RMSE),	98.95% accuracy	Future research must evaluate larger, more comprehensive data sets in order to develop a model that is more universal and all-encompassing.
[9]	Examine the variability of the solar resource at a suitable site using data from a geographic information system (GIS).	PSO-ANFIS, GA-ANFIS, GDALRNN, RBNN, ANFIS	The root means square error, mean absolute percentage error, computing time, robust coefficient of variation, and skill score are all related to this.	0.2951% for RMBE, 4.0998 for MAPE, 0.0403 for robust coefficient of variation, 10.7345 for skill score, and 9.71 seconds for computing time.	The findings indicate that the Western Cape Province has captured less solar energy than other provinces, and that greater utilisation of this resource is necessary to further South Africa's ambition for a low-carbon economy. The power corridors must also be updated to avoid anthropogenic activities related to utility-scale solar PV installation and operation threatening the avian habitat.
[10]	Almond kernel mass estimation based on shell characteristics	MLP-NN, RBF-NN, ANFIS, SVM	RMSE, MAPE	The RBF-NN classifier achieved an overall accuracy of 96.22%.	With 96.22% total accuracy, the RBF-NN classifier performed well. Given the precision of the prediction and processing efficiency, it is strongly advised for non-destructive evaluation of almond kernel mass.
[11]	Immature Pistacia vera genotypes' sex determination	SVM, SOM, ANFIS, KNN, and RBF	Accuracy	The best overall classification accuracy was attained by RBF (88.33%), ANFIS (66.3%), SVM (68.89%), and SOM (66.11%) among the five approaches.	These results could be corroborated by a larger collection of genetic biodiversity data from pistachios.
[12]	Analysis and Estimation of Foreign Exchange Reserves of India	BPNN	Correlation coefficients, MSE, relative error square, sensitivity	RMSE 1.30%, RMSRE 8.60%, Sensitivity from 0.11% to 0.86% for different output parameters,	Deep learning models can be used ti improve the performance
[13]	Stock Market Prediction	RF, Bagging, AdaBoost, DT, SVM, K-NN, and ANN	Precision, Recall, F-Score, Accuracy	Bagging and Random Forest with leaked datasets both perform above-averagely well, with accuracy of 93%.	Accuracy can be further improved by using Deep Neural Networks
[14]	Stock Market Investment	Fuzzy candlesticks pattern recognition	Sharpe ratio	SR 3.98	Exogenous inputs may be taken into account in stock market trading to improve prediction.
[15]	Stock Market Prediction	SVM- GA	Hit Rate is defined as the proportion of forecasts that were accurate during periods of 1-6 days.	The average of the methods 1 and 2's prediction accuracy is 63.84% and 64.5	A nonlinear autoregressive model can be used to get better results
[16]	Prediction of current market trends	Data stream mining and intelligent computing	Accuracy	Up to 73% accuracy as compared to other models	Using dynamic trading techniques and examining variations in transaction volume of intraday stock alerts in the long

4. FUTURE SCOPE

The fusion of soft computing methods with different forms of computational intelligence, such as granular computing and metaheuristic algorithms, must be investigated. Modern low-cost, lightning-fast microcontrollers can handle most soft computing tasks effectively. Many common household appliances, including refrigerators, cookers, and washing machines, already incorporate fuzzy logic, artificial neural networks, and expert systems. In addition to being widely used in daily life, soft computing has numerous industrial and commercial applications that are anticipated to expand during the next ten years. The study of word computing in fuzzy systems will grow, and evolutionary computation will also start to take shape. It is anticipated that they will be used in the development of more sophisticated, intelligent industrial systems. Given the possibilities, fuzzy logic is certainly a promising, perceptive, and focused tool for a range of therapeutic issues. Fuzzy logic can be successfully implemented into everyday clinical practise and has the potential to fundamentally alter medical diagnosis and care. Fuzzy controllers for blood pressure, ICP, and ventilator settings could be found in ICUs in the future, as could software for analysing MRI scans and planning neurosurgeries. If targeted research is carried out, it's also possible that neurophysiology labs will report EMGs and EEGs using fuzzy logic in the near future. In the foreseeable future, there is a huge market opportunity for portable goods based on fuzzy logic.

5. CONCLUSION

The paper continues with the descriptions on the fuzzy logic and the applications that benefits from them. Fuzzy logic is a component of soft computing, which is still in its early stages of development. A brief overview of fuzzy logic and its use in healthcare, agriculture, and financial systems was also introduced in this paper. Fuzzy models can be used in almost any domain, so in the coming era, soft computing will be a crucial instrument for actual analysis.

REFERENCES

1. Thukral, Sunny, and Vijay Rana. "Versatility of fuzzy logic in chronic diseases: A review." Medical hypotheses 122 (2019): 150-156.
2. Ozsahin, Dilber Uzun, et al. "Fuzzy logic in medicine." *Biomedical Signal Processing and Artificial Intelligence in Healthcare.* Academic Press, 2020. 153-182.
3. Lee, Miran, Tae-Geon Song, and Joo-Ho Lee. "Heartbeat classification using local transform pattern feature and hybrid neural fuzzy-logic system based on self-organizing map." *Biomedical Signal Processing and Control* 57 (2020): 101690.
4. Radhakrishnan, Sita, Suresh G. Nair, and Johney Isaac. "Analysis of parameters affecting blood oxygen saturation and modeling of fuzzy logic system for inspired oxygen prediction." *Computer methods and programs in biomedicine* 176 (2019): 43-49.
5. Sharma, Richa, and Anupam Kumar. "Optimal Interval Type-2 Fuzzy Logic Control Based Closed-Loop Regulation of Mean Arterial Blood Pressure Using the Controlled Drug Administration." *IEEE Sensors Journal* 22.7 (2022): 7195-7207.
6. Nwosu-Obieogu, Kenechi, et al. "Environmental sustenance via melon seed peel conversion to fermentable sugars using soft computing models." *Cleaner Engineering and Technology* 7 (2022): 100452.
7. Cadenas, José Manuel, et al. "Making decisions for frost prediction in agricultural crops in a soft computing framework." *Computers and Electronics in Agriculture* 175 (2020): 105587.
8. Heidari, Parviz, Mehdi Rezaei, and Abbas Rohani. "Soft computing-based approach on prediction promising pistachio seedling base on leaf characteristics." *Scientia Horticulturae* 274 (2020): 109647.
9. Adedeji, Paul A., et al. "Beyond site suitability: Investigating temporal variability for utility-scale solar-PV using soft computing techniques." *Renewable Energy Focus* 39 (2021): 72-89.
10. Ashtiani, Seyed-Hassan Miraei, Abbas Rohani, and Mohammad Hossein Aghkhani. "Soft computing-based method for estimation of almond kernel mass from its shell features." *Scientia Horticulturae* 262 (2020): 109071.
11. Rezaei, Mehdi, et al. "Using soft computing and leaf dimensions to determine sex in immature Pistacia vera genotypes." *Measurement* 174 (2021): 108988.
12. Chanda, Mriganka Mohan, Gautam Bandyopadhyay, and Neelotpaul Banerjee. "Analysis and estimation of India's foreign exchange reserves using soft computing techniques." *IIMB Management Review* 32.3 (2020): 280-290.
13. Subasi, Abdulhamit, et al. "Stock Market Prediction Using Machine Learning." *Procedia Computer Science* 194 (2021): 173-179.
14. Naranjo, Rodrigo, and Matilde Santos. "A fuzzy decision system for money investment in stock markets based on fuzzy candlesticks pattern recognition." *Expert Systems with Applications* 133 (2019): 34-48.
15. Ahmadi, Elham, et al. "New efficient hybrid candlestick technical analysis model for stock market timing on the basis of the Support Vector Machine and Heuristic Algorithms of Imperialist Competition and Genetic." *Expert Systems with Applications* 94 (2018): 21-31.
16. Lin, Chien-Cheng, Chun-Sheng Chen, and An-Pin Chen. "Using intelligent computing and data stream mining for behavioral finance associated with market profile and financial physics." *Applied Soft Computing* 68 (2018): 756-764.

Note: All the figures and tables in this chapter were made by the author.

Computational Intelligence and Mathematical Applications – Dr. Devendra Prasad et al. (eds)
© 2024 Taylor & Francis Group, London, ISBN 978-1-032-87721-1

34

Ocular Disease Detection in Retina Fundus Images Using Deep Learning: Methods, Datasets, and Performance Metrics

Gurpreet Kaur[1]

Research Scholar, Department of Computer Science and Engineering, M M Engg.College, MM(DU), Mullana, Ambala, Haryana, India

Amit Kumar Bindal[2]

Professor, Department of Computer Science and Engineering M M Engg.College, MM(DU), Mullana, Ambala, Haryana, India

Abstract: Detecting ocular pathology from retinal fundus images presents a crucial challenge in healthcare. Each pathology comprises different severity stages, which can be confirmed by identifying specific lesions. These lesions possess distinct morphological features. Additionally, there are instances where lesions from various pathologies share similarities in their features. Consequently, ocular pathology detection entails a multi-class classification challenge with complex resolution criteria. Numerous methods for detecting ocular pathologies from retinal fundus images have been proposed. Deep learning - based techniques, in particular, stands out for their enhanced detection performance, owing to their ability to adapt the network to the specific detection task. This article provides a comprehensive review of ocular disease detection methods employing various classification techniques. The review analyzes existing techniques for ocular disease classification. Subsequently, it delves into the extraction of reliable features and disease classification. The study encompasses simulation rules applied to evaluate the methods and the datasets utilized for training and testing, including ODIR and Kaggle EyePacs. Furthermore, it discusses ocular disease detection performance metrics such as accuracy and precision.

Keywords: Ocular disease detection, ODIR, Kaggle EyePacs dataset, Deep learning (DL), Morphological features

1. INTRODUCTION

According to the WHO, approximately 2.2 billion people worldwide undergo from visualization or eyesight loss impairments. The eye plays a key role in this statistic as one of the most important organs. One billion of these people are blind or have vision problems that range from mild to severe. The majority of these conditions is untreated and includes 88.4 million cases of blurred vision, ninety-four million cases of cataracts, 7.7 million cases of glaucoma, 4.2 million cases of corneal opacities, more than three million cases of diabetic retinopathy (DR), and 2 million cases of trachoma. An additional 826 million people worldwide experience near-vision loss due to uncorrected presbyopia. These ocular conditions are recognized globally as significant causes of visual impairment [1].

In Bangladesh, limited research has been conducted regarding the incidence of visual issues and blindness. Approximately 80% of urban residents require ophthalmological care, but access to such services remains inadequate. The Bangladesh National Blindness and Low Vision Survey found that twenty-one percent of the nation's people have low vision, which is characterized by visual acuity of no less than six inches. Non-communicable diseases like diabetes and smoking increase the risk of visual

[1]guri394.kaur@gmail.com, [2]amitbindal@mmumullana.org

DOI: 10.1201/9781003534112-34

impairments in Bangladesh. In 2013, the Urban Health Survey revealed that impoverished individuals living in shanty towns experience poor mental and physical health.

Deep learning (DL) methods are increasingly prevalent in image-based health analysis. Various eye diseases, including blindness, DR, and ocular diseases, can benefit from these advancements [2]

Any condition impairs the eye's proper function or affects visual acuity is categorized as an ocular disease. While most people experience optical subjects happened in their lives, some conditions require specialist care, while others can be managed at home. Globally, fundus problems are the leading cause of blindness. The most prevalent eye illnesses, which include as age-related macular degeneration (AMD), cataracts, glaucoma, and DR, are serious health problems. It is expected to impact over four hundred million individuals before 2030, making DR a critical worldwide medical hazard. This condition is particularly concerning because it is incurable and may lead to lifelong blindness [3].

1.1 Diabetic Retinopathy (DR)

DR disease characterized by a reduction in blood flow to the retina, results from the adverse effects of diabetes. This condition involves the development of lesions on the surface of the retina. As illustrated in Fig. 34.1(i), the symptoms associated with diabetic retinopathy are generally categorized into two types of lesions: red-colored lesions, which include small vessels and hemorrhages, and white lesions, which consist of fluids and cotton-wool regions.

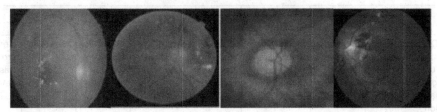

Fig. 34.1 Several diseases (i) Diabetic retinopathy (DR) (ii) Diabetic macular edema (iii) Glaucoma (iv) Age-related macular degeneration (AMD) [4]

1.2 Diabetic Macular Edema

As illustrated in Fig. 34.1(ii), a result is macular edema caused by diabetes that arises when capillaries in the retina's macular region are affected by fluid accumulation and the growth of exudates throughout various parts of the eye.

1.3 Glaucoma

Glaucoma is associated with the gradual loss fluctuations in the density of fibers in the optic nerve the optic nerve head. While eliminating glaucoma is impossible, medical treatment can help halt its progression. Therefore, early detection of this condition is crucial in preventing blindness. The identification of glaucoma relies on the manual examination of the optic disc using ophthalmoscopy, which involves assessing the morphological characteristics of the optic nerve, including its interior bright section and the neuroretinal edge, the outer region, as depicted in Figure 1(iii).

1.4 Age-related Macular Degeneration (AMD)

Both the central area of vision loss and an outward sector of vision distortion are effects associated with AMD. Drusen represents the most prevalent characteristic and clinical sign of dry AMD. Exudates, one of the primary symptoms of wet AMD, are depicted in Figure 1(iv). Various ML methods, such as the one proposed by **Akram et al. in 2022 [5],** utilize deep CNN and SVM to create an innovative automated model for recognizing eye diseases based on noticeable signs extracted from research findings. The DCNN model demonstrates superior performance compared to SVM models. **In 2019, Prasad et al. [6]** developed a deep neural network (DNN) model capable of classifying various syndromes, including diabetic retinopathy (DR). Their model offers improved early detection of glaucoma and DR, particularly when analyzing high-resolution retinal images acquired under diverse imaging conditions. Utilizing deep learning for screening eye diseases can expedite patient interactions with ophthalmologists. The resulting model, while having a relatively low complexity, achieved an accuracy rate of 80%. **Jain et al. in 2018 [7]** introduced a technique for the computerized classification of retinal fundus images using a DL model. They designed a model named "LCD Net," employing a convolutional neural network (CNN) capable of binary classification tasks.

2. RELATED WORK

Yaroub Elloumi et al. in 2023 [8] focused on recognizing ocular pathology from fundus images. They emphasized the need for morphological features to characterize different types of lesions. Additionally, they noted that multiple lesions related to pathologies can affect patients simultaneously. The study recommended various detection approaches for ocular pathologies from fundus images, with deep learning (DL) methods standing out due to their superior detection results. **Sushma K Sattigeri et al. in 2022 [9]** discussed eye-related diseases, which affect around fifteen million cases in India.

These conditions impair vision, including trachoma, corneal ulcers, and cataracts. Early-stage detection is crucial to prevent untreated cases, a significant cause of blindness in India. The authors utilized digital image processing technologies like segmentation, morphology, and DL methods like convolutional neural networks (CNNs). They successfully classified four types of eye diseases using CNNs, demonstrating that their deep neural network (DNN) model enables early detection of eye diseases, achieving a significant accuracy rate. **Abu Kowshir Bitto et al. in 2022 [1]** discussed difficulties related to human eyes and the conjunctiva caused by dust exposure. They emphasized the importance of visualization processes for early detection of eye diseases. The classification method was considered essential at various stages. The authors employed various CNN methods, including VGG-16, Inception-v3, and ResNet-50, to distinguish between eye conditions. Their recommended model achieved

Table 34.1 Analysis of existing methods

Author & Year	Title	Objective	Proposed Method	Problem/Gaps	Tool	Metrics	Findings
Yaroub Elloumi et al. [8] 2023	Ocular Diseases Diagnosis in Fundus Images using a Deep Learning: tools and Performance Evaluation	Diagnose ocular disease	Deep CNN, CNN, Fully and Partial DL methods	Disease detection is a complex task that needs various methods.	MATLAB Python C++	Acc SP SN	It enhanced the pre-processing steps and classified the disease detection.
Sushma K et al. [9] 2022	Eye Disease identification using DL	Eye disease identification	CNN DNN	Having diabetes as they age plays a central role in causing eye problems and diseases. Vision issue	-	Acc	Guide the patients to know their eye situation. DL model is cost-effective and easy to interface.
Abu Kowshir Bitto et al. [1] 2022	Categorical of standard eye disease detection Using convolutional neural network: a transfer learning approach	Disease detection	Inception-v3 VGG-16 RestNet-50	Limited detection process required more time for analysis.	-	Acc Precision F1 Score TPR FPR FNR TNR	It provides an accurate process of detection by using inception-V3
Borys Tymchenko et al. [10] 2022	Deep Learning Approach to Diabetic Retinopathy Detection	Diabetic retinopathy detection	EfficientNet-B4 EfficientNet-B5 SE-ResNeXt50	Not suitable for large datasets.	Python	SN SP Acc	The significant benefits of the proposed method are that it provides better simplification and moderates inconsistency with an ensemble of the networks and target dataset.
Akash Tayal et al. [11] 2021	DL-CNN-based approach with image processing techniques gnosis of retinal diseases	Diagnosis	3 CNN models such as five, seven and nine-layered CNN model	Only a single demographic dataset was used and did not include other images such as fundus photographs or angiographic pictures.	Python	SN SP Acc	Only a single demographic dataset was used and did not include other images such as fundus photographs or angiographic pictures.
Nadim Mahmud Dipu et al. [12] 2021	Ocular Disease Detection Advanced Neural Network Based Classification Algorithms	Detection	Resnet-34, EfficientNet, MobileNet V2 VGG-16	Complications in the detection process of disease.	Keras 2.3.1 Google Colab notebook	Acc Precision Recall F1-Score	This model provides better detection of ocular disease than the MobileNet model.

an impressive accuracy rate of 97.08% when using the Inception-v3 method. **Borys Tymchenko et al. 2020 [10]** addressed DR as a significant complication of diabetes leading to potential blindness. They stressed the significance of early detection for improved diagnosis and highlighted the efficacy of CNN methods in this context. The authors suggested an automatic DL- based technique for the phase detection of diabetic retinopathy through a single fundus photograph. They introduced a multi-stage transfer learning (TL) method using similar datasets with different labeling, achieving outstanding results with an accuracy of 0.99 on the APTOS 2019 blindness detection dataset. **Akash Tayal et al. in 2021 [11]** discussed how artificial intelligence (AI) could be employed for identifying problems and the challenges of consistency and interpretability in health data-based decision-making. Their proposed model utilized retina-based OCT images and was evaluated with three different CNN models. The model excelled in recognizing and extracting retinal layers, detecting abnormalities, and calculating eye defects. It achieved an impressive accuracy of 0.965, surpassing manual ophthalmological analysis. **Nadim Mahmud Dipu et al. in 2021 [12]** discussed the challenges ophthalmologists face in early screening and detecting ocular diseases using fundus images. They introduced four DL-based targeted ocular disease recognition models, utilizing state-of-the-art image classification methods and the ODIR dataset. The study required distinguishing eight classes representing various ocular diseases.

3. EXISTING ISSUES AND CHALLENGES

Ocular disease is considered a critical medical condition that significantly impacts human eyes. However, the process of detecting and diagnosing ocular diseases often yields only moderate results due to several issues and challenges, as outlined in [13]:

- One of the primary challenges in ocular disease detection involves identifying disease- promoting micro biota within the automatic ocular area or intraocular partitions. Understanding their essential roles is crucial.
- Another significant issue arises when surgery on diseased intraocular tissue is required to identify the causative agents responsible for vision-related problems.
- Biopsies or surgical procedures on intraocular tissue are undertaken during experimental preparations for studying eye diseases.
- For the early identification and identification of vision problems, limited devices and methods are offered.
- Traditional methods are time-consuming for ocular disease detection and diagnosis, emphasizing the need to develop more accurate models or tools for improved disease detection.
- Existing tools often exhibit limited performance and may not apply widely to large and small datasets. The tool or method's effectiveness often relies on the specific dataset used.

4. OCULAR DATASET

4.1 Disease Intelligent Recognition (ODIR)

It is a compilation of ophthalmic data comprising records from five thousand patients. These records include information about the patient's age, color fundus images of both eyes and diagnostic abbreviations provided by clinicians. The dataset comprises real-life data from *'Shanggong Medical Technology Co., Ltd.'* The images were captured using Canon, Kowa, and Zeiss high-resolution devices. The dataset is categorized into eight classes: normal (N), AMD (A), cataract (C), myopia (M), diabetes (D), hypertension (H), abnormalities (O), and glaucoma (G) [14], as illustrated in Fig. 34.2. A detailed description of the data utilized for the implemented method is explained in Table 34.2.

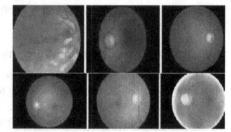

Fig. 34.2   ODIR dataset Image [14]

Table 34.2   Data division from ODIR dataset

Categories	Original Images	Train	Test
Normal	2096	1158	428
AMD	179	1158	386
DR	1356	838	256
Glaucoma	180	702	237
Others	1364	813	282

4.2 Kaggle EyePACS Dataset

It focused on diabetic retinopathy (DR), as depicted in Fig. 34.3, and comprises two subcategories: a training set with 35,126 images and an experiment collection with 53,576 images.

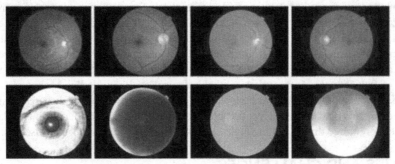

Fig. 34.3 Kaggle EyePacs dataset [15]

Various DR severity levels are represented by these photos, namely regular, relatively mild, severe enough, and regenerative DR. [15]. Table 34.3 defines the KaggleEyePACS data description.

Table 34.3 Kaggle EyePacs Data Detail

Original Grade	New Grade	Number of Images
No DR	Healthy	65300
Mild	Non-referable DR	19400
Moderate		
Severe	Referee	4000
Proliferation		

4.3 Eye Fundus Image (EFI) Dataset

It focuses on capturing the retinal layers of human eyes, as depicted in Fig. 34.4. It is designed to detect retinal alterations, such as variations in color saturation. The dataset includes attributes such as hemorrhages, lesions, and exudates associated with ocular disorders. These retinal descriptions are obtained using a Fundoscope. Additionally, the dataset incorporates the retina identification database (RIDB), which consists of retinal images with a resolution of 1504x1000 pixels, compressed *.jpeg format [16].

Fig. 34.4 EFI dataset image [16]

4.4 Optical Coherence Tomography Image Dataset(OCTID)

In clinical ophthalmology, it is a widely used non-invasive imaging technique. The primary goal is to capture cross-sectional photographs of retinal structures, allowing for early detection of a number of illnesses damaging the layer of the retina and its optic nerve head. As illustrated in Fig. 34.5, the OCTID is instrumental in assessing the functionality of nerves, blood circulation in the retina, and blood oxygenation [17].

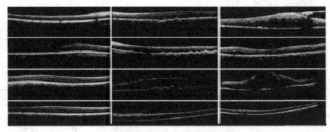

Fig. 34.5 OCTID [17]

5. SEVERAL OCULAR DISEASE DETECTION METHODS

Currently, deep learning (DL) and machine learning (ML) based models have presented capable outcomes in image classification and detection. Mostly convolutional neural network (CNN), support vector machine (SVM), Decision Tree (DT), and Visual Geometry Group (VGG)-16 Model-based representations have been widely considered for ocular disease recognition.

5.1 Multiple Instance Learning

It is an inadequately supervised ML method that gains a collection of labeled groups as an alternative to labeled orders. Each group covers various occurrences. A container is labeled with undesirable variables when all the occurrences are negative, whereas a container is constructive if various examples are positive [19]. In this, two variants of SVM methods, such as mi-SVM and MI-SVM. These methods are generalized as the powerful supervised SVM method.

5.2 Decision Tree (DT)

More than a few methods consist of this classifier multilayer perceptron (MLP) and Naïve Bayes classifier (NB). The productivity of a C4.5 DT classifier is an essential statistic in a hierarchical structure. This method is demonstrated as per requirement. This method is valid, well-organized, and widely expensive classification.

5.3 Naïve Bayes (NB)

This method was developed in 1950 and presented with various names in the early 1960s. It is an ML method, and it is based on Naïve Bayes' theorem. NB classifiers are very mountable, and restrictions are linear in different variables in a learning difficulty.

5.4 VGG-16 Model

This model is 16 16-layered CNN model. This model's convolutional layer (CL) has fixed dimensions such as 224*224 RGB resolutions. The ocular dataset image is transferred through a heap of CLs, where the filters are applied through a tiny receptive field. Three Connected layers (FCLs) are used in this heap of CLs. The concluding layer is called the soft-max layer. The formation of the FCLs is similar in all networks. All hidden layers are prepared using the rectification (ReLU) non-linearity [20].

5.5 CNN Model

It is a DL design that can be unsophisticated in allocating impact to more objects. The pre-handling needed in a ConvNet is inferior when distinguished using other alignment strategies. Although in unsophisticated approaches, networks are traditionally considered, with sufficient formulating, ConvNets are used to catch with these abilities [21].

5.6 Deep CNN Method

Deep NN, especially the convolutional method, proved its advantage in image classification. A lot of CNN-based techniques have been familiarized for constructing automatic DR analyses. The inception-v3 method is used to detect DR image construction for sensing. A twofold DCNN-based method distinguishes lesions in fundus imageries. It marks the severity of DR. A deep CNN model was also developed using a massive dataset for DR detection from fundus images [18]. Table 34.4 represents performance based on existing methods.

Table 34.4   Result analysis based on different methods

Methods	Accuracy (%)	Precision (%)
MIL (SVM) [19]	94.4	*
DT [22]	90	0.88
NB [22]	90	0.84
Deep CNN [18]	88.72	95.77
VGG-16 [20]	95	0.64
CNN [21]	94	*

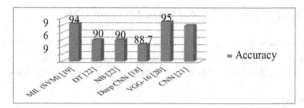

Fig. 34.6 Analysis of different methods: Accuracy Rate variables when all the occurrences are negative, whereas a container is constructive if various examples are positive [19]. In this, two variants of SVM methods, such as mi-SVM and MI-SVM. These methods are generalized as the powerful supervised SVM method.

Figure 34.6 represents ocular disease detection using existing methods such as DT, SVM, and NB. The SVM method reached 94.4% accuracy as compared to other methods. It represents ocular disease detection using DL methods such as deep CNN, VGG-16, and CNN. VGG-16 method reached an accuracy of 95% as compared to other methods.

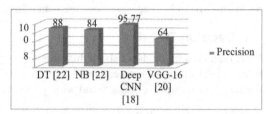

Fig. 34.7 Performance based on different methods: Precision

Figure 34.7 represents ocular disease detection using ML methods such as DT, SVM, and NB. The SVM method reached 0.88 of precision as compared to other methods. It represents ocular disease detection using DL methods such as deep CNN and VGG-16. VGG-16 method reached a precision of 98% as compared to another method.

6. CONCLUSION AND FUTURE SCOPE

This article provides an overview of ocular diseases, their categories, signs and symptoms, as well as Machine Learning (ML) and Deep Learning (DL) methods for ocular disease detection. It conducts an analysis of different classifiers such as VGG-16, SVM, NB, and DT, along with performance metrics including accuracy and precision rate. The discussion revolves around diverse classifiers employed in the diagnosis of various ocular diseases from retinal fundus images, using various databases like ODIR, Kaggle EyePacs, and more. The analysis demonstrates that the VGG-16 model achieves a remarkable accuracy of 95%, which is the highest among the compared methods. Additionally, various categories of ocular disease datasets are presented, fostering novel approaches for improved detection and classification. Future research should explore the practical application of DL and ML methods in clinical settings.

REFERENCES

1. K. Bitto., & I. Mahmud,. Multi categorical of common eye disease detect using convolutional neural network: a transfer learning approach. *Bulletin of Electrical Engineering and Informatics*, *11*(4), 2378-2387, (2022).
2. M. S. Khan., N. Tafshir., K. N. Alam,. Dhruba, A. R., Khan, M. M., Albraikan, A. A., & Almalki, F. A.. Deep learning for ocular disease recognition: an inner-class balance. *Computational Intelligence and Neuroscience*, (2022).
3. O. Adio,. A. Alikor., & E. Awoyesuku., Survey of pediatric ophthalmic diagnoses in a teaching hospital in Nigeria. *Nigerianjournal of medicine: journal of the National Association of Resident Doctors of Nigeria*, *20*(1), 105-108, (2011).
4. O. J. Perdomo Charry., & F. A. González. A systematic reviewof deeplearning methods applied to ocular images. *Ciencia e Ingenieria Neogranadina*, *30*(1), 9-26, (2020).
5. Akram & R. Debnath, An automated eye disease recognition system from visual content of facial imagesusing machine learning techniques. *Turkish Journal of Electrical Engineering and Computer Sciences*, *28*(2), 917-932, (2020).
6. K.. Prasad., P. S. Sajith, Neema, M., Madhu, L., & Priya, P. N. Multiple eye disease detection using Deep Neural Network. In *TENCON 2019-2019 IEEE Region 10 Conference (TENCON)* (pp. 2148-2153). IEEE, (2019).
7. L. Jain, Murthy, H. S., Patel, C., & D. Bansal.. Retinal eye disease detection using deep learning. In *2018 Fourteenth International Conference on Information Processing (ICINPRO)* (pp. 1-6). IEEE, (2018).
8. Y. Elloumi., M. Akil, & H. Boudegga, Ocular diseases diagnosis in fundus images using a deep learning: approaches, tools and performance evaluation. In *Real-Time Image Processing and Deep Learning 2019* (Vol. 10996, pp. 221-228). SPIE, (2019).
9. S. Neupane, J. Ables., W. Anderson, S. Mittal., S. Rahimi, I. Banicescu., & M. Seale. Explainable intrusion detection systems (x-ids): A survey of current methods, challenges, and opportunities. *IEEE Access*, *10*, 112392-112415, (2022).
10. Tymchenko, P. Marchenko, & D. Spodarets., Deep learning approach to diabetic retinopathy detection. *arXiv preprint arXiv:2003.02261*, (2020).

11. Tayal, J. Gupta, J., A. Solanki, K. Bisht, , A. Nayyar, & Masud, M. DL-CNN-based approach with image processing techniques for diagnosis of retinal diseases. *Multimedia Systems*, 1-22, (2021).

12. N. M. Dipu, S. A. Shohan, & K. M. A. Salam., Ocular disease detection using advanced neural network based classification algorithms. *Asian Journal For Convergence In Technology (AJCT) ISSN-2350- 1146*, 7(2), 91-99, . (2021).

13. H. Xu, H., & N. A. Rao. Grand challenges in ocular inflammatory diseases. *Frontiers in Ophthalmology*, 2, 756689, (2022).

14. Smitha, & P.Jidesh, Classification of multiple retinal disorders from enhanced fundus images using semi-supervised GAN. *SN Computer Science*, 3, 1-11, (2022).

15. Xie, L., Yang, S., Squirrell, D., & Vaghefi, E. (2020). Towards implementation of AI in New Zealand national diabetic screening program: Cloud-based, robust, and bespoke. *Plos one*, 15(4), e0225015.

16. M. U. Akram, A. A. Salam, S. G. Khawaja, S. G. H. Naqvi, & S. A Khan, RIDB: a dataset of fundus images for retina based person identification. *Data in Brief*, 33, 106433, (2020).

17. P. Gholami,., P. Roy, M. K. Parthasarathy, & V. Lakshminarayanan, V. OCTID: Optical coherence tomography image database. *Computers & Electrical Engineering*, 81, 106532, (2020).

18. Z. Gao, J. Li., Y. Chen, Yi, Z., & J. Zhong, J. Diagnosis of diabetic retinopathy using deep neural networks. *IEEE Access*, 7, 3360-3370, (2018).

19. W. Sun, X. Liu, , & Z. Yang, Z. Automated detectionof age-relatedmacular degenerationin OCT images using multiple instance learning. In *Ninth International Conference on Digital Image Processing (ICDIP 2017)* (Vol. 10420, pp. 822-826). SPIE.

20. V. Lakshmi., L. Monisha., L., Vinay., M. S. Mounesh., Ocular Disease Recognition and Detection using VGG Algorithm. International Journal of Advanced Research in Science, Communication and Technology (IJARSCT) ISSN- 2581-9429, (2022).

21. R. Devanaboina., S. Badri, R. M. Depa., S. Bhutada., Ocular Eye Disease Prediction Using Machine Learning. IJCRT, ISSN: 2320-2882, (2021).

22. P. R. Aruna Ramanan., K. R. Gowtham S. Balaji., & P. Nilesh., Eye Disease Prediction Using Machine Learning Techniques. *Networks*, 24, 255-262, (2020)

Computational Intelligence and Mathematical Applications – Dr. Devendra Prasad et al. (eds)
© 2024 Taylor & Francis Group, London, ISBN 978-1-032-87721-1

35 | Comparative Analysis of Classical and Quantum Machine Learning for Breast Cancer Classification

Manaswini De*, Arunima Jaiswal, Tanya Dixit

Department of Computer Science and Engineering, Indira Gandhi Delhi Technical University for Women, Kashmere Gate, Delhi

Abstract: Breast cancer, the unruly development of abnormal cells in breasts, is a pressing public health issue and needs early detection for improving survival rates. Machine learning algorithms have shown promising results in breast cancer classification and detection. This study evaluates the performances of traditional and quantum machine learning techniques for data classification of breast cancer, emphasizing their effectiveness in discerning patterns using reduced set of features. An approach is proposed to reduce 30-dimensional features to 2-dimensional features using widely applied Principal Component Analysis (PCA), Incremental PCA, Truncated Singular Value Decomposition and Kernel PCA techniques. This study aims to provide an insight on the impact of reduction of features on the classification of breast cancer using classical and quantum machine learning techniques.

Keywords: Machine learning, Breast cancer classification, Feature reduction, Variational quantum classifier, Quantum support vector classifier, KNN, Decision tree, Logistic regression model, Classical support vector classifier

1. INTRODUCTION

Breast cancer (BC) is the uncontrolled development of abnormal and atypical cells in breasts. The World Health Organization has reported that BC is the type of cancer that occurs most frequently among women. The biennial report 2020-2021 by International Agency for Research on Cancer pointed out the gravity of female breast cancer in recent years (IARC 2021). Breast cancer is one of the primary concerns in India also and the present scenario and challenges ahead have been highlighted in literature (Mehrotra & Yadav 2022). BC may develop in lobules or ducts present in the breast as well as fatty tissue known as the adipose tissue. There are mainly two types of BC: 'malignant' and 'benign'. Malignant tumours are cancerous in nature and can spread to other tissues and may also travel to lymph nodes in body. Benign tumours are non-cancerous and do not necessarily require treatment. Although there is no definitive method for diagnosing breast cancer, early detection serves as the initial stage in treatment, risk evaluation and stop further growth.

In recent years, machine learning as well as deep learning models are being studied and utilized for early detection of breast cancer. Machine learning is a technique for analysing data that instructs a computer on predicting outcomes using various algorithms. Prevalent algorithms in machine learning for classification including K-Nearest Neighbours, Decision Tree, Support Vector Machine, Logistic Regression etc. can be used for classifying breast cancer into malignant and benign cases. Recent studies have shown very promising results to classify breast cancer cases using machine learning algorithms. These studies indicate that the success rates vary from ~93% to 99% using different machine learning algorithms in classical classifiers (Hazra et al. 2020, Tahmooresi et al. 2018, Allugunti 2022, Roslidar et al. 2020, Khandezamin et al. 2020, Strelcenia & Prakoonwit 2023, Mashudi et al. 2021).

*Corresponding author: manaswini080btcse20@igdtuw.ac.in

DOI: 10.1201/9781003534112-35

Quantum computing has risen as a promising approach in the field of data analysis and classification. The utilization of quantum computing techniques holds promise for enhancing the processing and analysis of data. Its computational power and capability to process large data have attracted significant attention in the recent years. Quantum machine learning algorithms, when applied to breast cancer data, hold the promise of enhancing accuracy, reducing computational resources, and expediting the diagnosis and prognosis of breast cancer patients (Musaddiq et al. 2022, Adebiyi et al. 2023, Mishra et al. 2020). Maheshwari et al. (2022) reviewed the applications of quantum machine learning algorithms in the health care domain and commented on limitations and future prospects. Currently, classical machine learning holds a slight edge over quantum machine learning in terms of accuracy and computational time. However, it is anticipated that quantum computing will prove more efficient for handling intricate data in the near future.

Machine learning algorithms utilize the power of artificial intelligence to analyse a given problem in an efficient way. In recent time, Support Vector Machine (SVM), a supervised machine learning model, is widely utilized in classification as well as regression problems (Brereton & Lloyd 2010, Fadime & Erkan 2022). SVM's main objective is to obtain an optimal hyper plane for classifying a given dataset into classes.

KNN is a supervised learning technique employed in clustering and regression. It assigns a label to a data point by considering the predominant class among its k-nearest neighbours. Logistic Regression is employed in problems involving binary and multiclass classification. It assumes a decision boundary and works well when the classes are separable by a hyper plane. Decision tree classifier is a fast and versatile technique for capturing non-linear relationships and interactions between features. They recursively split the data based on features to create a tree structure and can be used for classification (Muhammet 2020).

Quantum computing utilizes the concepts of quantum mechanics like quantum bits (qubits), entanglement and superposition for enhancing the data processing capability. Quantum Machine Learning (QML) involves applying the principles of quantum computing to improve machine learning algorithms. Some well-known QML algorithms include Variational Quantum Classifier (VQC) and Quantum Support Vector Classifier (QSVC). VQC is a QML algorithm that utilizes quantum circuits to approximate classical functions. VQCs work by employing a quantum circuit to learn a function that maps input data to desired outputs. A classical optimization algorithm is utilized to train the quantum circuit by minimizing a loss function, which quantifies the disparity between the predicted and actual outputs. VQCs have shown promising results in a variety of applications (Mashudi et al. 2021, Santos & Exposito 2023, Maheshwari et al. 2020).

The computational efficiency of a model can greatly be enhanced by eliminating the redundant features in any dataset (Park et al. 2020, Zhu et al. 2005). Dimensionality reduction is a fundamental method in data analysis and machine learning aiming to reduce the number of features (or dimensions) in a dataset while preserving as much valuable information as possible. The primary goal of dimensionality reduction is to simplify complex datasets, making them more manageable, interpretable, and computationally efficient (Mert et al. 2015). Various dimensionality reduction techniques such as Principal Component Analysis (PCA), Incremental PCA (IPCA), Kernal PCA, Truncated Singular Value Decomposition (TSVD), etc. are widely used techniques for reduction of dimensionality in a data set for classification (Ibrahim et al. 2021, Petinrin et al. 2023). All these techniques assume linear relationships between the features in a data set and unable to handle complex non-linear data patterns.

Data pre-processing to extract useful information from original raw observations plays an important role in machine learning models. In general, with the increase in the number of input features in quantum classifier, the computational time increases considerably. The main objective of this study is to obtain a reduced number of features (i.e., 2 feature subsets) for a quantum classifier which could provide comparable results with that obtained from a classical classifier. Therefore, in this study, an approach is proposed to reduce 30-dimensional (30D) features to a 2-dimensional (2D) features for efficient handling of data in classification using quantum classifiers. Furthermore, the impact of reduction of features on the success rates of classical as well as quantum classifier has been investigated in this study.

In the Qiskit documentation (Qiskit 0.44 Documentation), a comparison example is given to explain the process of training and testing the classical and quantum model using Iris data set. Similar methodology has been adopted in this study for classification of Wisconsin Breast Cancer data set using Variational Quantum Classifier (VQC) as well as Quantum Support Vector Classifier (QSVC). Different classifiers e.g., Decision Tree (DTC), Logistic Regression (LR), and K-Nearest Neighbour (KNN) have been employed as classical classifiers for comparative study.

The Section 2 of this paper provides a brief on some of the related studies being carried out in this field. Section 3 gives information on the data set employed in this study and the necessary pre-processing required before utilizing the processed data for classification. In Section 4, an approach for reduction of features and architectures of classical and quantum classifiers are presented. The comparison of the results obtained from this study with that from the published literature is considered in Section 5. Eventually, the conclusions and inferences drawn from this study are summarized in Section 6.

2. RELATED WORK

In the previous few years, a diverse range of studies and investigations have been performed on the usage of machine learning for BC detection and classification. Hazra et al. (2020) have employed Artificial Neural Networks (ANN) along with Decision Tree Classifier (DTC) to classify breast cancer tumours as benign and malignant. The success rates of ANN and DTC were 98.55% and 96.23% respectively. Performances of both models were compared based on several metrics such as sensitivity, accuracy, precision, specificity and recall. Tahmooresi et al. (2018) reviewed and examined the role of machine learning algorithms in early detection of BC. The performances of four ML models i.e., ANN, DTC, KNN, and SVM have been investigated in this research.

Allugunti (2022) discussed the computer-aided diagnosis method, which can classify patients into three classes (cancerous, non-cancerous, and normal) based on thermo-graphic images of the breast. The performances of Convolutional Neural Network (CNN) and SVM have been compared. A data set of 200 thermo-graphic images of breast from the Breast Cancer Digital Repository (BCDR) has been used in the study. The paper reports that CNN classifier achieved a stellar accuracy of 97.5%, thereafter by the SVM classifier with 95% and RF classifier with 92.5%. Roslidar et al. (2020) reviewed the potential of thermography for early detection of BC and discussed use of deep learning (DL) models, especially CNN for classifying breast thermo grams into normal and abnormal categories.

In the past few years, quantum-inspired machine learning techniques have gained significant interest in resolving complex problems, classifications, and real-time decision making processes (Adebiyi et al. 2023, Zeguendry et al. 2023, Garcia et al. 2022). Lloyd et al. (2013) demonstrated methods to enhance the speed of quantum machine learning over classical computers for high-dimensional input vectors. Innan et al. (2023) proposed an approach for binary classification of data using quantum SVMs to enhance the accuracy.

Musaddiq et al. (2022) demonstrated cancer cell detection using the Quantum Neural Networks (QNN). The quantum neural networks are employed to classify cancerous cells into malignant and benign. The findings suggest that the QNNs can outperform CNNs in terms of accuracy within a shorter computational timeframe. QNN exhibits superior performance compared to CNN in successfully training and generating a valid model, particularly when dealing with smaller datasets. The accuracy achieved by QNN is 97.5%, whereas CNN achieves an accuracy of 92.5%.

Premanand et al. (2023) employed Support Vector Classifiers (SVC) together with Quantum Support Vector Classifiers (QSVC) using the Qiskit framework. The findings of that study reveal that the SVC attained an accuracy of 97.3% within 0.008 seconds, whereas the QSVC achieved 93.8% accuracy but required significantly more time, specifically 893.55 seconds. Additionally, the VQC method yielded an accuracy of 81%. Havenstein et al. (2018) evaluated the performances of the Quantum Variational Support vector Machines (QSVMs) and the classical Support Vector Machines (SVMs) on three multi-class classification problems, including the Wisconsin Breast Cancer dataset. They used different quantum feature maps, such as ZZFeatureMap and PauliFeatureMap, and different classical kernels, such as linear and radial basis function kernels.

Santos & Exposito (2023) investigated the effect of two sets of features on the performances of classical SVC and VQC classifiers. Two sets of most relevant features namely Set 1: perimeter_worst, concave point_worst, concave points_mean, and concave points_se, and Set 2: perimeter_worst and concave points_worst are used in classical SVC and VQC for categorization of Wisconsin BC dataset. Within this study, SVC and VQC models were experimented with these 4 features and 2 features sets and the train/test scores were used as comparison metrics. Fadime & Erkan (2022) investigated the impact of optimization of hyper-parameters such as feature dimensions and Feature Maps (ZFeatureMap, ZZFeatureMap, and PauliFeatureMap) on the performances of Linear SVM, Non-linear SVM, and QSVM using the Wisconsin breast cancer data.

Mert et al. (2015) investigated the effects of dimensionality reduction on various classifiers like K-Nearest Neighbours (KNN), Artificial Neural Network (ANN), Radial Basis Function Neural Network (RBFNN), and Support Vector Machine (SVM) using independent component analysis applied to Wisconsin dataset. The effectiveness of quantum convolutional neural networks for detection of BC has been demonstrated by Mishra et al. (2020).

3. DATA

The Wisconsin Breast Cancer (Diagnostic) data, a benchmark data set, publicly accessible from the Kaggle website as well as Scikit-learn module in Python (available under liberal licence) has been used in this study (Kaggle website, Scikit-learn website). The data contains 569 observations with a distribution of 357 benign plus 212 malignant cases. There are 30 distinct characteristic features that are quantified from the digitized image of the breast masses procured through Fine Needle Aspiration (FNA). These features are the numerical measures of ten attributes of the cell nuclei, including means, standard errors and worst values of radius, texture, perimeter, area, smoothness, compactness, concavity, concave points, symmetry, and fractal dimension. The target variable can be either benign (non-cancerous), denoted by value 0 or malignant (cancerous), denoted by the value 1.

3.1 Pre-Processing

The pre-processing of input data could play an important role for classification problems using quantum machine learning (Mancilla & Pere 2022). The pre-processing steps for input data include data cleaning, normalization, feature selection together with splitting the data into training and testing segments. In Wisconsin breast cancer data, two columns namely 'Unnamed: 32' and 'id' are irrelevant for the present study and hence removed before further processing. In this data set, there is no null or missing value. The objective of data normalization is to standardize or transform data into a consistent range or distribution, influencing the encoding and computation of quantum models. Here, the complete data set is subjected to dimension reduction techniques to reduce the original 30 dimensional features to 2 dimensional features. The reduced features are subsequently separated into a training set and a testing set. The training set is utilized for training classifier models, whereas the test set is used to evaluate the performance of these models.

4. METHODOLOGY

4.1 Proposed Feature Selection Approach

The pre-processed cleaned data set with 30 features (i.e., 30-dimensions) was subjected to normalization using standard techniques available in Scikit-learn library (Scikit-learn website) [34]. The normalized complete data set with 30 features are subjected to standard linear dimensionality reduction techniques i.e., Principal Component Analysis (PCA), Incremental PCA (IPCA), and Truncated Singular Value Decomposition (TSVD) techniques for reduction of dimension of dataset from 30-dimension to 6-dimension. It is observed that the first six principal components obtained from PCA, TSVD and IPCA could explain ~88% of total variances of the data set. However, the dominance of original individual features in each of these 6 principal components varies with the technique employed for dimensionality reduction. Therefore, a subset of 14 dominant features (which are common in these 6 components) contributing most to the calculated 6 principal components is identified. Further, these 6-dimensional features are again reduced to 2-dimensional features using Kernel-PCA (KPCA). The variances explained by the 2 principal components obtained from KPCA are [0.61, 0.39]. Kernel PCA has the ability to capture underlying non-linear patterns in a dataset, a capability lacking in conventional linear dimensionality reduction techniques like PCA, TSVD and IPCA. Further, linear dimensionality reduction techniques are in general sensitive to data outliers. Hence, non-linear dimensionality reduction technique KPCA is used for reduction of dimensionality to 2 dimensions. Within this study, the performances of traditional and quantum classifiers are investigated using the reduced 2-dimensional features obtained from KPCA. These results are compared with the published results given in Santos & Exposito (2023) and Premanand et al. (2023).

The first 2 principal components out of 6 components obtained from PCA, TSVD and IPCA could explain only 70%, 64%, and 70% respectively of the total variances in the data set. Here, it is worth to mention that initially experiments were conducted using only the first 2 principal components; however, the performances of quantum classifiers were not encouraging. Hence, the 14 dominant features common in the 6 principal components obtained from PCA, TSVD, and IPCA are further subjected to KPCA technique for reduction of dimensionality. KPCA with RBF transformation function has been used in this study.

4.2 Classification using Classical Machine Learning

At first, all the original features available in Wisconsin Breast Cancer dataset are normalized and used as input features to the classical models. The types of cases (benign and malignant) are used as labelled data for fitting the models. The complete data set is segregated as 80% training and 20% testing data. Once the training data is fitted into the mode, the identical model is assessed using the test data to compute accuracies, sensitivity and F1-score. Next, the reduced 2-dimensional features obtained from the proposed feature reduction approach are used to train and test the classifiers.

A wide range of in-built functions and algorithms are available in Scikit-learn library for implementing and evaluating various machine learning classifiers. All the classifiers required labelled training data for decision-making. Training a Decision Tree Classifiers (DTC) can be fast. Decision tree additionally requires decision criteria (e.g. Gini impurity) and stopping criteria. It operates by recursively splitting the dataset according to features, constructing a tree structure where leaf nodes represent the predicted outcome for new data.

Logistic Regression (LR) operates by creating a model for the probability of the target variable being part of a specific class, utilizing a logistic function. It estimates coefficients through optimization, making it suitable for binary or multiclass classification. LR is suitable when the correlation between the dependent and independent variables is approximately linear (Muhammet 2020). K-Nearest Neighbours (KNN) requires specifying the number of neighbours (k) and categorizes new data points by determining the predominant class among their K-nearest neighbours. Decision boundaries are defined by the closeness of all data points present in the feature space. The selection of k is crucial, as opting for a value too small renders the model sensitive to noise, while on the other hand, choosing a large value can result in smoothing of decision boundaries. These are the classical models that have been used in the present study for classification of breast cancer.

4.3 Classification using Quantum Machine Learning Models

Quantum classifier is an advanced form of classical SVC that utilizes the advantages of quantum states for complex machine learning tasks using the supervised learning approach. Conventional data needs to be converted to a format suitable for quantum processing. Quantum embedding translates classical data points to quantum data points using the given quantum circuit. The challenge lies in designing efficient quantum circuits that can perform this translation without requiring an excessive number of qubits or operations. More number of qubits makes the quantum processing comparatively slower. Classical data can be prepared for quantum models using different quantum feature maps.

Qiskit library has in-built functions like ZZFeatureMap and PauliFeatureMap for encoding classical data into quantum states (Qiskit website, Scikit-learn website). The ZZFeatureMap uses second-order Pauli-Z evolution circuits to transform the data by applying a Hadamard gate to each qubit, followed by a U1 gate with a nonlinear function of the data as the parameter. Then, it applies a series of controlled-Z gates between pairs of qubits, followed by another layer of U1 gates with a nonlinear function of the data and the entanglement as the parameter. The in-built ansatz namely RealAmplitude and EfficientSU2 are available in Qiskit module. The efficiencies of different combinations of configuration for encoding data and parameterization of quantum circuit (ansatz) are analyzed and finally ZZFeatureMap and EfficientSU2 are used with quantum classifier VQC in the present study.

A variational quantum circuit is formed by utilizing quantum gates and qubits. This circuit includes parameterized gates, which can be adjusted during training. An optimization algorithm is then employed for adjusting the parameters of the quantum circuit in a way that minimizes or maximizes the objective function, depending on the problem. However, the challenge lies in finding the optimal variational parameters and the trade-off between accuracy and computational efficiency. Various optimization algorithms have been developed for training of VQCs such as gradient-based, gradient-free, hybrid algorithm, etc. There is no one-size-fits-all solution, and different algorithms may suit different scenarios and objectives. In this study, COBYLA (Constrained Optimization By Linear Approximation), a gradient-free optimization technique, is used as the optimizer in VQC.

QSVC is a kernel-based approach that employs the quantum kernel trick to transform classical data into a high-dimensional quantum feature space (Qiskit website). In such a feature space, the QSVC algorithm can then learn a linear decision boundary that segregates the data into different classes. The QSVC model is implemented with fidelity quantum kernel for quantum data transformation and kernel creation respectively. Fidelity quantum kernel focuses on capturing the fidelity, or the similarity, between the different quantum states. In the present study, ZZFeatureMap encoding has been used in QSVC model.

5. RESULTS

In this study, four comparison metrics namely accuracy, sensitivity, F1-score, and execution time are used for comparison of results between classical models and quantum ML models. Accuracy is the ratio of correctly classified instances to the aggregate instances and can be expressed as (Khandezamin et al. 2020):

$$\text{Accuracy} = \frac{\text{TN} + \text{TP}}{\text{TP} + \text{TN} + \text{FP} + \text{FN}} \times 100$$

Here TP (True Positive) denotes count of accurately classified positive occurrences; TN (True Negative) represents count of accurately classified negative occurrences; FP (False Positive) signifies count of negative occurrences falsely classified as positive; and FN (False Negative) indicates the count of positive occurrences incorrectly classified as negative. It measures how well the classifier can predict the correct class labels for new data. Higher accuracy means better performance. Sensitivity and F1-score are two measures of performance of a classifier on a given dataset, especially when the data is imbalanced or the cost of false negatives is high. Sensitivity, also called recall, is the proportion of true positives (TP) relative to the aggregate count of positive occurrences and is expressed as (Khandezamin et al. 2020):

$$\text{Sensitivity} = \frac{\text{TP}}{\text{TP} + \text{FN}} \times 100$$

It measures how well the classifier can identify a positive class of instances. F1-score represents the harmonic mean of precision and recall, which are metrics that evaluate how well the classifier can classify data into positive and negative classes and is expressed as (Khandezamin et al. 2020):

$$\text{F1 score} = \frac{2 * \text{precision} * \text{recall}}{\text{precision} + \text{recall}}, \text{ where Precision} = \frac{\text{TP}}{\text{FP} + \text{TP}} \times 100$$

Execution time is the computational time to train and test the classifier on a given dataset. It measures the efficiency of the classifier in terms of computational resources. Lower execution time means better performance of the model (Havenstein 2018, Ozpolat & Karabatak 2023). The results obtained from classical and quantum classifiers have been depicted in the following tables:

Table 35.1   Classical machine learning

Model	Features*	Accuracy		Sensitivity		F1-Score		Comp Time (s)
		Train	Test	Train	Test	Train	Test	
SVC	All	98.4	97.7	96.3	95.8	97.9	96.9	0.01
	F1	93.8	92.6	90.9	91.2	91.6	89.8	0
DTC	All	98.2	93.0	96.4	91.5	97.5	91.5	111
	F1	94.7	90.4	92.1	87.2	92.7	88.2	45
LG	All	98.9	96.5	97.6	95.7	98.5	95.7	0.06
	F1	93.1	93.0	90.3	93.6	90.6	91.7	0
KNN	All	97.8	95.6	95.2	89.4	96.9	94.4	0.04
	F1	94.5	92.1	92.7	93.6	92.4	90.7	0

*Features used as input data: All – Total (30) Nos of Features; F1 –Reduced 2D features.

Table 35.2   Quantum machine learning

Model	Accuracy		Sensitivity		F1-Score		Comp Time (s)
	Train	Test	Train	Test	Train	Test	
VQC	93.6	92.6	90.2	87.5	91.5	89.2	388
QSVC	94.0	92.1	94.0	92.1	91.7	89.1	2796

The train score and test score were used as performance indicator by Santos & Exposito (2023). The study indicates that the maximum train/test-score of VQC model with RealAmplitude as ansatz was 88% when specific 2-features were used as input; whereas, 90% was achieved when specific 4-features were used with EfficientSU2 as ansatz. Further, the SVC model gives maximum train-test-score as 95% (Santos & Exposito 2023). It is also noticed from their studies that a large difference in results were not obtained when RealAmplitude or EfficientSU2 functions are used in the model. Compared to this study, the present study with reduced sets of 2D features give a better performance in case of two qubit VQC classifier with EfficientSU2 ansatz.

In this study, it is observed that the results obtained from classical models degrade slightly when reduced sets of features are used as compared to the results obtained with all features. The accuracies achieved from classical models vary from 93% to

97.7% with the test data (when all features are used), whereas, the accuracies vary from 90.4% to 93% with the reduced 2-D feature subsets. It is also observed that the computational time for executing classical classifier is negligible as compared to quantum classifiers.

Further, the performances of quantum models with reduced sets of features are comparable to that obtained from classical models. This indicates that quantum models have the ability to extract the relevant and hidden patterns from a reduced set of data for classification purpose. The confusion matrix for train and test data set of one of the typical experimental cases obtained with VQC classifier is shown in Fig. 35.1.

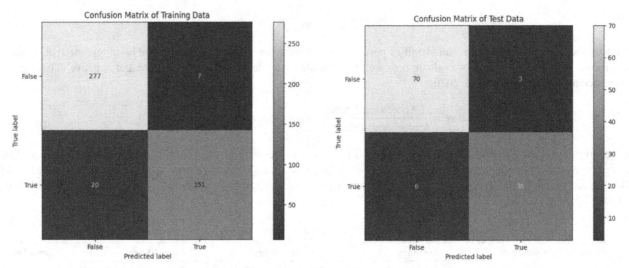

Fig. 35.1 Confusion matrix for train and test data set of one of the typical experimental cases

Santos & Exposito (2023) in their study demonstrated that the results with classical approach using specific 4 features and 2 features are comparable. In the present study the classical models show comparatively better results with all features than that obtained using reduced features. However, the results obtained with quantum classifiers with reduced sets of features are better than that presented in Santos & Exposito (2023) and Premanand et al. (2023).

6. CONCLUSIONS

This study proposes an approach for feature reduction and investigates the performances of both classical and quantum classifier models using the metrics namely accuracy, sensitivity, F1-score and computational time. The results obtained from quantum classifiers using reduced set of feature subset are promising. In the current study, when comparing quantum and classical models and finding similar outcomes, it is evident that as quantum hardware advances, quantum classifiers will inevitably outperform classical classifiers in terms of both computational speed and accuracy. In future, the efficiency of the proposed feature reduction technique will be further experimented using other types of data sets.

In the present study, one qubit per feature is used in quantum classifier models. With the increase in number of input features, the requirement of more qubits in the quantum models and hence the computational time increases considerably. Therefore, the entire dataset was not experimented with the quantum classifier models. With the increase in capability and efficiency of machine learning algorithms in terms of processing power using more number of qubits, quantum classifiers could be used with large input data set in future. Overall, this study indicates the potential and capability of QML techniques to classify the benign along with malignant cases in breast cancer data with fair level of accuracies using reduced 2-dimensional features subset.

REFERENCES

1. IARC. 2021. IARC Biennial Report 2020-2021. IARC: Lyon, France.
2. Mehrotra, R. & Yadav, K. 2022. Breast cancer in India: Present scenario and the challenges ahead. *World Journal of Clinical Oncology* 13(3): 209–218. https://www.ncbi.nlm.nih.gov/pmc/articles /PMC8966510/.

3. Hazra, R., Banerjee, M. & Badia, L. 2020. Machine learning for breast cancer classification with ANN and decision tree. *11th IEEE Annual Information Technology, Electronics and Mobile Communication Conference (IEMCON), Vancouver, BC, Canada, 2020*: 0522-0527. doi: 10.1109/IEMCON51383. 2020.9284936.

4. Tahmooresi, M., Afshar, A., Bashari Rad, B., Nowshath, K.B., & Bamiah, M.A. 2018. Early detection of breast cancer using machine learning techniques. *Journal of Telecommunication Electronic and Computer Engineering* 10: 21-27. www.researchgate.net/publication/327974742_Early_Detection _of_Breast_Cancer_Using_ Machine_Learning_Techniques.

5. Allugunti, V.R. 2022. Breast cancer detection based on thermographic images using machine learning and deep learning algorithms. *International Journal of Engineering in Computer Science* 4: 49–56. https://doi.org/10.33545/26633582.2022.v4.i1a.68.

6. Roslidar, R., Rahman, A., Muharar, R., Syahputra, M.R., Arnia, F., Syukri, M., Pradhan, B. & Munadi, K. 2020. A review of recent progress in thermal imaging and deep learning approaches for breast cancer detection. *IEEE Access* 8: 116176-116194. doi: 10.1109/ACCESS.2020.3004056.

7. Khandezamin, Z., Naderan, M. & Rashti, M.J. 2020. Detection and classificationof breast cancer using logistic regression feature selection and gmdh classifier. *Journal of Biomedical Informatics* 111: 103591. https://doi.org/10.1016/j.jbi.2020.103591.

8. Strelcenia, E. & Prakoonwit, S. 2023. Effective feature engineering and classification of breast cancer diagnosis: a comparative study. *BioMedInformatics* 3: 616-631. https://doi.org/10.3390/ biomedinformatics3030042.

9. Mashudi, N. A., Rossli, S.A., Ahmad, N. & Noor, N.M. 2021. Breast cancer classification: feature investigation using machine learning approaches. *International Journal of Integrated Engineering* 13: 107-118.

10. Musaddiq, A.A., Amjad, S. & Muazaz, A.A. 2022. Investigation of early-stage breast cancer detection using quantum neural network. https://arxiv.org/ftp/arxiv/papers/ 2210/2210.03882.pdf

11. Adebiyi, M.O., Fatinikun-Olaniyan, D., Osang, F. & Adebiyi, A.A. 2023. Quantum theory approach to performance enhancement in machine learning. *International Conference on Science, Engineering and Business for Sustainable Development Goals (SEB-SDG), Omu-Aran, Nigeria, 2023*: 1-7. doi: 10.1109/SEB-SDG57117.2023.10124582.

12. Mishra, N., Bisarya, A., Kumar, S., Maouaki, W.E., Mukhopadhyay, S., Behra, B.K., Panigrahi, O.K. & De, D. 2020. Breast cancer detection using quantum convolution neural networks: a demonstration on a quantum computer. medRxiv preprint. doi: https://doi.org/10.1101/2020.06.21.20136655.

13. Maheshwari, D., Garcia-Zapirain, B. & Sierra-Sosa, D. 2022. Quantum machine learning applications in the biomedical domain: a systematic review. *IEEE Access* 10. doi: 10.1109/ACCESS.2022.3195044.

14. Brereton, R.G. & Lloyd, G.R. 2010. Support vector machines for classification and regression. *The Analyst* 135: 230-267. doi 10.1039/b918972f.

15. Fadime, D. & Erkan, T. 2022. Hyper-parameter tuning for quantum support vector machine. *Advances in Electrical and Computer Engineering; Suceava* 22: 47-54. doi 10.4316/AECE. 2022.04006.

16. Muhammet Faith Ak 2020. A comparative analysis of breast cancer detection and diagnosis using data visualization and machine learning application. *Healthcare* 8(2): 111. https://doi.org/10.3390/ healthcare8020111.

17. Santos, S.D. & Exposito, D.E. 2023. Classical vs. quantum machine learning for breast cancer detection. *19th International Conference on the Design of Reliable Communication Networks (DRCN), Vilanova i la Geltru, Spain, 2023*: 1-5. doi: 10.1109/DRCN57075.2023.10108230.

18. Maheshwari, D., Garcia-Zapirain, B. & Sierra-Sosa, D. 2020. Machine learning applied to diabetes dataset using quantum versus classical computation. *IEEE International Symposium on Signal Processing and Information Technology (ISSPIT), Louisville, KY, USA, 2020*: 1-6. doi: 10.1109/ISSPIT51521.2020.9408944.

19. Park, J.E., Quanz, B., Wood, S., Higgins, H. & Ray, H. 2020. Practical application improvement to quantum svm: theory to practice. arXiv:2012.07725 [quant-ph]. https://doi.org/10.48550 /arXiv.2012.07725.

20. Zhu, M., Zhu, J. & Chen, W. 2005. Effect analysis of dimension reduction on support vector machines. *International Conference on Natural Language Processing and Knowledge Engineering, Wuhan, China, 2005*: 592-596. doi: 10.1109/NLPKE.2005. 1598806.

21. Mert, A., Kilic, N., Bilgili, E. & Akan, A. 2015. Breast cancer detection with reduced feature set. *Computational and Mathematical Methods in Medicine*: Article ID 265138. doi 10.1155/2015/265138.

22. Ibrahim, S., Nazir, S. & Velastin, S.A. 2021. Feature selection using correlation analysis and principal component analysis for accurate breast cancer diagnosis. *Journal of Imaging* 7(11): 225. https://doi.org/10.3390/ jimaging7110225.

23. Petinrin, O.O., Saeed, F., Salim, N., Toseef, M., Liu, Z. & Muyide, I.O. 2023. Dimension reduction and classifier-based feature selection for oversampled gene expression data and cancer classification. *Processes* 11: 1940. https://doi.org/10.3390/ pr11071940.

24. Qiskit 0.44 Documentation: https://qiskit.org/documentation/ (Accessed on 10 Sept 2023)

25. Zeguendry, A., Zahi, J. & Mohamed, Q. 2023. Quantum machine learning: a review and case studies. *Entropy* 25(2): 287. https://doi.org/10.3390/e25020287

26. García, D.P., Benito, J.C. & García-Peñalvo, F.J. 2022. Systematic literature review: quantum machine learning and its applications. arXiv:2201.04093 [quant-ph]. https://doi.org/10.48550/arXiv.2201. 04093.

27. Lloyd, S., Mohseni, M. & Rebentrost, P. 2013. Quantum algorithms for supervised and unsupervised machine learning. arXiv:1307.0411V2 [quant-ph]. https://arxiv.org/pdf/1307.0411.pdf.

28. Innan, N., Al-Zafar Khan, M., Panda, B. & Bennai, M. 2023. Enhancing quantum support vector machines through variational kernel training. arXiv:2305.06063 [quant-ph]. https://doi.org/ 10.48550/arXiv.2305.06063.

29. Premanand, V., Snavya Sai, M. B., Srinivas, S. & Reddy, S. 2023. Quantum machine learning for breast cancer detection: a comparative study with conventional machine learning methods. *Indian Journal of Natural Science* 14(78): 57728-57736.

30. Havenstein, C., Thomas, D., & Chandrasekaran, S. 2018. Comparisons of performance between quantum and classical machine learning. *SMU Data Science Review* 1(4): Article No. 11.

31. Kaggle website: https://www.kaggle.com/code/buddhiniw/breast-cancer-prediction.

32. Scikit-learn website: http://scikit-learn.org/stable/modules /generated/sklearn.datasets.

33. Mancilla, J. & Pere, C. 2022. A preprocessing perspective for quantum machine learning classification advantage in finance using NISQ algorithms. arXiv:2208.13251 [quant-ph]. https://doi.org/ 10.48550/arXiv.2208.13251.

34. Ozpolat, Z. & Karabatak, M. 2023. Performance evaluation of quantum-based machine learning algorithms for cardiac arrhythmia classification. *Diagnostics (Basel)* 13(6): 1099. doi: 10.3390/diagnostics13061099. PMID: 36980406; PMCID: PMC10047100.

Note: All the figures and tables in this chapter were made by the author.

Computational Intelligence and Mathematical Applications – Dr. Devendra Prasad et al. (eds)
© 2024 Taylor & Francis Group, London, ISBN 978-1-032-87721-1

36

Comparative Analysis of Machine Learning and Deep Learning Algorithms for the Detection of Hate Speech

Arpana Jha[1], Arunima Jaiswal[2], Anshika Singh[3], Sampurnna Swain[4], Eshika Aggarwal[5]
Computer Science & Engineering Department, Indira Gandhi Delhi Technical University for Women, Delhi, India

Abstract: Hate speech on social media platforms poses significant challenges, necessitating advanced approaches for effective detection and mitigation. The study focuses on a dataset comprising approximately 25,000 tweets classified into 3 classes: hate speech, offensive, and neither, collected from Twitter, reflecting a diverse range of linguistic expressions and sentiments. This research conducts a comparative analysis of machine learning, ensemble learning, and deep learning algorithms such as KNN, Logistic Regression, Random Forest, Support Vector Classifier, Naive Bayes, Decision Tree, Neural Network LSTM, CatBoost, Stacking, AdaBoost, XGboost, Pasting, and Bagging, in discerning hate speech from non-hateful content within the ETHOS dataset. The findings of this research provide valuable insights for the development of effective and reliable hate speech detection systems, aiding in the creation of safer online spaces and promoting healthier online discourse.

Keywords: Hate speech, ETHOS dataset, Machine learning, Deep learning, Ensemble learning, Count-vectorization

1. INTRODUCTION

Hate speech has become a pervasive issue in online platforms, with social media being a significant breeding ground for the propagation of harmful and offensive language. As online communities continue to grow, the need for effective hate speech detection mechanisms becomes crucial to maintaining a safe and inclusive digital environment [1]. A significant gap exists in the current research landscape- an absence of a comprehensive comparative study that systematically evaluates the strengths and weaknesses of these ML and DL models [2] This study addresses the pervasive issue of online hate speech by conducting a thorough comparison of machine learning algorithms for effective detection. The evaluation focuses on algorithms including KNN, Decision Trees, Random Forest, SVM, Logistic Regression, Naïve Bayes, Catboost, Stacking, XGBoost, AdaBoost, Pasting, Bagging, Neural Networks, and LSTM, with performance measured using accuracy, precision, recall, and F1-score.[3]

2. RELATED WORK

The identification of trolls and hate speech has garnered substantial attention from researchers. Further efforts include a semi-supervised approach and statistical topic modeling for offensive content detection on Twitter [4] and a supervised machine learning text classifier focusing on race, ethnicity, and religion [5]. Neural language models have also been leveraged for learning distributed low-dimensional representations of comments [6]. Ensemble methods garnered attention for their ability to improve overall model performance. Ensemble methods, such as combining typed dependencies and bag-of-words features, increase robustness and accuracy in distinguishing hate speech utterances.

[1]arpana018btcse20@igdtuw.ac.in, [2]arunimajaiswal@igdtuw.ac.in, [3]anshika022btcse20@igdtuw.ac.in, [4]sampurnna029btcse20@igdtuw.ac.in, [5]eshika030btcse20@igdtuw.ac.in

DOI: 10.1201/9781003534112-36

3. METHODOLOGY

The proposed methodology for the detection of hate speech involves choosing a suitable and authentic dataset, performing preprocessing of data, extracting suitable features, and applying various ML, DL, and Ensemble classification models, and evaluating their performances as shown in Fig. 36.1. The methodology for hate speech detection involves using the ETHOS dataset, derived from YouTube and Reddit comments, and a dataset of 25k tweets from Twitter [7]. The ETHOS dataset, publicly available on papers with code, undergoes preprocessing and feature extraction. It comprises two subsets for binary and multi-label classification, containing 998 and 433 comments, respectively. With 24,784 rows and 5 columns, each representing a tweet, the dataset includes counts of specific content types, overall tweet categories, and text content. The columns for "hate speech," "offensive language," and "neither" occurrences range from 0-6. The target variable is categorized as "hate speech" (0), "offensive speech" (1), or "neither" (2), facilitating diverse linguistic analysis for hate speech detection using ML, DL, and Ensemble models.

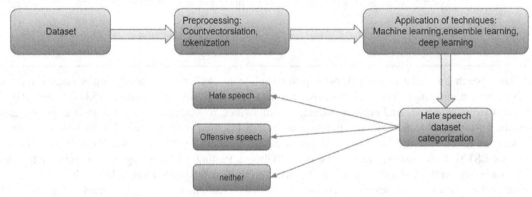

Fig. 36.1   The above figure describes the generalized approach to our study of hate speech detection using the stated machine learning, deep learning, and ensemble techniques

3.1 Preprocessing

Data preprocessing in the Ethos dataset involves cleaning by removing rows with missing values, dropping irrelevant columns ('Unnamed: 0' and 'count'), and categorizing data into three classes ('0', '1', '2'). The dataset is then split into training and testing sets. For Classic Machine Learning, the training set is used for model training, and the test set assesses performance. Deep Learning Algorithms undergo text preprocessing, including tokenization, conversion to numerical sequences, and padding. Target labels are encoded using a label encoder in this process.

4. PROPOSED APPROACH

4.1 Feature Extraction

The preprocessed and cleaned dataset undergoes normalization using standard techniques from the scikit-learn library. Subsequently, a CountVectorizer instance is created, serving as a text feature extraction method. This CountVectorizer is fitted to the training data, transforming text data into a sparse matrix of token counts. Each row in the matrix corresponds to a tweet, and each column represents a unique word in the vocabulary, with entries indicating the count of each word in the respective document. This process ensures consistent features (words) in both training and test data representations [8].

4.2 Classification Using Classical Machine Learning

The text outlines the use of classical machine learning models for hate speech detection, including Naïve Bayes, Support Vector Machine (SVM), Logistic Regression, Random Forest, Decision Tree, and K-Nearest Neighbours (KNN). Naïve Bayes excels in analyzing individual words or phrases. SVM and Logistic Regression serve as adept linear classifiers, with SVM finding optimal hyperplanes and Logistic Regression being computationally efficient. Random Forest, an ensemble of decision trees, is versatile for handling complex textual data. Decision trees capture complex relationships, and KNN is adaptable for localized patterns in textual analysis. These models, implemented in sci-kit learn[9], form a foundational architecture for hate speech detection, combining accessibility and adaptability.

4.3 Classification using Deep Learning

The evaluation of deep learning techniques, specifically Neural Networks and LSTM, for hate speech detection involves robust methodologies. Neural Networks demonstrate proficiency in learning intricate patterns through dataset preprocessing, word embedding, and fine-tuned training, showcasing effectiveness in discerning hate speech features. LSTM networks, a type of recurrent neural network, excel in capturing contextual dependencies in sequential data. Their application in hate speech detection involves preprocessing, embedding, and iterative fine-tuning, resulting in significant efficacy. Both techniques showcase the capability of deep learning in addressing the complex task of hate speech detection by understanding nuanced language patterns and features [10].

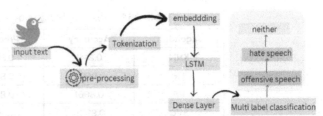

Fig. 36.2 LSTM algorithm which involves tokenization after preprocessing the data, creating an LSTM model with embedding and dense layers, compiling, training, and evaluating the model to produce multi-label classification

4.4 Classification Using Ensemble Algorithms

Ensemble Techniques are majorly classified into three categories Bagging, Stacking, and Boosting. This study employs multiple Ensemble Techniques falling under the above categories such as CatBoost, AdaBoost, XGBoost, Bagging, Pasting Ensemble, and Stacking.

The study discusses the application of various ensemble learning algorithms for hate speech detection. CatBoost employs a sequential decision tree ensemble approach, known for its effectiveness in capturing complex relationships within high-dimensional data. It is highlighted for its innate ability to handle imbalanced datasets, assigning higher weights to minority classes during training. The study emphasizes CatBoost's scalability achieved through a specialized algorithm for efficient handling of large datasets.

Extreme Gradient Boost (XGBoost) is presented as a powerful tool, utilizing sequential decision tree ensembles to learn from the errors of previous trees. Its strength lies in capturing subtle features indicative of hate speech, making it well-suited for tasks demanding interpretability and model transparency. The algorithm's efficiency in handling large datasets and adaptability to various types of features contributes to its versatility in modeling diverse hate speech data.[11]

AdaBoost, another sequential ensemble learner, is discussed for its unique approach of assigning more weight to misclassified instances in each iteration. This adaptability and focus on challenging examples are emphasized as crucial factors contributing to AdaBoost's ability to achieve high accuracy in hate speech detection.

The stacking ensemble approach is introduced as a method enhancing robustness to noisy data, improving generalization capabilities, and adapting well to varying data distributions. By combining the strengths of Support Vector Machine (SVM) and Logistic Regression (LR) with the ensemble learning power of Random Forest, the stacking ensemble is positioned as an effective strategy for addressing the complexities of hate speech classification.[12]

Bagging, involving creating subsets of the dataset through random sampling with replacement, is discussed as a technique that reduces overfitting and improves model generalization. Its stability and robustness are highlighted, particularly in mitigating the impact of outliers or noisy instances in the training data. The effectiveness of bagging in handling imbalanced datasets common in hate speech detection is underscored through the training of models on various balanced subsets of the data.

Pasting Bagging, a variation of the Bagging Classifier, is briefly explored through experimentation with certain parameters and repeated stratified K-fold cross-validation. The focus here is on assessing its performance and showcasing its effectiveness in ensemble learning.

5. RESULT AND DISCUSSION

In hate speech detection evaluation, key metrics include accuracy, precision, recall, and F1 score. Accuracy gives a general measure but can be misleading with imbalanced classes. Precision ensures accuracy of positive predictions, crucial in high false positive cost scenarios. Recall measures the model's ability to capture all positive instances, important in minimizing false negatives. F1 score, the harmonic mean of precision and recall, balances these metrics, providing a comprehensive understanding of algorithmic performance for informed decision-making[13].

The results are as follows:

Table 36.1 Results obtained by applying machine learning algorithms

Algorithm	Accuracy	Precision	F1 score	Recall
KNN	0.8141	0.8030	0.7981	0.8141
LR	0.9000	0.8882	0.8916	0.9000
RF	0.8661	0.8490	0.8454	0.8661
SVC	0.8996	0.8791	0.8759	0.8996
NB	0.8494	0.8242	0.8164	0.8494
DT	0.8905	0.8800	0.8842	0.8893

Table 36.1 details the outcomes of traditional machine learning models, showcasing their respective strengths in identifying instances of hate speech. Here Logistic Regression yields the highest accuracy. Among the classifiers (KNN, SVM, Decision Tree, Random Forest, Naive Bayes), logistic regression yielded the highest accuracy of 90.0006%. This indicates that logistic regression's probabilistic approach to classification outperformed the other models in capturing the underlying patterns in the dataset. Logistic regression is well-suited for binary classification tasks and relies on a straightforward optimization process, making it effective when the relationships between features and the target.

Moving on to, Table 36.2 presents the results for neural network models, including a traditional neural network and LSTM. These models exhibited competitive performance metrics, showcasing the potential of deep learning approaches in hate speech detection tasks.

Table 36.2 Results obtained by applying deep learning algorithms

Algorithm	Accuracy	Precision	F1 score	Recall
Neural Network	0.7700	0.9296	0.9635	1.0
LSTM	0.7675	0.6144	0.6768	0.7675

Additionally, Table 36.3, which outlines the performance of ensemble models, we observed compelling results from various algorithms, with the Pasting Ensemble standing out as the top performer. This reinforces the robustness and generalizability of ensemble techniques in mitigating the complexities associated with hate speech classification.

Table 36.3 Results obtained by applying ensemble learning algorithms

Algorithm	Accuracy	Precision	F1 score	Recall
Catboost	0.8594	0.8423	0.8301	0.8594
Stacking	0.9014	0.8994	0.8930	0.9014
XGBoost	0.8921	0.8868	0.8781	0.8921
Adaboost	0.8879	0.8827	0.8756	0.8879
Pasting	0.9040	0.8881	0.8929	0.9022
Bagging	0.9039	0.8903	0.8950	0.9039

The findings suggest that while traditional machine learning models contribute valuable insights, ensemble methods, particularly the Pasting Ensemble, excel in achieving superior accuracy. The observed superior accuracy of Pasting Ensemble in hate speech detection compared to other algorithms can be attributed to several factors. Firstly, its ability to mitigate overfitting by aggregating predictions from multiple models, each trained on a different subset of the dataset, is crucial. This ensemble approach helps create a more generalized model that performs well on new, unseen instances. Additionally, the robustness of pasting bagging to noisy data and outliers contributes to improved performance, especially in scenarios where hate speech patterns might exhibit diverse and nuanced linguistic features. Moreover, neural network models display promising capabilities in capturing intricate patterns within hate speech data. The results highlight the importance of considering a diverse range of algorithms when developing hate speech detection systems, emphasizing the nuanced nature of the task.

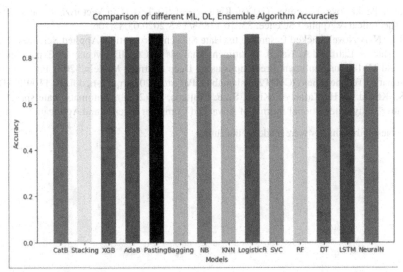

Fig. 36.3 The above bar graph compares the accuracies of different ml, dl, and ensemble techniques for hate speech detection

6. CONCLUSION AND FUTURE SCOPE

This research compares hate speech detection algorithms, including Classic ML, Ensemble, and Deep Learning. Ensemble techniques, particularly Pasting outperforms individual models with the highest accuracy at 0.904. Logistic regression excels among Classic ML algorithms, achieving an accuracy of 0.900. In Deep Learning, the LSTM model exhibits an accuracy of 0.767, showcasing its ability to capture temporal dependencies. The study emphasizes the effectiveness of Ensemble techniques, highlighting the significance of Pasting in superior hate speech detection.

The future scope involves integrating advanced technologies such as recurrent neural networks and transformer models to capture intricate linguistic nuances. Exploration of quantum learning algorithms aims for unprecedented speed in large-scale data analysis. Additionally, the research looks into multimodal approaches, combining image and text analysis, to address non-textual elements. Overall, the goal is to contribute to the development of a robust hate speech detection framework capable of mitigating linguistic variations and contextual complexities across diverse languages and cultures.

REFERENCES

1. O. de Gibert, N. Perez, A. García-Pablos, and M. Cuadros, "Hate speech dataset from a white supremacy forum," in Proc. 2nd Workshop Abusive Lang. Online (ALW2), 2018, pp. 11–20.
2. Liu, H., Burnap, P., Alorainy, W., & Williams, M. L. (2019). Fuzzy multi-task learning for hate speech type identifcation. In The World Wide Web conference on—WWW '19 (pp. 3006–3012). ACM Press.
3. Scikit-learn: Machine Learning in Python, Pedregosa et al., JMLR 12, pp. 2825-2830, 2011.
4. Peter Burnap and Matthew Leighton Williams. Hate speech, machine classification and statistical modelling of information flows on Twitter. In Internet, Policy and Politics, 2014.
5. Nemanja Djuric et al. Hate speech detection with comment embeddings. In 24th International Conference on World Wide Web, pages 29–30. ACM, 2015.
6. M. Chau and J. Xu, "Mining Communities and Their Relationships in Blogs: A Study of Online Hate Groups, (2007), International Journal of Human-Computer Studies, pp. 57–70.
7. ETHOS Dataset | Papers With Code
8. Pedregosa F, Varoquaux G, Gramfort A, Michel V, Thirion B, Grisel O, et al. Scikit-learn: MachineLearning in Python. JMLR. 2011; 12:2825–2830.
9. Martín Abadi, Ashish Agarwal, Paul Barham, Eugene Brevdo,Zhifeng Chen, Craig Citro, Greg S. Corrado, Andy Davis,Jeffrey Dean, Matthieu Devin, Sanjay Ghemawat, Ian Goodfellow,Andrew Harp, Geoffrey Irving, Michael Isard, Rafal Jozefowicz, Yangqing Jia,Lukasz Kaiser, Manjunath Kudlur, Josh Levenberg, Dan Mané, Mike Schuster,Rajat Monga, Sherry Moore, Derek Murray, Chris Olah, Jonathon Shlens,TensorFlow: Large-scalemachine learning on heterogeneous systems,2015. Software available from tensorflow.org.

10. Kozhevnikov, Vadim & Pankratova, Evgeniya. (2020). Research of the Text Data Vectorization and Classification Algorithms of Machine Learning. Theoretical & Applied Science. 85. 10.15863/TAS.2020.05.85.106.

11. Aljero MKA, Dimililer N. A Novel Stacked Ensemble for Hate Speech Recognition. Applied Sciences. 2021; 11(24):11684.

12. [8] G. M. Raza, Z. S. Butt, S. Latif and A. Wahid, "Sentiment Analysis on COVID Tweets: An Experimental Analysis on the Impact of Count Vectorizer and TF-IDF on Sentiment Predictions using Deep Learning Models," 2021 International Conference on Digital Futures and Transformative Technologies (ICoDT2), Islamabad, Pakistan, 2021, pp. 1-6, doi: 10.1109/ICoDT252288.2021.9441508.

13. Abro, S., Shaikh, S., Khand, Z. H., Zafar, A., Khan, S., & Mujtaba, G. (2020). Automatic hate speech detection using machine learning: A comparative study. International Journal of Advanced Computer Science and Applications, 11(8).

Note: All the figures and tables in this chapter were made by the author.

Computational Intelligence and Mathematical Applications – Dr. Devendra Prasad et al. (eds)
© 2024 Taylor & Francis Group, London, ISBN 978-1-032-87721-1

37 | Quantum Inspired Anomaly Detection for Industrial Machine Sensor Data

Tanisha Kindo[1], Arunima Jaiswal[2], Kirti Singh[3] and Vanshika[4]

Department of Computer Science and Engineering, Indira Gandhi Delhi Technical University for Women, Delhi, India

Abstract: Anomaly detection is a crucial task in data analysis to recognize patterns or instances that deviate significantly from the standard within a dataset. Industrial machine anomaly detection is a specialized application of anomaly detection techniques in industrial automation and machinery. This study compares the performance of classical and quantum machine learning algorithms in detecting anomalies in industrial machines. A method for identifying outliers is proposed that makes use of machine learning algorithms such as Isolation Forest, Hotelling's T2, Local Outlier Factor (LOF), One-class Support Vector Machine, and Quantum-Inspired SVM to detect anomalies in industrial machine sensor data.

Keywords: Machine learning algorithm, Machine temperature data, Quantum-inspired support vector machine, Isolation forest classifier, Hotelling's t-squared, Local outlier factor, One-class SVM

1. INTRODUCTION

Anomaly detection is the identification of patterns that deviate significantly from the expected behavior within a dataset. Detecting anomalies is a crucial challenge that has gained attention across various research fields and application domains. When applied to industrial machine data, anomaly detection becomes a foundation for ensuring the smooth operation of machinery, reducing disruptions, and preventing failures. However, machine sensor data complexities, arising from malfunctions, environmental changes, or variations, pose challenges. If undetected, anomalies can lead to inefficiencies and failures. This study focuses on detecting anomalies in the "machine_temperature_system_failure" subset of the NAB dataset [1], using unsupervised learning due to limited labeled anomaly data [2]. Quantum-inspired SVM is employed for improved classification accuracy, expecting it to outperform classical algorithms [3].

The study's primary contribution lies in applying diverse unsupervised algorithms and exploring the potential of quantum-inspired techniques for anomaly detection in machine sensors, concluding with a comprehensive results comparison.

2. RELATED WORK

One of the studies addressed the critical problem of identifying anomalous subsequence in time series data, that applies to a variety of industries including finance, healthcare, and manufacturing [4]. Another study examined 20 univariate statistical, machine learning, and deep learning methods for anomaly detection in time-series data. The study demonstrated that statistical methods detect point and collective anomalies with lower computation times by assessing the accuracy and computation time on publicly available datasets [5].

[1]tanisha163btcse20@igdtuw.ac.in, [2]arunimajaiswal@igdtuw.ac.in, [3]kirti162btcse20@igdtuw.ac.in, [4]vanshika181btcse20@igdtuw.ac.in,

DOI: 10.1201/9781003534112-37

The review by Hu et al. [6] employed Hotelling's T2 techniques to distinguish between sensor and aircraft engine faults. One of the studies presented a LOF-based algorithm that was applied for efficiently detecting anomalies [7]. Liu et al. [8] introduced Isolation Forest, a novel model-based anomaly detection method that focuses on isolating anomalies rather than profiling normal instances.

The study by Widodo et al. [9] reviewed the use of support vector machines (SVM) in fault diagnosis and machine condition monitoring. Another study investigated the potential of quantum machine learning (QML), which uses quantum systems to process classical data in a variety of fields [10].

3. METHODOLOGY

The employed methodology serves as a roadmap explaining the steps and procedures taken to achieve the study's objectives and highlights the systematic approach employed throughout the research. The given block diagram visually represents the key steps in the proposed methodology.

The block diagram (Fig. 37.1) begins with the NAB Machine Temperature System Failure dataset selection. It follows with data preprocessing, incorporating EDA and Time Series Analysis for pattern and statistical insight. Unsupervised algorithms (Isolation Forest, LOF, One-class SVM, Hotelling's T2) identify anomalies in temperature readings. A quantum-inspired approach using Quantum SVM enhances anomaly detection. The final step checks and compares results, providing a comprehensive understanding of each method's effectiveness. This systematic process helps efficiently uncover unusual events in industrial machine temperature data.

3.1 Data Selection

The dataset selected, "machine_temperature_system_failure," is a subset of the Numenta Anomaly Benchmark (NAB) [11]. This dataset captures temperature sensor data from an internal component of an industrial machine. It forms the basis of this anomaly detection study and encapsulates crucial timestamp and univariate value information. The variables within this dataset are complicatedly structured to capture the evolution of machine temperature. Given the univariate nature of the data, the focus of the study lies in comprehending the temporal patterns and fluctuations within the machine's temperature over time.

The selection of this dataset aligns with the study's objective of addressing anomalies in industrial machine behavior, emphasizing the importance of accurate detection in preventing operational disruptions and equipment failures.

3.2 Preprocessing

Normalization and scaling are essential preprocessing techniques that were applied to the temperature values, ensuring a consistent range that aids in algorithm convergence. The data undergoes inspection, addressing null and duplicate values to maintain data integrity. The original univariate format is transformed into a more versatile multivariate structure, extracting relevant features, such as year, month, day, hour, and minute from timestamps using scikit learn library techniques [12].

These extracted features enable the algorithms to recognize trends over time and greatly contribute to the anomaly detection process. Furthermore, the dataset is divided into 80% for training and 20% for testing, allowing for a thorough evaluation of algorithm performance [13].

3.2 Exploratory Data Analysis (EDA) and Time Series Analysis

In the EDA phase, industrial machine data was thoroughly analyzed using the HoloViews library [14] for meaningful insights. Various visualizations, including bar charts and histograms, unveiled temperature distribution patterns across different time

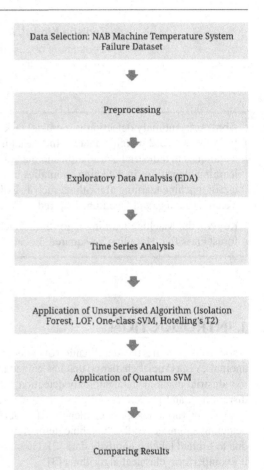

Fig. 37.1 Industrial machine anomaly detection workflow

dimensions [15]. Bar charts depicted temperature reading counts, offering insights into monthly and yearly trends, and aiding anomaly identification. Histograms detailed temperature distribution, and breakdowns by year and month allowed nuanced variation examination. Time series analysis used curve plots [16] to visualize temperature fluctuations, emphasizing anomalies. Daily trends were captured by computing mean daily temperatures, creating a curve highlighting daily variations.

3.3 Application of Unsupervised Algorithms

This research paper employs the application and comparative analysis of classical unsupervised algorithms—Hotelling's T2, Isolation Forest, One-class SVM, and LOF.

Hotelling's T2 is a statistical method for detecting outliers in data that has a multivariate distribution [17]. This method was used in particular because of the interdependence of various features in the machine temperature dataset. The use of Hotelling's T2 provided a reliable method for detecting anomalies that account for the complex relationships between various temperature-related parameters.

The Isolation Forest algorithm isolates anomalies by randomly partitioning the dataset [18]. Anomalies are identified as instances requiring fewer partitions to isolate, making it an effective algorithm for the high-dimensional machine temperature dataset. The algorithm's ability to isolate anomalies quickly and effectively contributes to the system's overall speed and accuracy.

SVMs are designed for binary classification with the assumption that the majority of the data belongs to the normal class [18, 20]. This makes it suitable for detecting anomalies in machine temperature readings as deviations from the expected norm. The use of One-Class SVM created a binary classification framework tailored to our specific anomaly detection requirements. This model excels at detecting deviations from normal behavior, allowing for a more precise approach to anomaly detection.

The Local Outlier Factor (LOF) algorithm detects anomalies by comparing the local density of instances to the density of their neighbors [19]. It is effective at detecting anomalies that are not always obvious, making it useful for the proposed anomaly detection methodology. By taking into account local data density, this algorithm excels at capturing anomalies that can be significant, adding depth to the anomaly detection results.

3.4 Application of Quantum-Inspired SVM Classifier

This study introduces the application of a Quantum-inspired Support Vector Machine (SVM), using a unique quantum-inspired feature map. This method transforms the dataset, capturing detailed patterns that classical SVMs might miss. Applying this quantum-inspired feature map (Eq. 1) to the machine temperature dataset, we employ a quantum-inspired SVM model for training [21,22].

$$np.sin(x[0]), np.cos(x[1]), np.sin(x[2]) \qquad \cdot (1)$$

This quantum-inspired SVM model uses a radial basis function (RBF) kernel [22] and scales the gamma parameter of the data to ensure reliable classification.

4. RESULTS/DISCUSSION

The evaluation of anomaly detection algorithms involves the careful consideration of multiple performance metrics, including accuracy, precision, F1 score, and recall.

Accuracy [23] is a fundamental metric that measures the ratio of correctly identified instances to the total number of instances. In this study, the Hotelling's T2 algorithm demonstrated the highest accuracy at 92.5%. This implies that when this algorithm identifies an anomaly, it is highly likely to be a genuine anomaly. In this study, while comparing Recall [23] across the anomaly detection models, Quantum-inspired SVM outperformed other algorithms with a recall value of 100% and gave the highest Precision [23] of 89.3%. This means that the Quantum-inspired SVM identified all true anomalies present in the dataset. On the other hand, One-Class SVM, while exhibiting reasonable precision and accuracy, showed a lower recall value of 63.4%, indicating that it missed a significant number of true anomalies. F1 score [23], the harmonic mean of recall and precision, provides a balanced measure that considers both false positives and false negatives. Quantum-inspired SVM excelled in F1 score with 94.4%, showcasing its balance between precision and recall, making it an effective anomaly detection model. To clearly understand the performance of each machine learning algorithm, the comparison for each model is shown below.

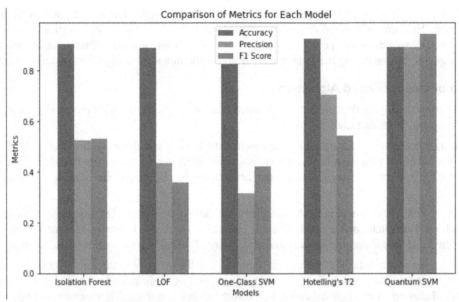

Fig. 37.2 Comparison of anomaly detection metrics

To understand the difference between the performance metric of the unsupervised One-Class SVM and Quantum-Inspired SVM, a bar graph is plotted below.

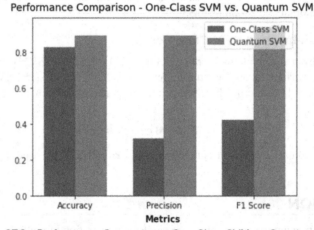

Fig. 37.3 Performance Comparison - One-Class SVM vs. Quantum SVM

The model Accuracy, F1 Score, Precision, and Recall of unsupervised models and quantum-inspired SVM are tabulated in Tables 37.1 and 37.2 respectively.

Table 37.1 Results obtained from unsupervised machine learning

Model	Isolation Forest	LOF	One-class SVM	Hotelling's T2
Accuracy	90.5	89.0	82.6	92.5
Precision	52.5	43.4	31.6	70.3
F1 score	52.5	35.7	42.2	54.4
Recall	52.8	11.1	63.4	44.3

Table 37.2 Results obtained from Quantum machine learning

Model	Quantum Inspired SVM
Accuracy	89.3
Precision	89.3
F1 score	94.4
Recall	100

5. CONCLUSION AND FUTURE SCOPE

This study thoroughly evaluates the performances of unsupervised and quantum-inspired machine learning algorithms on metrics such as accuracy, precision, F1 score, and recall, in detecting anomalies in industrial machine temperature sensors. Quantum-inspired SVM achieved 89.3% accuracy and outperformed the classical machine learning algorithms, providing a 94.4% F1 Score and a 100% Recall value.

Although this study provides insight into the detection of industrial machine anomalies using quantum machine learning, there is still scope for improvement in the accuracy. In the future, the quantum-inspired SVM model will be improved further to achieve even better accuracy in detecting the anomalies in the industrial machine temperature sensor data.

REFERENCES

1. Numenta/NAB: The numenta anomaly benchmark, https://github.com/numenta/NAB.
2. Varun Chandola, Arindam Banerjee, and Vipin Kumar. 2009. Anomaly detection: A survey. ACM Comput. Surv. 41, 3, Article 15 (July 2009), 58 pages. https://doi.org/10.1145/1541880.1541882.
3. C. Ding, T. -Y. Bao and H. -L. Huang, "Quantum-Inspired Support Vector Machine," in IEEE Transactions on Neural Networks and Learning Systems, vol. 33, no. 12, pp. 7210-7222, Dec. 2022, doi: 10.1109/TNNLS.2021.3084467.
4. Sebastian Schmidl, Phillip Wenig, and Thorsten Papenbrock. 2022. Anomaly detection in time series: a comprehensive evaluation. Proc. VLDB Endow. 15, 9 (May 2022), 1779–1797. https://doi.org/10.14778/3538598.3538602.
5. Braei, M., & Wagner, S. (2020). Anomaly Detection in Univariate Time-series: A Survey on the State-of-the-Art. ArXiv. / abs/2004.00433.
6. Xiao Hu, Raj Subbu, P. Bonissone, Hai Qiu and N. Iyer, "Multivariate anomaly detection in real-world industrial systems," 2008 IEEE International Joint Conference on Neural Networks (IEEE World Congress on Computational Intelligence), Hong Kong, China, 2008, pp. 2766-2771, doi: 10.1109/IJCNN.2008.4634187.
7. Lin Xu, Yi-Ren Yeh, Yuh-Jye Lee, Jing Li, A Hierarchical Framework Using Approximated Local Outlier Factor for Efficient Anomaly Detection, Procedia Computer Science, Volume 19, 2013, Pages 1174-1181, ISSN 1877-0509, https://doi.org/10.1016/j.procs.2013.06.168.
8. F. T. Liu, K. M. Ting and Z. -H. Zhou, "Isolation Forest," 2008 Eighth IEEE International Conference on Data Mining, Pisa, Italy, 2008, pp. 413-422, doi: 10.1109/ICDM.2008.17.
9. Achmad Widodo, Bo-Suk Yang, Support vector machine in machine condition monitoring and fault diagnosis, Mechanical Systems and Signal Processing, Volume 21, Issue 6, 2007, Pages 2560-2574, ISSN 0888-3270, https://doi.org/10.1016/j.ymssp.2006.12.007.
10. T. M. Khan and A. Robles-Kelly, "Machine Learning: Quantum vs Classical," in IEEE Access, vol. 8, pp. 219275-219294, 2020, doi: 10.1109 /ACCESS.2020. 3041719.
11. BoltzmannBrain. "Numenta Anomaly Benchmark (NAB)." Kaggle, 19 Aug. 2016, www.kaggle.com/datasets/boltzmannbrain/nab.
12. Scikit-learn: Machine Learning in Python, Pedregosa et al., JMLR 12, pp. 2825-2830, 2011.
13. T. Zito, N. Wilbert, L. Wiskott, and P. Berkes. Modular toolkit for data processing (MDP): A Python data processing framework. Frontiers in Neuroinformatics, 2, 2008.
14. Stevens, Jean-Luc & Rudiger, Philipp & Bednar, James. (2015). HoloViews: Building Complex Visualizations Easily for Reproducible Science. 10.25080/Majora-7b98e3ed-00a.
15. Komorowski, Matthieu & Marshall, Dominic & Salciccioli, Justin & Crutain, Yves. (2016). Exploratory Data Analysis. 10.1007/978-3-319-43742-2_15.
16. Velicer, Wayne & Fava, Joseph. (2003). Time Series Analysis. 10.1002/0471264385.wei0223.
17. Muhammad Ahsan, Muhammad Mashuri, Muhammad Hisyam Lee, Heri Kuswanto, Dedy Dwi Prastyo, Robust adaptive multivariate Hotelling's T2 control chart based on kernel density estimation for intrusion detection system, Expert Systems with Applications, Volume 145, 2020, 113105, ISSN 0957-4174, https://doi.org/10.1016/j.eswa.2019.113105.
18. Victorambonati. "Unsupervised Anomaly Detection." Kaggle, Kaggle, 4 Aug. 2017, www.kaggle.com/code/victorambonati/unsupervised-anomaly-detection.
19. Zhangyu Cheng, Chengming Zou, and Jianwei Dong. 2019. Outlier detection using isolation forest and local outlier factor. In Proceedings of the Conference on Research in Adaptive and Convergent Systems (RACS '19). Association for Computing Machinery, New York, NY, USA, 161–168. https://doi.org/10.1145/3338840.3355641.
20. C.C. Chang and C.J. Lin. LIBSVM: a library for support vector machines. http://www.csie. ntu.edu.tw/cjlin/libsvm, 2001.
21. S. Van der Walt, S.C Colbert, and G. Varoquaux. The NumPy array: A structure for efficient numerical computation. Computing in Science and Engineering, 11, 2011.
22. Chen, B., & Chern, J. (2022). Generating quantum feature maps for SVM classifier. ArXiv. /abs/2207.11449.
23. Ajay Kulkarni, Deri Chong, Feras A. Batarseh, 5 - Foundations of data imbalance and solutions for a data democracy, Editor(s): Feras A. Batarseh, Ruixin Yang, Data Democracy, Academic Press, 2020, Pages 83-106, ISBN 9780128183663, https://doi.org/10.1016/B978-0-12-818366-3.00005-8.

Note: All the figures and tables in this chapter were made by the author.

Computational Intelligence and Mathematical Applications – Dr. Devendra Prasad et al. (eds)
© 2024 Taylor & Francis Group, London, ISBN 978-1-032-87721-1

38 | Challenges and Opportunities in IoT-based Smart Farming: A Survey Report

Isha Chopra[1]

Research Scholar, Department of CSE, Maharishi Markandeshwar Engineering College, MM(DU), Mullana, Ambala, Haryana

Amit Kumar Bindal[2]

Professor, Department of CSE, Maharishi Markandeshwar Engineering College, MM(DU), Mullana, Ambala, Haryana, India

Zatin Gupta[3]

Assistant Professor, Department of Computer Science & Engineering, School of Computing Science & Engineering, Galgotias University, Greater Noida, Uttar Pradesh, India

Abstract: The advancement of technology has facilitated different industries in different ways. The agricultural sector is one of the industries that have benefited from the implication of IoT technology for crop cultivation. Thus, the following review paper has presented the impact of IoT technology in smart agriculture. The paper's primary aim is to give a review paper based on smart agriculture systems using IoT devices. To develop the results, the report has discussed various factors of smart agriculture and the benefits of IoT technology. Additionally, a close discussion related to the components of IoT technology is presented in the review paper. Certain issues, such as energy consumption, high cost, and the knowledge gap, were found to hinder the application. Moreover, solutions to the problems are discussed coherently. At the same time, secondary qualitative methods have facilitated the discussion of the requirements of smart agriculture. The overall paper reflects tangible knowledge regarding smart agriculture using IoT systems.

Keywords: IoT, Smart agriculture, Precision farming

1. INTRODUCTION

Advances in technology have boosted industry growth and productivity. The methodology of an empirical analysis undertakes all the steps undertaken to develop results. Moreover, the method investigates the different elements that contributed to the development of the results based on the objectives of a study [3]. Therefore, a secondary qualitative methodology was employed to present a review paper. Moreover, information from past literature was gathered and coherently discussed to comprehend IoT devices' use for precise agriculture. Furthermore, the influence of IoT devices is analysed based on the qualitative and quantitative information gathered from the past analysis literature. Additionally, the secondary data is pre-verified; hence it can be contemplated that the outcome of the study is reliable [5]. Moreover, With the implication of secondary qualitative methods, a detailed perspective is gained regarding the use of IoT devices in agriculture.

[1]isshchopra08@gmail.com, [2]amitbindal@mmumullana.org, [3]zatin.gupta2000@gmail.com

DOI: 10.1201/9781003534112-38

Table 38.1   The difference in the technologies used in IoT-based farming and traditional farming

IoT-based farming	Traditional farming
It uses a modern technique to increase the moisture level of the water	This uses traditional irrigation technology
This system included different sensors to inform the moisture level and other conditions of the soil	Farmers used to visit for checking the moisture level of the soil
The ingredients used in the soil are artificial	Here natural and artificial components used in the soil
Monitoring systems control soil fertility through the modern system	The traditional argricultural system is based on the manual system

2. UNDERSTANDING AND MANAGING SOIL HEALTH

Through this understanding of soil quality farmers can understand soil moisture, nutrient levels, and pH which are essential for improved agriculture. Therefore, through the utilization of sensors soil conditions can be obsessed with a data-backed method [9]. Moreover, smart farmers depend on traditional manners and achieve additional information. Hence, with this knowledge, they can modify their fertilization and other soil management techniques, resulting in optimal plant development and increased yields. Moreover, understanding the soil quality is essential to understand the corps' growth [10]. The growth of crops thrives when the soil is in good condition, and at the same time nutritional value of the crop can be preserved. Plants receive all their necessary nutrients, and water, from quality soil. In addition, the structural support of the plants comes from the soil. Therefore, with the understanding of the soil farmers can guarantee the best conditions for plant growth, and greater harvests can be intended. Moreover, it can be estimated that with a greater knowledge of soil quality, maintaining soil health can be done systematically with all the knowledge gathered with IoT devices. Having this information also helps when thinking about applying fertiliser to the soil, which boosts the nutritional value of the end product [11]. Soil quality knowledge is obviously useful for fertiliser management, which is critical for crop yield and quality control. Depending on the plantation, it is also crucial to know the soil's water-holding capacity [11]. Soil that can retain water is ideal for growing rice because it creates a more humid growing environment [12]. Therefore, this kind of cultivation must think about the sand-holding capacity. Soil management affects crop nutritional value, so it's crucial to know how soil quality is measured.

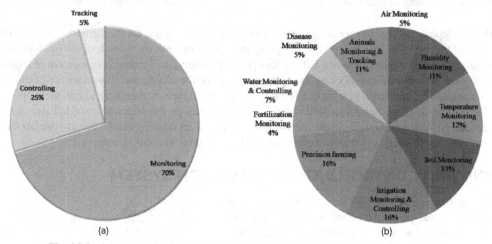

Fig. 38.1   (a) and (b) Role of IoT devices in different domains of agriculture [8] [9]

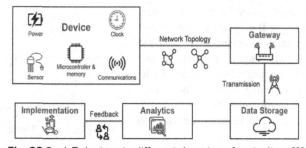

Fig. 38.2   IoT devices in different domains of agriculture [8]

3. MONITORING OF CROPS FOR RISK ANALYSIS AND MANAGEMENT

During the analysis of past literature, it was noted that there are different strata associated with the process of agriculture. Moreover, farmers must go through different phases from planting to achieving the final yield [13]. Additionally, it was noted that certain risks are associated with each agriculture stratum. Such risk can be harmful to the process and can degrade the quality and quantity of the yield. The risk for cultivating such factors can differ depending on the cultivation season. Additionally, there might be natural risk factors that cannot be predicted precisely. For instance, natural calamities are one of the risk factors that cannot be predicted precisely with manual calculations [14]. However, with the use of IoT technology, appropriate evidence can be calculated with appropriate pieces of evidence. Additionally, with the appropriate use of such evidence, intercity can be measured, and appropriate precautions can be taken.

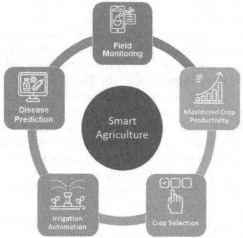

Fig. 38.3   Elements of crop monitoring [12]

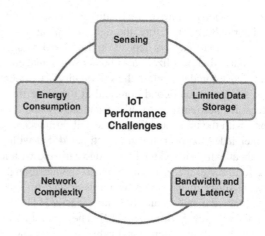

Fig. 38.4   Challenges in smart agriculture [19]

Furthermore, the development of monitoring the growth of crops can be of great use for the farmers. There are different factors, such as the quality of the produce which can aid in analysing the risk along with the growth of the produce. The above figure illustrates the elements of crop monitoring using IoT devices [17]. The selection of a crop is an element of monitoring the crop. Additionally, it can be contemplated that choosing the right crop based on the quality and characteristics of the plant is essential. In addition, environmental factors play a major role in crop development. Hence in smart agriculture, the crops are chosen based on the evident factors that aid in the holistic growth of the plant. Thus, with the implementation of IoT devices, better seeds for plantation can be chosen [18]. Additionally, at the time of growth, regular monitoring can be done to predict the disease for the plants. An in-depth disease prediction that allows one to act at the right time can also be achieved.

4. PLANNING AND CONTROLLING THE IRRIGATION SYSTEM

For cultivation, irrigation and watering of the plant are subjective factors that depend on the plant. Additionally, there are some plants where the quality of water and quantity of these major factors are. For instance, *Oryza sativa, commonly known as rice,* requires different levels of water depending on the stage. Therefore, the basin irrigation method is used in cultivating rice. For "basin irrigation," making shallow, flat basins in the ground to store water for agricultural irrigation is essential [17]. At the same time, regulating the water level is essential for water crops. This technique is frequently used to irrigate crops with a high-water need, such as rice, and it is particularly effective for regions with relatively flat terrain. Hence, with the implementation of IoT-based devices, a better irrigation system can be achieved for such a cultivation method.

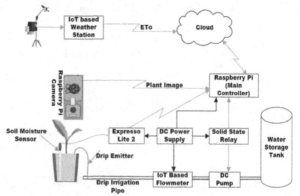

Fig. 38.5   Model for the irrigation system [11]

5. CONCLUSION

The review paper explores smart agriculture using IoT-based systems, highlighting their potential to improve crop yield and automate processes like irrigation and fertilisation despite the knowledge gap affecting their initial scalability.

REFERENCES

1. Abdullah, N., Durani, N. A. B., Shari, M. F. B., Siong, K. S., Hau, V. K. W., Siong, W. N., & Ahmad, I. K. A. (2020). Towards smart agriculture monitoring using fuzzy systems. Ieee Access, 9, 4097-4111.
2. Abhiram, M. S. D., Kuppili, J., & Manga, N. A. (2020, February). Smart farming system using IoT for efficient crop growth. In 2020 IEEE International Students' Conference on Electrical, Electronics and Computer Science (SCEECS) (pp. 1-4). IEEE.
3. Adi, P. D. P., Prasetya, D. A., Arifuddin, R., Sari, A. P., Mukti, F. S., Sihombing, V., ... & Yanris, G. J. (2021, October). Application of IoT-LoRa Technology and Design in irrigation canals to improve the quality of agricultural products in Batu Indonesia. In 2021 2nd International Conference On Smart Cities, Automation & Intelligent Computing Systems (ICON-SONICS) (pp. 88-94). IEEE.
4. Ali, A. H., Chisab, R. F., & Mnati, M. J. (2019). Smart monitoring and controlling for agricultural pumps using LoRa IOT technology. Indonesian Journal of Electrical Engineering and Computer Science, 13(1), 286-292.
5. Alreshidi, E. (2019). Smart sustainable agriculture (SSA) solution underpinned by internet of things (IoT) and artificial intelligence (AI). arXiv preprint arXiv:1906.03106.
6. Amirova, E. F., Kirillova, O. V., Kuznetsov, M. G., Gazetdinov, S. M., & Gumerova, G. H. (2020). Internet of things as a digital tool for the development of agricultural economy. In BIO Web of Conferences (Vol. 17, p. 00050). EDP Sciences.
7. Angin, P., Anisi, M. H., Göksel, F., Gürsoy, C., & Büyükgülcü, A. (2020). AgriLoRa: a digital twin framework for smart agriculture. J. Wirel. Mob. Networks Ubiquitous Comput. Dependable Appl., 11(4), 77-96.
8. Antony, A. P., Leith, K., Jolley, C., Lu, J., & Sweeney, D. J. (2020). A review of practice and implementation of the internet of things (IoT) for smallholder agriculture. Sustainability, 12(9), 3750.
9. Araby, A. A., Abd Elhameed, M. M., Magdy, N. M., Abdelaal, N., Abd Allah, Y. T., Darweesh, M. S., ... & Mostafa, H. (2019, May). Smart iot monitoring system for agriculture with predictive analysis. In 2019 8th International Conference on Modern Circuits and Systems Technologies (MOCAST) (pp. 1-4). IEEE.
10. Ardiansah, I., Bafdal, N., Suryadi, E., & Bono, A. (2020). Greenhouse monitoring and automation using Arduino: a review on precision farming and internet of things (IoT). Int. J. Adv. Sci. Eng. Inf. Technol, 10(2), 703-709.
11. Atmaja, A. P., El Hakim, A., Wibowo, A. P. A., & Pratama, L. A. (2021). Communication systems of smart agriculture based on wireless sensor networks in IoT. Journal of Robotics and Control (JRC), 2(4), 297-301.
12. Awan, S. H., Ahmed, S., Safwan, N., Najam, Z., Hashim, M. Z., & Safdar, T. (2019). Role of internet of things (IoT) with blockchain technology for the development of smart farming. Journal of Mechanics of Continua and Mathematical Sciences, 14(5), 170-188.
13. Ayaz, M., Ammad-Uddin, M., Sharif, Z., Mansour, A., & Aggoune, E. H. M. (2019). Internet-of-Things (IoT)-based smart agriculture: Toward making the fields talk. IEEE access, 7, 129551-129583.
14. Bouali, E. T., Abid, M. R., Boufounas, E. M., Hamed, T. A., & Benhaddou, D. (2021). Renewable energy integration into cloud & IoT-based smart agriculture. IEEE Access, 10, 1175-1191.
15. Boursianis, A. D., Papadopoulou, M. S., Diamantoulakis, P., Liopa-Tsakalidi, A., Barouchas, P., Salahas, G., ... & Goudos, S. K. (2022). Internet of things (IoT) and agricultural unmanned aerial vehicles (UAVs) in smart farming: A comprehensive review. Internet of Things, 18, 100187.
16. Deepa, B., Anusha, C., & Chaya Devi, P. (2021). Smart agriculture using iot. In Intelligent System Design: Proceedings of Intelligent System Design: INDIA 2019 (pp. 11-19). Springer Singapore.
17. Devanand, W. A., Raghunath, R. D., Baliram, A. S., & Kazi, K. (2019). Smart agriculture system using IoT. Int. J. Innov. Res. Technol, 5(10).
18. Farooq, M. S., Riaz, S., Abid, A., Abid, K., & Naeem, M. A. (2019). A Survey on the Role of IoT in Agriculture for the Implementation of Smart Farming. Ieee Access, 7, 156237-156271.
19. Farooq, M. S., Riaz, S., Abid, A., Umer, T., & Zikria, Y. B. (2020). Role of IoT technology in agriculture: A systematic literature review. Electronics, 9(2), 319.
20. Ferrag, M. A., Shu, L., Yang, X., Derhab, A., & Maglaras, L. (2020). Security and privacy for green IoT-based agriculture: Review, blockchain solutions, and challenges. IEEE access, 8, 32031-32053.
21. Gao, D., Sun, Q., Hu, B., & Zhang, S. (2020). A framework for agricultural pest and disease monitoring based on internet-of-things and unmanned aerial vehicles. Sensors, 20(5), 1487.
22. García, L., Parra, L., Jimenez, J. M., Lloret, J., & Lorenz, P. (2020). IoT-based smart irrigation systems: An overview on the recent trends on sensors and IoT systems for irrigation in precision agriculture. Sensors, 20(4), 1042.
23. Hu, W. J., Fan, J., Du, Y. X., Li, B. S., Xiong, N., & Bekkering, E. (2020). MDFC–ResNet: an agricultural IoT system to accurately recognize crop diseases. IEEE Access, 8, 115287-115298.

24. Jaiswal, S. P., Bhadoria, V. S., Agrawal, A., & Ahuja, H. (2019). Internet of Things (IoT) for smart agriculture and farming in developing nations. International Journal of Scientific & Technology Research, 8(12), 1049-1056.

25. Khan, N., Ray, R. L., Sargani, G. R., Ihtisham, M., Khayyam, M., & Ismail, S. (2021). Current progress and future prospects of agriculture technology: Gateway to sustainable agriculture. Sustainability, 13(9), 4883.

26. Khoa, T. A., Man, M. M., Nguyen, T. Y., Nguyen, V., & Nam, N. H. (2019). Smart agriculture using IoT multi-sensors: A novel watering management system. Journal of Sensor and Actuator Networks, 8(3), 45.

27. Kim, W. S., Lee, W. S., & Kim, Y. J. (2020). A review of the applications of the internet of things (IoT) for agricultural automation. Journal of Biosystems Engineering, 45, 385-400.

28. Kour, V. P., & Arora, S. (2020). Recent developments of the internet of things in agriculture: a survey. Ieee Access, 8, 129924-129957.

29. Kuthadi, V. M., Selvaraj, R., Rao, Y. V., Kumar, P. S., Mustafa, M., Phasinam, K., & Okoronkwo, E. (2023). Towards security and privacy concerns in the internet of things in the agriculture sector. Turkish Journal of Physiotherapy and Rehabilitation, 32(3).

30. Lakhwani, K., Gianey, H., Agarwal, N., & Gupta, S. (2019). Development of IoT for smart agriculture a review. In Emerging Trends in Expert Applications and Security: Proceedings of ICETEAS 2018 (pp. 425-432). Springer Singapore.

31. Li, C., & Niu, B. (2020). Design of smart agriculture based on big data and Internet of things. International Journal of Distributed Sensor Networks, 16(5), 1550147720917065.

32. Gupta, Z., Gupta, Z. & Bindal, A. K. (2023). Delay-Conscious Service Quality Constraint in IoT Sensor Networks for Smart Farming. Journal of Computer Science, 19(8), 1029-1049. https://doi.org/10.3844/jcssp.2023.1029.1049

33. Gupta, Z. ., Bindal, A. K. ., Shukla, S. ., Chopra, I. ., Tiwari, V. ., & Srivastava, S. . (2023). Energy Efficient IoT-Sensors Network for Smart Farming. International Journal on Recent and Innovation Trends in Computing and Communication, 11(5), 255–265. https://doi.org/10.17762/ijritcc.v11i5.6612

34. R. Raman, Z. Gupta, R. Singh, U. Joshi, S. C. P and M. Kalyan Chakravarthi, "Potential of Vehicular Ad-Hoc Network for use in Developing a Drone-Based Delivery System," 2023 3rd International Conference on Advance Computing and Innovative Technologies in Engineering (ICACITE), Greater Noida, India, 2023, pp. 897-902, doi: 10.1109/ICACITE57410.2023.10182733.

35. R. Raman, Z. Gupta, A. Gupta, S. V. Akram, D. Saini and J. Bhalani, "Internet-Of-Things Wireless Communication with a Focus on the Protection of User Privacy and the Delivery of Relevant Facts," 2023 International Conference on Artificial Intelligence and Smart Communication (AISC), Greater Noida, India, 2023, pp. 656-660, doi: 10.1109/AISC56616.2023.10084975.

36. R. Raman, D. Buddhi, G. Lakhera, Z. Gupta, A. Joshi and D. Saini, "An investigation on the role of artificial intelligence in scalable visual data analytics," 2023 International Conference on Artificial Intelligence and Smart Communication (AISC), Greater Noida, India, 2023, pp. 666-670, doi: 10.1109/AISC56616.2023.10085495.

37. R. Raman, Z. Gupta, S. V. Akram, D. Lalit Thakur, B. G. Pillai and M. K. Chakravarthi, "Network Security Concerns for Designing Robotic Systems: A Review," 2023 International Conference on Artificial Intelligence and Smart Communication (AISC), Greater Noida, India, 2023, pp. 661-665, doi: 10.1109/AISC56616.2023.10085453.

38. R. Raman, R. Singh, Z. Gupta, S. Verma, A. Rajput and S. M. Parikh, "Wireless Communication With Extreme Reliability and Low Latency: Tail, Risk and Scale," 2022 5th International Conference on Contemporary Computing and Informatics (IC3I), Uttar Pradesh, India, 2022, pp. 1699-1703, doi: 10.1109/IC3I56241.2022.10073096.

39. K. Goel and A. K. Bindal, "Wireless Sensor Network in Precision Agriculture: A Survey Report," 2018 Fifth International Conference on Parallel, Distributed and Grid Computing (PDGC), 2018, pp. 176-181, doi: 10.1109/PDGC.2018.8745854.

40. Z. Gupta and A. Bindal, "Comprehensive Survey on Sustainable Smart Agriculture using IOT Technologies," 2022 2nd International Conference on Advance Computing and Innovative Technologies in Engineering (ICACITE), Greater Noida, India, 2022, pp. 2640-2645, doi: 10.1109/ICACITE53722.2022.9823837.

Computational Intelligence and Mathematical Applications – Dr. Devendra Prasad et al. (eds)
© 2024 Taylor & Francis Group, London, ISBN 978-1-032-87721-1

39

An Advanced Internet of Things-Based Health Monitoring System Assisted by Machine Learning

Venkateswaran Radhakrishnan[1]

Faculty-Networking and Information Security, College of Computing & Information Sciences, University of Technology and Applied Sciences, Salalah, Oman

Husna Tabassum[2]

Assistant Professor, Department of CSE, HKBK College of Engineering, Nagavara, Bangalore

Zatin Gupta[3]

Assistant Professor, School of Computing Science & Engineering, Galgotias University, Greater Noida, Uttar Pradesh

Abstract: In the realm of community health, there is a mountain of information and a wide variety of approaches to processing it. Systems for processing data are widely used. The results of CVD are forecasted using this method. This technique can diagnose heart disease. Healthcare applications of IoT and ML will be explored in this research. Part of the job entails controlling the Internet of Things (IoT) with a pulse sensor and Arduino, displaying data in a time-stamped format. IFTTT can analyse CSV sensor data from Google Sheets. The datasets are sorted into groups based on the data treatment parameters employed in their development and evaluation. To evaluate the parameters, this technique uses artificial intelligence calculations and classifications to compile data. Machine Learning Algorithms, in particular the Decision Tree Algorithm and the Random Backwoods Classification Algorithm, are used in Python to parse, inspect, and screen the dataset before processing. SVMs excel at identifying cardiovascular disorders (SVM). A highly accurate method of cardiac disease prediction is the one that has been proposed. In the early stages of cardiac disease, early detection is aided by both hardware and software. In the future, this information will be used to anticipate health management issues.

Keywords: IoT, AI, ML, HMS, Viability

1. INTRODUCTION

Humans necessitate medical care. Diseases can be cured or prevented if they are treated early on. CT scans allow for speedy diagnosis of internal or external bodily problems; PET scans allow for early detection of heart attacks, strokes, and other diseases. Disorders of ageing that cannot be predicted The increasing number of people in the world puts a strain on healthcare systems everywhere, necessitating more food and medical professionals if people are to receive high-quality healthcare [1]. The Internet of Things may help reduce stress in hospitals and other medical facilities. Before that time, infected patients were carefully watched. The goal of many scientists is disease prevention. The data used in assisted rehabilitation is plentiful; it is essential, but the importance of analysing and using it is rarely discussed. Various wearable frameworks are in place to protect data in transit. Data privacy and management of Internet of Things devices are significant concerns for medical clinic administrators and IT staff. [2]

[1]r.venkateswaran2020@gmail.com, [2]husna.tabbu786@gmail.com, [3]zatin.gupta2000@gmail.com

DOI: 10.1201/9781003534112-39

Artificial intelligence (AI) has been proposed as a shorthand for a set of practices unique to human culture (AI). We can create a more peaceful and prosperous tomorrow using our collective intellect. Because it can learn new things on its own without being specifically taught, AI has the potential to outperform even the most brilliant humans. We avoid the need for a formal justification for our results by supplementing traditional computation with actual data rather than writing code and providing reason based on the data.

There are many real-world benefits to learning about artificial intelligence (AI). Access to this information can lead to breakthroughs in research and development that ultimately benefit our daily lives and experiences as a whole. The Internet of Things (IoT) concept revolves around utilising the Internet to enable remote coordination and validation of physical objects. According to the McKinsey Global Institute's research, AI will be the driving force behind the next period of explosive expansion. Today, we talk about the Internet of Things (IoT) in the context of wireless sensing and data collection related to the health monitoring system.

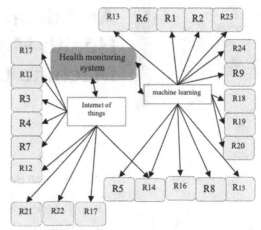

Fig. 39.1 System engineering methodology of a cutting-edge machine learning-based IoT-based health monitoring system

The safety of any new tool created by humans is always an issue. Health insurance is more important than ever in light of the recent economic damage caused by the Coronavirus. [3] Current solution: Internet of Things-based healthcare monitoring system, which is most effective in areas where the virus has already spread. Human services can now reach more people at a lower cost thanks to advancements in remote patient monitoring. With the help of sensors and the Internet to relay alerts, this project intends to create an intelligent patient health monitoring system. Reducing the need for emergency room visits, hospital stays, and diagnostic procedures is one way to build monitoring systems that can save money [4]. Sensors in the body's sheath communicate with a microcontroller, which communicates with an LCD and a remote alert connection, revealing temperature and precognition.

Let us say the system notices some temperature fluctuations in the brain. In that case, the Internet of Things sends real-time alerts to the client regarding the patient's heart rate and temperature. [5] We used an online T-set tolerances well-being monitoring system to speed up the screening process while maintaining a high standard of care. Unnecessary medical care is one way in which people can waste money. Remotely monitored smart sensors could save many lives in an area where doctors cannot predict the spread of an epidemic. [6].

Table 39.1 Displays the diagnostic criteria for standard health examinations

Pulse rate	Body temperature with patient ID		
	Low	Normal	High
Low	Diagnosis required	Unhealthy	Diagnosis required
Normal	Hypothermia	Healthy	Hyperthermia
High	Diagnosis required	Unhealthy	Diagnosis required

2. COMBINING AI WITH HEALTH TRACKING TECHNOLOGY

The medical field has been profoundly affected by technological developments. Access to healthcare is being revolutionised by the increasing interoperability of electronic health records (EHRs), the Internet of Things (IoT), telemedicine, and other information technologies. The field of machine learning is one of the most dynamic in the technological world today. As a result of machine learning's ability to streamline disease diagnosis and provide clinicians with tools to create more targeted treatment plans, the healthcare artificial intelligence market is expected to grow to more than USD 40.2 billion by 2020, expanding at a compound annual growth rate of 49.7 per cent [7].

More quickly and accurately than a neural network, it analyses massive amounts of patient data using complex algorithms. Machine learning management systems will partially replace human physicians, but these systems will make the doctors' jobs easier and help them make better decisions. Let us take a look at how ML is enhancing healthcare delivery. [8]

3. SERVER FOR IOT DEVICES

Sensors in a medical facility can convert a patient's vital signs into electrical signals as soon as they walk through the door. The electrical flag is then digitalised by the RFID chip and stored. Using a standardised format, the Zigbee Protocol transfers information between the embedded devices and a local server. [9] The Zigbee communication standard is compatible with this layout. These days, everyone wants their tech to be compact and efficient. Wireless LAN (Wi-Fi) transmits information between treatment facilities and therapeutics servers. There is a lot of data saved on the medical server. The server will append the new information to the patient's medical file before sending it to the specialist if one already exists on the therapeutic website. If the client has no therapy records in the server's database, the server will generate a unique identifier and save it there. The data is sent to the expert, who then reviews it. [10]

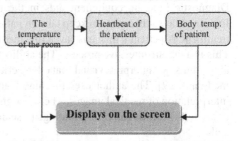

Fig. 39.2   The Monitoring System of Health

4. IMPLICATIONS OF IMPLEMENTING IOT IN HEALTHCARE

IoT technology enhances patient monitoring in healthcare by enabling real-time remote diagnosis, treatment, and cost-effectiveness. It facilitates precise diagnosis, electronic health records, and medication administration. However, privacy concerns and hacking risks remain, and a unified set of IoT standards is needed for full integration. Despite potential costs, IoT can ultimately lower healthcare costs [11-16].

5. PROMISING FUTURE OF IOT FOR HEALTH-RELATED MONITORING SYSTEMS

There will be a $billion market for Internet of Things (IoT) health technologies by 2022, reports Business Insider. Demand growth, improvements in 5G connectivity and IoT technology, and rising use of health IT software are all factors that will fuel this expansion. Apple, Google, and Samsung, the gap between fitness tracking apps and alternative medicine, will all play a role in this [17]. The Internet of Things (IoT) and the prospect of a further digital medical revolution will continue to fascinate and reshape the healthcare industry regardless of the drawbacks. Therefore, it is time to stop worrying about the barriers and move forward with the journey toward connected healthcare equipment [18].

6. USES FOR MACHINE LEARNING IN THE MEDICAL SURVEILLANCE SYSTEM

6.1 Maintaining Health Records

Hospitals began implementing electronic healthcare software as personal computers, and the Internet became more widespread in the early 1990s. According to CDC data [19], 85% of MDs and DOs use EMRs (Electronic Medical Records). Digital health records improve the availability of information. It is challenging to locate trustworthy sources of information. Tools for quickly processing data are within reach, thanks to machine learning—natural language processing and optical character recognition aid data collection and management. More than 700,000 doctors, nurses, and other medical professionals in the United States use Ciox [20]. The purpose of developing Healthsource was to enhance network coordination. The distribution of health information on a national scale can be streamlined with the help of AI and machine learning solutions, which also help manage the chaos of unorganised records and data. Data collection can also make use of NLP and handwriting recognition technology.

6.2 Accuracy improvements in treatment plans

Providing effective treatment frequently requires substantial effort from medical professionals. Cancer treatment is complex. Machine learning algorithms can analyse patient records and generate individualised treatment plans using databases of effective strategies to kill cancer cells while sparing healthy tissue. In 88% of radiation therapy for bladder cancer cases, treatment options based on machine learning were selected [21]. Research Laboratory is used in over 2,600 different medical facilities. One product with the intention of bettering medical care is Ray Station. The software generates personalised therapy planners using machine learning algorithms based on patient geometry, dosages, and treatment duration. With the help of privacy-preserving tools, medical facilities can compare notes and refine their practices.

6.3 Analysing medical images

Diagnostic imaging equipment aids in the treatment of illness. Imaging allows therapists to keep tabs on patient's progress during treatment. A patient's image is saved in a medical imaging system. This is a labour-intensive process. The ability of image processing algorithms to interpret visual data has critical applications in medicine [27]. The author uses machine learning to enhance the interpretation of medical images. Their recognition of multiple cell types will be invaluable to pathologists searching for malignant cells.

Table 39.2 The table represents the pulse rate and states of the pulse rate of the patient

heartbeat or pulse rate of the patient	States
60BPM – 120BPM	Normal
>120BPM	High
100BPM	Normal
<60BPM	Low

7. AREAS WHERE MACHINE LEARNING AND IOT CAN HELP IMPROVE HEALTHCARE

Without timely and appropriate therapeutic maintenance, patients with cardiac issues or who have been assaulted face an uncertain prognosis. The elderly, the young, the lighting experts, family members, and friends of a patient can all be easily identified in this way. Using sensors and the Internet to keep track of patient and family issues, we predict that Patient Health Monitoring will have similarly impressive results. ML methods (IoT) have enhanced many categories of IoT devices. AI is now commonplace, especially in the field of online commerce. Patterns in large datasets are what data mining is all about. Data mining is extracting valuable knowledge from large data sets or streams. Information preparation involves the collection of private information from numerous databases. In this article, we use a database of patients and related information to predict coronary illness in a patient population [23]. Classifiers that use parameters like temperature and heart rate can effectively summarise these summaries.

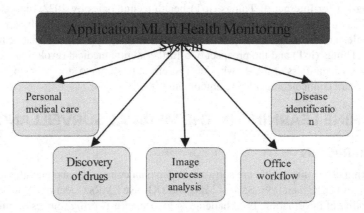

Fig. 39.3 Application of machine learning in the health monitoring system

For the sake of argument, assume that both the heart rate and the body temperatures (Low and minimal) are Low and High, High and Low, or Low and High (High and high). In such a scenario, the patient needs immediate medical attention. Patients are considered unwell if their pulse rate and internal temperature are regular and low. A patient is considered hypothermic if their heart rate is low and their body temperature is also common. Patients are believed to have a fever if their temperature falls between (Normal and high). A healthy patient has a temperature between 37 and 39 degrees Celsius [24].

8. CONCLUSION

Internet of Things devices, particularly for health monitoring, are quickly gaining acceptance as viable alternatives for remote value tracking. It paves the way for people to be kept in the cloud, where data on their financial well-being can be safely stored, and doctors worldwide can access it to make diagnoses and keep tabs on patients. This paper presents a healthcare monitoring system that uses the Internet of Things. The LCD screen showed the user's internal temperature, the heart rate monitor's readings, and the room's temperature and humidity. These sensor readings are sent to a medical server through a wireless connection. The data is transmitted to the intended recipient via an Internet of Things-enabled mobile device. Despite the results, the doctor may decide to proceed with a diagnosis.

REFERENCES

1. R. R, "Sewage Environment and Workers Health Monitoring System Using IOT and ML", International Journal for Research in Applied Science and Engineering Technology, vol. 10, no. 7, pp. 1688-1694, 2022. Available: 10.22214/ijraset.2022.45572.

2. "IoT-based Animal Health Monitoring and Crop Monitoring System", International Journal of Recent Trends in Engineering and Research, vol. 3, no. 2, pp. 252-257, 2017. Available: 10.23883/ijrter.2017.3035.qudpb.

3. "Human Health Monitoring System using IOT", International Journal of Recent Trends in Engineering and Research, vol. 4, no. 4, pp. 425-432, 2018. Available: 10.23883/ijrter.2018.4256.fsht9.

4. T. Patel, P. Sharma, V. Yadav, and S. Suman, "IoT Based Remote Health Monitoring System", SSRN Electronic Journal, 2020. Available: 10.2139/ssrn.3536149.

5. "Implementation of low-cost Automated Wheelchair with Healthcare Monitoring System using IoT and ML", International Journal of Pharmaceutical Research, vol. 12, no. 03, 2020. Available: 10.31838/ijpr/2020.12.03.193.

6. "A Health Care Monitoring System with Wireless Body Area Network using IoT", International Journal of Recent Trends in Engineering and Research, vol. 3, no. 11, pp. 112-117, 2017. Available: 10.23883/ijrter.2017.3499.edo9r.

7. D. S., "Cryptography-based Security Solutions to IoT Enabled Health Care Monitoring System", Journal of Advanced Research in Dynamical and Control Systems, vol. 12, no. 7, pp. 265-272, 2020. Available: 10.5373/yards/v12i7/20202008.

8. "IOT BASED HEALTH MONITORING SYSTEM", Journal of Critical Reviews, vol. 7, no. 04, 2020. Available: 10.31838/jcr.07.04.137.

9. "Health Care Monitoring System Using Internet of Things (IoT)", International Journal of Pharmaceutical Research, vol. 12, no. 03, 2020. Available: 10.31838/ijpr/2020.12.03.194.

10. S, A. Phaniraj, G. K, J. Thomas, and V. Mahale, "SMS: Soil Monitoring System using IoT and ML", Journal of Web Development and Web Designing, vol. 05, no. 02, pp. 10-19, 2020. Available: 10.46610/jowdwd.2020.v05i02.003.

11. P. Soam, P. Sharma and N. Joshi, "Health Monitoring System Using IoT: A Review", SSRN Electronic Journal, 2020. Available: 10.2139/ssrn.3606060.

12. S. Arora and A. Goel, "IoT Smart Health Monitoring System", SSRN Electronic Journal, 2020. Available: 10.2139/ssrn.3602545.

13. "Patient Health Monitoring System Using Iot Gsm", Journal of critical reviews, vol. 7, no. 05, 2020. Available: 10.31838/jcr.07.05.340.

14. . P, A T, A. A and D. Devi, "IoT and ML Based Monitoring of Urban Wastewater System", International Journal of Electronics and Communication Engineering, vol. 7, no. 6, pp. 17-22, 2020. Available: 10.14445/23488549/piece-v7i6p103.

15. M. Taştan, "IoT Based Wearable Smart Health Monitoring System", Celal Bayar Üniversitesi Fen Bilimleri Dergisi, pp. 343-350, 2018. Available: 10.18466/cbayarfbe.451076.

16. M. Kulkarni and M. Kulkarni, "Secure Health Monitoring of Soldiers with Tracking System using IoT: A Survey", International Journal of Trend in Scientific Research and Development, vol. -3, no. -4, pp. 693-696, 2019. Available: 10.31142/ijtsrd23834.

17. P. Gupta* and A. Bisht, "IOT-based Patient Health Monitoring System using ML", International Journal of Engineering and Advanced Technology, vol. 9, no. 1, pp. 5086-5091, 2019. Available: 10.35940/ijeat.a2148.109119.

18. Suman Mann, Amit Kumar Bindal, Archana Balyan, Vijay Shukla, Zatin Gupta, Vivek Tomar, Shahajan Miah, "Multiresolution-Based Singular Value Decomposition Approach for Breast Cancer Image Classification", BioMed Research International, vol. 2022, Article ID 6392206, 11 pages, 2022. https://doi.org/10.1155/2022/6392206

19. Mann, Suman, et al. "Artificial Intelligence-based Blockchain Technology for Skin Cancer Investigation Complemented with Dietary Assessment and Recommendation using Correlation Analysis in Elder Individuals." Journal of Food Quality 2022 (2022)

20. R. Gopi, S. Veena, S. Balasubramanian, D. Ramya, P. Ilanchezhian et al., "IoT based disease prediction using MapReduce and lsqn3 techniques," Intelligent Automation & Soft Computing, vol. 34, no.2, pp. 1215–1230, 2022.

21. Z. Gupta and A. Bindal, "Comprehensive Survey on Sustainable Smart Agriculture using IOT Technologies," 2022 2nd International Conference on Advance Computing and Innovative Technologies in Engineering (ICACITE), 2022, pp. 2640-2645, DOI: 10.1109/ICACITE53722.2022.9823837.

22. Bansal, P. Saxena, R. Garg and Z. Gupta, "Data Analysis based Digital Step Towards Waste Management," 2022 6th International Conference on Trends in Electronics and Informatics (ICOEI), 2022, pp. 430-435, doi:10.1109/ICOEI53556.2022.9777149.

23. R. Raman, R. Singh, Z. Gupta, S. Verma, A. Rajput and S. M. Parikh, "Wireless Communication With Extreme Reliability and Low Latency: Tail, Risk and Scale," 2022 5th International Conference on Contemporary Computing and Informatics (IC3I), Uttar Pradesh, India, 2022, pp. 1699-1703, doi: 10.1109/IC3I56241.2022.10073096.

24. SK Hasane Ahammad, D. K. J.Saini, Phishing URL detection using machine learning methods, Advances in Engineering Software, Volume 173, 2022,103288, ISSN 0965-9978, https://doi.org/10.1016/j.advengsoft.2022.103288

25. Yadav, P, S. Kumar, and D. K. J.Saini. "A Novel Method of Butterfly Optimization Algorithm for Load Balancing in Cloud Computing". International Journal on Recent and Innovation Trends in Computing and Communication, vol. 10, no. 8, Aug. 2022, pp. 110-5, doi:10.17762/ijritcc.v10i8.5683.

26. R. Raman, D. Buddhi, G. Lakhera, Z. Gupta, A. Joshi and D. Saini, "An investigation on the role of artificial intelligence in scalable visual data analytics," 2023 International Conference on Artificial Intelligence and Smart Communication (AISC), Greater Noida, India, 2023, pp. 666-670, doi: 10.1109/AISC56616.2023.10085495.

27. R. Raman, Z. Gupta, A. Gupta, S. V. Akram, D. Saini and J. Bhalani, "Internet-Of-Things Wireless Communication with a Focus on the Protection of User Privacy and the Delivery of Relevant Facts," 2023 International Conference on Artificial Intelligence and Smart Communication (AISC), Greater Noida, India, 2023, pp. 656-660, doi: 10.1109/AISC56616.2023.10084975.

28. Shailesh Kamble, Dilip Kumar J. Saini, Vinay Kumar, Arun Kumar Gautam, Shikha Verma, Ashish Tiwari & Dinesh Goyal (2022) Detection and tracking of moving cloud services from video using saliency map model, Journal of Discrete Mathematical Sciences and Cryptography, 25:4, 1083-1092, DOI: 10.1080/09720529.2022.2072436

29. G. S. Kumar, D. De la Cruz-Cámaco, M. Ravichand, K. Joshi, Z. Gupta and S. Gupta, "Monitoring and Predicting Performance of Students in Degree Programs using Machine Learning," 2023 10th International Conference on Computing for Sustainable Global Development (INDIACom), New Delhi, India, 2023, pp. 1311-1315.

30. Jang Bahadur, D. K., and L.Lakshmanan, "Wireless Sensor Network Optimization for Multi-Sensor Analytics in Smart Healthcare System" Specialusis Ugdymas, ISSN NO: 1392-5369 Vol. 1 No. 43 (2022)

31. R. Raman, Z. Gupta, S. V. Akram, D. Lalit Thakur, B. G. Pillai and M. K. Chakravarthi, "Network Security Concerns for Designing Robotic Systems: A Review," 2023 International Conference on Artificial Intelligence and Smart Communication (AISC), Greater Noida, India, 2023, pp. 661-665, Doi: 10.1109/AISC56616.2023.10085453.

Note: All the figures and tables in this chapter were made by the author.

Computational Intelligence and Mathematical Applications – Dr. Devendra Prasad et al. (eds)
© 2024 Taylor & Francis Group, London, ISBN 978-1-032-87721-1

Artificial Intelligence and Machine Learning in Women's Health: Unveiling PCOS with Image Analysis

Rishit Aggarwal[1]

Department of Computer Science and Engineering, Graphic Era University, Dehradun, India

Shweta Jain[2], Anju Bhandari Gandhi[3], Drishti Dhingra[4]

Department of Artificial Intelligence and Machine Learning, Panipat Institute of Engineering and Technology, Samalkha, Panipat, India

S.C. Gupta[5]

Department of Computer Science and Engineering, Panipat Institute of Engineering and Technology, Samalkha, Panipat, India

Abstract: One common endro-endocrinological dysfunction among women of reproductive age is Polycystic Ovary Syndrome (PCOS). A group of sex hormones called androgens are overproduced in women, and this can lead to a variety of syndromes, including PCOS. PCOS is responsible for a number of syndromes, such as oligo-ovulation, hirsutism, alopecia, acne, and hyperandrogenemia. It also contributes significantly to female infertility. Worldwide, PCOS affects 15% of women who are of reproductive age. One cannot stress how important it is to identify PCOS early on because of the severity of its harmful effects. PCONet, a Convolutional Neural Networking based (CNN)model designed to recognize infected polycystic ovaries from fit ovarian ultrasound imaging part. Moreover, it contains improved InceptionV3, a 45-layer pre-trained convolutional neural network, by applying the transfer learning technique to the classification of ultrasound pictures of polycystic ovarian cancer. We compared these two models using a range of quantitative performance evaluation metrics and found that, with an outstanding performance of 98.12%, PCONet is the better model, while the optimised InceptionV3 showed an accuracy of 96.56% on testing-based imaging dataset.

Keywords: Efficient machine learning (ML), Bioengineering, Polycystic ovary syndrome, Ultrasound images, Image classification, Convolutional neural networks

1. INTRODUCTION

PCOS is a complex disorder with a wide range of clinical symptoms, including hyperandrogenism (hirsutism, hyperandrogenaemia), ovarian dysfunction (oligo-ovulation, PCOM), and hyperandrogenism. Currently used and managed by the majority of scientific communities and healthcare based associations, the Rotterdam criteria is the most widely accepted classification for PCOS. By definition, if a patient exhibits at least two of the three symptoms, they are diagnosed with PCOS. Ovulation disorder, ovulation dysfunction, and hyperandrogenism are these symptoms. Figure 40.1 illustrates the distinction between a polycystic ovary and a normal ovary.[1-4]

PCOS is a complex disorder that comprises a number of clinical symptoms, ovarian dysfunction (oligo- ovulation, PCOM), and hyperandrogenism. The Rotterdam definition, which is now the most widely acknowledged categorization for PCOS, is properly

[1]rishitaggarwal1@gmail.com, [2]jainshweta290204@gmail.com, [3]anjugandhi.cse@piet.co.in, [4]drishtidhingra24@gmail.com, [5]hod.cse@piet.co.in

DOI: 10.1201/9781003534112-40

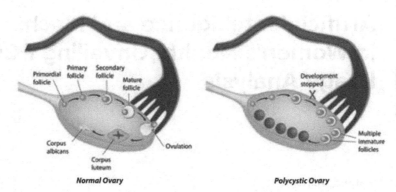

Fig. 40.1 Normal ovary vs polycystic ovary

Source: https://www.news-medical.net/health/Polycystic-Ovary-Syndrome-%28PCOS%29-Irregular-Function-of-Ovaries.aspx

managed by the vast scientific groups and health bodies. By definition, if a patient exhibits at least two of the three symptoms, they are diagnosed with PCOS. Chronic oligomenorrhea and ovulation disorders leading to endometrial hyperplasia, infertility, and carcinoma; overabundance of androgen potentially resulting in cutaneous manifestations; and PCOM potentially. This raises the risk of ovarian hyperstimulation syndrome,which can have significant effects.

Our contributions to this paper-

PCOS detection from ultrasound images required substantially less work than PCOS categorization utilising data tables containing both numerical and categorical information. This was discovered by reviewing prior studies on deep learning-based PCOS detection.

(a) Developed a CNN model for detecting ovarian cysts using ultrasound pictures of the ovaries. Our approach surpasses the pre-trained model in identifying the ultrasound pictures in our test dataset, with an ovarian cyst detection accuracy of 96.40%.

(b) Refined and tailored the convolutional neural network (CNN) model to detect ovarian cysts in ultrasound pictures of the ovaries using our dataset.

2. RELATED WORK AND LITERATURE REVIEW

The following methods were used to diagnose PCOS: k-nearest neighbor (KNN), decision tree classifier (DT's Classifier), naive bayes(NB), logistic regression (LR), and support vector machine(SVM). It was evaluated that the Decision Tree was the most efficient and applicable strategy in their specific situation. To recognize PCOS disease from the patient's clinical informational data table, classifiers including logistic regression, random forest, naive Bayes classification, support vector machines, and classification and regression trees (CART) were utilised.[2,3]

With an accuracy of 96%, the random forest classifier was the most accurate. Kumari utilized DenseNet-121, VGG-19, DenseNet-121, InceptionV3, and ResNet-50 to classify PCOS based on an image dataset. Her findings revealed that VGG-19 had the highest accuracy among the four, scoring 70%. Using the efficient architecture of Generative Adversarial Networks (GANs) & supplementing the data information, the author of this study was able to overcome the limitations imposed by the dataset. There were 94 images in the collection; 50 of those showed PCOS, while the other 44 were ultrasonic images without any PCOS imagery.[5-8]

3. METHODOLOGY AND PROPOSED SYSTEM

Two models of convolutional neural networks were employed in this investigation. We have tuned the InceptionV3 convolutional neural network to match the requirements of our dataset by optimizing the model using the transfer learning technique. Furthermore, Using our dataset, we created a convolutional neural network (Lightweight CNN), properly evaluating ovarian cysts.

3.1 Dataset Description

Two distinct datasets—Dataset A and B—made up our database for this project. Our models were trained and validated using Dataset A. However, to test the models objectively, we employed a discrete dataset B rather than separating a test set from dataset A. The description of the dataset is:

(a) *Dataset A:* We downloaded dataset A from Kaggle, which served as our first dataset and was used to train and evaluate both of our models. At first, Dataset A contained 1,143 training set images that were not infected and 781 images that were infected. Ultrasound images of infected cysts on the ovary were labelled as "infected," whereas those of a fit normal ovary were labelled as "not infected." Figure 40.2 shows representative ultrasound pictures of both healthy and cystic ovarian tissues.

(b) *Dataset B:* Now, taking the test dataset, Dataset B, we acquired an ultrasound imaging dataset of cystic and healthy ovarian tissues is shown in Fig. 40.3.

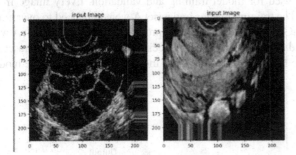

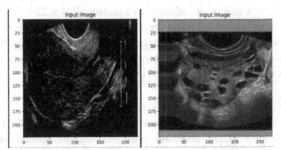

Fig. 40.2 Dataset A (First input is a polycystic ovarian ultrasound image and second input is a healthy fit ovarian ultrasound image)

Fig. 40.3 Dataset B (First input is a healthy fit ovarian ultrasound image and Second input is a infected polycystic ovarian ultrasound image)

Source: https://www.kaggle.com/datasets/anaghachoudhari/pcos-detection-using-ultrasound-images

The training workflow of a deep learning model for Polycystic Ovary Syndrome (PCOS) includes preprocessing and dataset collection. Medical photos or other relevant features might be included in the dataset. Following the creation of the model architecture, which is depicted in Fig. 40.4., The dataset is divided into training , validation sets. Subsequently, the model is trained on the training set, in order to minimize error, backpropagation is used to adjust its weights. Hyperparameters can

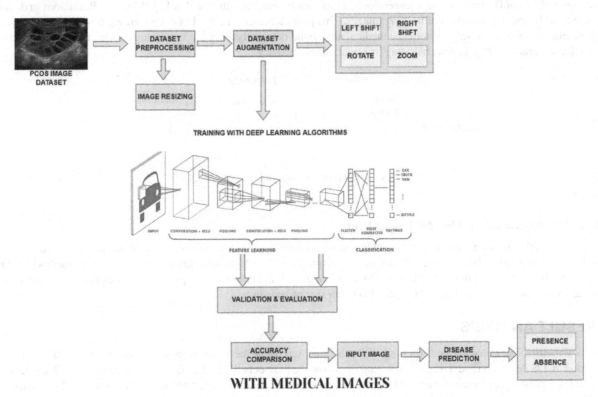

Fig. 40.4 Workflow for training and evaluating our models

Source: Author

be adjusted to improve performance. After training, the model is evaluated using an independent test set to determine its generalization capabilities. To assess how well the model detects PCOS, performance metrics such as specificity, sensitivity, and accuracy are computed. This allows the deep learning model to be optimized for PCOS detection. It is possible to improve model accuracy through iterative refinement. For robustness, the final model is validated on fresh, untested data. The procedure includes an ongoing feedback loop of refinement and testing to optimize the deep-learning model for PCOS detection.

3.2 Image Preprocessing

Images of various sizes were included in dataset A, which was used for model training and validation. Every image from datasets A and B was adjusted to a uniform 224 x 224-pixel size. The training images in dataset A were enhanced to overcome constraints after the image data was normalized by the ImageDataGenerator from Keras. Approximately 80% of dataset A was used in the training set, with the remaining 20% used for the validation set. The test set, specifically designated as Dataset B. Since each set included 16 batches, variations in the dataset were taken into account and a thorough evaluation of the model was achieved on standardized images.

3.3 Model Description

In the research paper, a convolutional neural network (CNN) model for PCOS detection using medical ultrasound images is presented. The model, which makes use of the MobileNet architecture, is started with weights that have already been trained, giving effective support for the extraction of features from images. To ensure that features learned during training are preserved, the base MobileNet's layers are all frozen. According to Fig. 40.5, the output that has been flattened is fed into a dense layer that has a sigmoid activation, making it easier to classify PCOS cases as binary. The binary cross-entropy loss and RMSprop optimizer

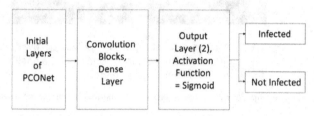

Fig. 40.5 Simple architecture of PCONet

Source: Author

are used to compile the model, with accuracy being the primary evaluation metric. Resizing to a consistent 224x224 pixels solves the problem of fluctuating image dimensions in this creative way, as illustrated in Fig. 40.6. The limitations in dataset A are countered by the augmentation of training images. This paper advances the field of deep learning based PCOS diagnostics by presenting a viable approach to automated medical ultrasound data analysis. After 30 training epochs, the model achieves an astounding accuracy of 96.4%, demonstrating its remarkable efficiency.

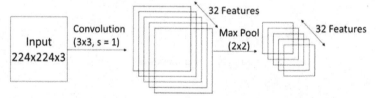

Fig. 40.6 The first convolutional block of PCONeT

Source: Author

3.4 Training Dataset and Model Evaluation

The dataset used was heterogeneous and included images of different sizes, so careful preprocessing was required. Both models were subjected to the same training set and image preprocessing technique, guaranteeing a uniform method for model comparison. Both models underwent thirty epochs of training, with the number of steps in each epoch being calculated as ,the total number of images was divided by the selected batch size of sixteen.

4. RESULT ANALYSIS

Accuracy metrics calculated on both training and validation sets were the main focus of the quantitative evaluation of our suggested PCOS detection model. Figure 40.7 illustrates how the dataset facilitated a binary classification task. It was labelled - 'infected' for polycystic ovarian ultrasound images; 'not infected' for healthy ovarian ultrasound images. PCONet's impressive 96.4% accuracy on the training set showed off its ability to discriminate between ultrasound images of healthy ovaries and those with polycystic ovarian cysts.

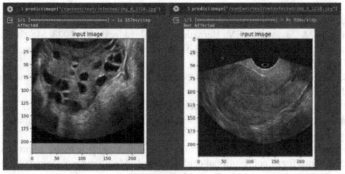

Fig. 40.7 Outputs

Source: https://www.kaggle.com/datasets/anaghachoudhari/pcos-detection-using-ultrasound-images

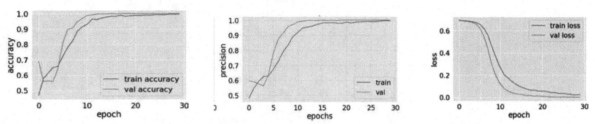

Fig. 40.8 Training and validation accuracy, precision, and loss

Source: Results from the code

We used dataset B, which consists of actual hospital images, to test the accuracy of our models, and the results showed 96.4% accuracy. In our test set, there were 1143 healthy ovarian images and 781 infected images. Despite being a significantly lighter model than InceptionV3, PCONet has an accuracy of 98.12% on our test set, demonstrating 1.56% higher accuracy than InceptionV3, which has an accuracy of 96.56% on our test set. When it comes to detecting cystic ovarian cancer, PCONet outperforms InceptionV3 by 2%.

5. CONCLUSIONS

To sum up, this study presents a CNN model designed especially for polycystic ovarian ultrasound image classification. The work is extended to optimise a pre-trained InceptionV3 model, and a thorough quantitative analysis shows that PCONet achieves higher accuracy (98.56%) than InceptionV3 (98.12%). PCONet's strong performance is further validated by analyses of precision, recall, and f1 score. The impartial assessment, carried out on a separate test set, guarantees the validity of the comparative outcomes. PCONet's implementation in healthcare facilities is expected to result in a notable progress in the early identification of Polycystic Ovary Syndrome (PCOS), which will facilitate prompt interventions aimed at reducing its detrimental effects. Subsequent efforts will focus on improving PCONet's accuracy and investigating how well ovarian image detection works with different imaging modalities, including magnetic resonance imaging, ultrasound, and photoacoustic imaging. This study demonstrates the potential of PCONet to transform PCOS detection and contribute to all-encompassing healthcare solutions, laying the groundwork for innovative advances in reproductive health diagnostics.

REFERENCES

1. Suha, S.A. and Islam, M.N., 2022. An extended machine learning technique for polycystic ovary syndrome detection using ovarian ultrasound image. *Scientific Reports, 12(1), pp. 17123.*
2. Gopalakrishnan, C. and Iyapparaja, M., 2021. Multilevel thresholding-based follicle detection and classification of polycystic ovary syndrome from ultrasound images using machine learning. *International Journal of System Assurance Engineering and Management, pp. 1-8.*
3. Bhosale, S., Joshi, L. and Shivsharanan, A., 2022. PCOS (polycystic ovarian syndrome) detection using deep learning. *International Research Journal of Modernization in Engineering Technology and Science, 4(01).*

4. Ahmed, S., Rahman, M.S., Jahan, I., Kaiser, M.S., Hosen, A.S., Ghirime, D. and Kim, S.H., 2023. A review on the Detection techniques of polycystic ovary syndrome using Machine Learning. *IEEE Access.*

5. Chitra, P., Srilatha, K., Sumathi, M., Jayasudha, F.V., Bernatin, T. and Jagadeesh, M., 2023, March. Classification of Ultrasound PCOS image using Deep learning-based hybrid models. In 2023 *Second International Conference on Electronics and Renewable Systems (ICEARS) (pp. 1389-1394). IEEE.*

6. GÜLHAN, P.G., ÖZMEN, G. and ALPTEKİN, H., 2023. CNN based Determination of polycystic ovarian syndrome using Automatic Follicle Detection Methods. *Politeknik Dergisi, p.1-1.*

7. Abouhawwash, M., Sridevi, S., Sundararajan, S.C.M., Pachlor, R., Karim, F.K. and Khafaga, D.S., 2023. Automatic diagnosis of polycystic ovarian syndrome using Wrapper methodology with deep learning techniques. *Computer Systems Science & Engineering, 47(1).*

8. Ravishankar, T.N., Jadhav, H.M., Kumar, N.S. and Ambala, S., 2023. A deep learning approach for ovarian cyst detection and classification (OCD-FCNN) using a fuzzy convolutional neural network. *Measurements: Sensors, 27, pp. 100797.*

9. Jain, S., Gandhi, A., Singla, S., Garg, L. and Mehla, S., 2022, December. Quantum Machine Learning and Quantum Communication Networks: The 2030s and the Future. In *2022 International Conference on Computational Modelling, Simulation and Optimization (ICCMSO) (pp. 59-66). IEEE*

Computational Intelligence and Mathematical Applications – Dr. Devendra Prasad et al. (eds)
© 2024 Taylor & Francis Group, London, ISBN 978-1-032-87721-1

41 | Hybrid FUCOM-TOPSIS based Matching of Users' Requirements with Life Insurance Policy Features

Asha Rani*, Kavita Taneja
Department of Computer Science and Applications, Panjab University, Chandigarh, India

Harmunish Taneja
Department of Computer Science and IT, DAV College, Sector 10, Chandigarh, India

Abstract: Selecting a life insurance policy has become quite challenging due to the extensive options resulting from increased competition in the insurance sector. There is extreme need for an intelligent life insurance recommender system that considers not only demographic features but also personalized priorities for policy criteria. This paper proposes hybrid FUCOM-TOPSIS based matching of users' requirements with life insurance policy features using Full Consistency Method (FUCOM) and Techniques for Order of Preference by Similarity to Ideal Solution (TOPSIS). The experimentation involves assessing 165 potential life insurance policies based on 14 preference criteria derived after analysing user demographic information. The stability of the results is established through sensitivity analysis. The correlation coefficient analysis demonstrates the correctness of proposed matching method for effective life insurance recommendations.

Keywords: Weight elicitation, Preference elicitation, Life insurance, Multi criteria decision making (MCDM), FUCOM, TOPSIS, Recommender system

1. INTRODUCTION

Life insurance is a crucial financial instrument for those who have dependents or financial responsibilities, since it provides a method to support their loved ones. It might be difficult to choose the best insurance provider due to the complexities and wide range of preferences involved while buying life insurance (Pattnaik et al., 2021). Online life insurance recommendations integrate efficiency, convenience, and thorough research to help people choose the best suited life insurance policy. MCDM is found to be mapping well to the domain due to higher complexities, absence of ratings, lesser frequency of visitors and preferences of the user (Rani et al., 2021a).

Weight elicitation and preference elicitation are the two major steps in MCDM approaches. Scaling, weight ranking, point allocation methods etc. are some direct weighting methods while the indirect approach include weight generated from theories and mathematical models (Ginevicius & Podvezko, 2004).

By integrating the Full Consistency Method (FUCOM), a pair-wise comparison methodology, with the methodology for Order of Preference by Similarity to Ideal Solution (TOPSIS), a group decision-making strategy, in life insurance selection domain, the proposed paper demonstrates a new hybrid approach for required matching method.

This matching method includes FUCOM and TOPSIS methods for the life insurance policy selection by the users in order to provide them the best insurance cover which matches their personalized requirements and priorities. The novelty of the proposed method is its suggested application in the life insurance domain with 165 life insurance policies (alternatives) and

*Corresponding author: asharana20@gmail.com

DOI: 10.1201/9781003534112-41

further 14 criteria. An efficient matching method in the domain of life insurance will lead to increase in insurance penetration and will also increase policy holder safeguards.

Organization of the paper is as follows. Section II shows literature review on weight elicitation methods and various MCDM methods employed in life insurance domain. Section III illustrates hybrid FUCOM-TOPSIS method for matching user's requirements life insurance policy features. Section IV provides the experimentation and results and discussion are discussed in section V. Finally, Section VI concludes the paper.

2. LITERATURE SURVEY

MCDM has several uses in a variety of fields and professions, including finance, education, supply chain management, airlines, healthcare, engineering production, economics, etc. The choice of a weighting mechanism is an important step in MCDM problems since it facilitates the DM process and helps to identify the relative relevance of each criterion (Ayan et al., 2023). Literature shows various methods of weight elicitation (Aldian, 2005)(Diakoulaki et al., 1995)(Pamucar & Ecer, 2020)(Pamučar et al., 2018)(Ginevicius & Podvezko, 2004)(Odu, 2019)(Kao, 2010)(Hung & Chen, 2009).

G.O. Odu, (Odu, 2019) establish the fact that subjective weighing methods are simpler and more straightforward to calculate than objective weighting approaches, which use a mathematical relation to calculate the weights without the customer's involvement. However, they provide a clearer explanation of the elicitation process and are more frequently employed in practice(Singh & Pant, 2021).

FUCOM, AHP, ANP and BWM cover the major domain of subjective methods. The criteria weights in these methods are dependent of the priorities for each criterion supplied by the customer (Do & Nguyen, 2022). The number of comparison steps and the constraints defined while determining the criteria values in FUCOM weight method makes it a better choice than that of other methods including AHP weight, ANP and BWM weight (Pamučar et al., 2018). By computing the error value (DFC), for the generated weight vectors, FUCOM offers capability to validate the model. FUCOM provides more accurate findings in terms of consistency and is considerably more versatile and less complicated (Pamučar et al., 2018).

A few MCDM methods such as, GRA (Pandey & Pandey, 2017), AHP (Hinduja & Pandey, 2018) and MAUT (Abbas et al., 2015) are studied and employed for recommending life insurance products. Lack of end-to-end solution for life insurance recommendation is observed in literature study due to inefficient matching of user's requirements with policy features. TOPSIS is found to be the most widely used in MCDM methods (Rani et al., 2021b)(Trung et al., 2021). Research comparing TOPSIS with methods like MOORA, SAW, WASPAS, VIKOR, and COPRAS concludes TOPSIS yields comparable, effective results while maintaining simplicity and processing efficiency. However, using TOPSIS alone may lead to uncertainties in reliability and relevant outcomes (Ali et al., 2020). Recognizing these limitations, a new hybrid method integrating FUCOM and TOPSIS that could address matching issues and provide consistent outcomes with reduced mathematical complexity is need of the hour.

3. HYBRID FUCOM-TOPSIS METHOD

1. Store user comparative priority score of life insurance policy criteria in an array to form an objective function.
2. Sort life insurance policy criteria with respect to comparative priority score of criteria.
3. Calculate performance matrix of life insurance policies
4. Perform matching of user's requirement with life insurance policy features
    4.1 Obtain weight coefficients using FUCOM
    4.2 Calculate weighted normalized decision matrix by using weight coefficients and performance matrix of life insurance policies
    4.3 Calculate rank of recommendation preference order by applying TOPSIS method.
5. Sort ranking of policies with respect to closeness coefficient.

4. EXPERIMENTATION

The foremost requirement for the matching method is the dataset. Lack of availability of standardized dataset in life insurance domain prompted the creation of policy data model and user data model. The policy brochures of 165 life insurance policies from 5 life insurance companies (Life Insurance Corporation of India (LIC), SBI Life Insurance Co. Ltd., HDFC Standard Life Insurance Co. Ltd., ICICI Prudential Life Insurance Co. Ltd. and Bajaj Allianz Life Insurance Co. Ltd.) are used for creating the policy database categorizing the details of each policy in 25 different policy criteria. Data comprising of demographic details and preferences for policy features is collected from 800 perspective users through a questionnaire and user data model is created.

Hybrid FUCOM-TOPSIS method is applied for matching of users' requirements with life insurance policy features.

Recommendation of the alternatives: Based on the matching score obtained from step (2) ranking of alternatives is performed.

Proposed matching method: User data model and policy data model

A total of 14 criteria of life insurance policies have been chosen for the present study. These features/criteria are selected after a literature review and consultation with domain experts. Life insurance policy criteria are categorized into policy constraints (checks the eligibility of the customer for the particular policy) and policy features (takes comparative score from customer). The proposed matching method requires 04 policy constraints, i.e., Entry Age (P1), Term (P2), Sum Assured (P3) and Maturity Age (P4). Also, 10 policy criteria are identified as Premium Amount (C1), Premium payment flexibility (C2), Tax benefits (C3), Benefits on death (C4), Benefits on survival (C5), Policy Loan (C6), Bonus (C7), Riders availability (C8), Premium rebate (C9) and Claim Settlement ratio (C10) for supporting proposed matching based decision making which are to be assessed for selecting a preferred personalized policy by the user. The user preferences on these criteria are compared and matched with the performance of different policies on each criteria and the best policy is recommended. The policy data model contains the quantized score of each policy over each criterion.

5. RESULTS AND DISCUSSION

The proposed hybrid FUCOM-TOPSIS method is utilized to analyze the given data and match with the optimal policy option.

The proposed method first takes demographic features as listed in Table 41.1 from the user through the interface. The policy constraints previously kept in the database are compared to these features. In this scenario, this user-provided demographic information is found to have a match with 10 life insurance policies from the database.

Table 41.1 User inputs- demographic features

Features	Criterion Name
Age	45
Sum Assured	600000
Annual Income	500000

The quantitative benefits that a policy generates for its client are obtained by normalizing its performance on each criterion and is already estimated and recorded in the knowledge repository. In next step, user's preferences for different criteria of the life insurance policies are taken. User is asked to select the most important criterion and assign it the rank one. Rest all 10 criteria are compared with this criteria and given a comparable relative priority on the scale [1- 9]. Table 41.2 shows the ranking and criteria priorities. FUCOM method is applied on the inputs and the weights for the 10 criteria are obtained.

Table 41.2 Priority score assigned to criteria with respect to the most preferred criteria C10

Criteria	C10	C2	C1	C7	C9	C3	C4	C5	C8	C6
Comparative Score	1	2	2	4	5	6	6	7	8	9
Weights generated through FUCOM	0.158	0.158	0.052	0.052	0.045	0.035	0.0790	0.039	0.063	0.316

The proposed method then matches each criterion of 10 policies to the preferences specified by the user. TOPSIS method is applied for preference elicitation and providing the ranking of the policies on the basis of users' unique personalized requirements. In addition to keeping its simplicity, TOPSIS is effective at describing each choice in mathematical form and has strong processing efficiency. This approach is based on the fact that the best solution should be the one that is closest to the ideal solution.

The alternative whose proximity coefficient (Ci) is close to 1 is chosen as the best option in the last phase. In Table 41.3, closeness coefficient of Policy no. 8 is 1 and hence it is found to be the best match policy for the given preferences and constraints. Policy no. 9 is the 2nd preferred matched option. The graph in Fig. 41.1 depicts the distance for each policy from ideal solution and anti-ideal solution using FUCOM-TOPSIS approach.

5.1 Sensitivity Analysis

Sensitivity analysis assesses the ranking's consistency and stability concerning criteria weights. Ten test runs were conducted, altering weight coefficients to observe changes in alternative ranks. Throughout the ten runs, Policy No. 8 consistently secures

Table 41.3 Result of hybrid FUCOM-TOPSIS method (Value of closeness coefficient of each life insurance policy)

Policy name	Closeness coefficient C_i
Policy 1	0.15437530
Policy 2	0.15437530
Policy 3	0.14334115
Policy 4	0.28914181
Policy 5	0.32457852
Policy 6	0.12390398
Policy 7	0.41491107
Policy 8	1.00000000
Policy 9	0.48202238
Policy 10	0.57919406

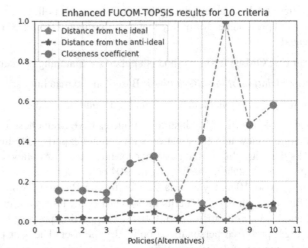

Fig. 41.1 Ideal solution, anti-ideal Solution and closeness coefficient obtained from FUCOM-TOPSIS approach

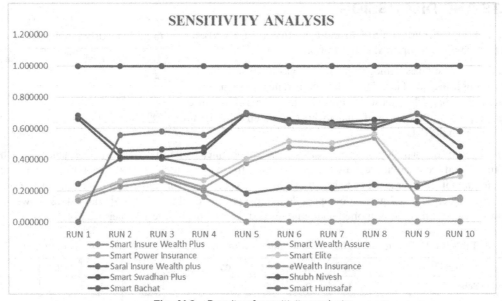

Fig. 41.2 Results of sensitivity analysis

the top matched rank as per user's requirement. Policy No. 10 emerges as the second-best option in 6 out of 10 runs. If premium payment flexibility (C6) is prioritized, Policy No. 7 becomes the optimal choice due to its highest score in this criterion. The sensitivity analysis in Fig. 41.2 demonstrates that variations in priority ranking and scores correspond with changes in our hybrid FUCOM-TOPSIS method, making it well-suited for recommending personalized, unbiased policies to users. The matching method exhibits dynamic adaptability to the policy criteria weight change. The computed coefficient of correlation shown in Fig. 41.3 ranging from 0.88 to 1 corresponds with the effective matching of user's requirements with life insurance policy criteria.

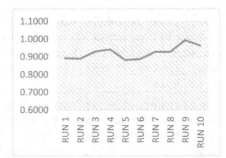

Fig. 41.3 Correlation coefficient between various test runs of the hybrid FUCOM-TOPSIS method

6. CONCLUSION

People often find insurance policies complex and rely on agents for selection. Suggestions from agents and web aggregators may be biased and lack personalization. An intelligent recommender system in the insurance domain is essential for recommending individual life insurance policies. The proposed hybrid FUCOM-TOPSISMCDM based matching method offers end-to-end solutions with personalized policy recommendations based on user demographics and priorities. The paper identifies features and criteria of users and life insurance policy and policy and user data models with dataset from 800 life insurance policy seekers in India. Sensitivity analysis for 10 test runs shows dynamic adaptation in priority rankings and scores of the proposed matching method. Also, the correlation coefficient, ranging from 0.88 to 1, indicates direct matching with life insurance policy preference scores and user-provided criteria preferences through the proposed matching method.

REFERENCES

1. Abbas, A., Bilal, K., Zhang, L., & Khan, S. U. (2015). A cloud based health insurance plan recommendation system: A user centered approach. *Future Generation Computer Systems*, *43–44*, 99–109. https://doi.org/10.1016/j.future.2014.08.010

2. Aldian, A. (2005). A Consistent Method To Determine Flexible Criteria Weights for Multicriteria Transport Project Evaluation in Developing Countries. *Journal of the Eastern Asia Society for Transportation Studies*, *6*, 3948–3963.

3. Ali, Y., Mehmood, B., Huzaifa, M., Yasir, U., & Khan, A. U. (2020). Development of a new hybrid multi criteria decision-making method for a car selection scenario. *Facta Universitatis, Series: Mechanical Engineering*, *18*(3), 357–373. https://doi.org/10.22190/FUME200305031A

4. Ayan, B., Abacıoğlu, S., & Basilio, M. P. (2023). A Comprehensive Review of the Novel Weighting Methods for Multi-Criteria Decision-Making. *Information (Switzerland)*, *14*(5). https://doi.org/10.3390/info14050285

5. Diakoulaki, D., Mavrotas, G., & Papayannakis, L. (1995). Determining objective weights in multiple criteria problems: The critic method. *Computers and Operations Research*, *22*(7), 763–770. https://doi.org/10.1016/0305-0548(94)00059-H

6. Do, D. T., & Nguyen, N. T. (2022). Applying Cocoso, Mabac, Mairca, Eamr, Topsis and Weight Determination Methods for Multi-Criteria Decision Making in Hole Turning Process. *Strojnicky Casopis*, *72*(2), 15–40. https://doi.org/10.2478/scjme-2022-0014

7. Ginevicius, R., & Podvezko, V. (2004). Objective and Subjective Approaches to Determining the Criterion Weight in Multicriteria Models. *Proceeding of International Conference RelStat 2004*, *1*, 133–139.

8. Hinduja, A., & Pandey, M. (2018). An Intuitionistic Fuzzy AHP based Multi Criteria Recommender System for Life Insurance Products. *International Journal of Advanced Studies in Computer Science and Engineering*, *7*(1), 1–8. https://doi.org/https://www.researchgate.net/publication/323143943

9. Hung, C., & Chen, L. (2009). A Fuzzy TOPSIS Decision Making Model with Entropy Weight under Intuitionistic Fuzzy Environment. *International MultiConference of Engineers and Computer Scientists 2009*, *I*, 18–21.

10. Kao, C. (2010). Weight determination for consistently ranking alternatives in multiple criteria decision analysis. *Applied Mathematical Modelling*, *34*(7), 1779–1787. https://doi.org/10.1016/j.apm.2009.09.022

11. Odu, G. O. (2019). Weighting methods for multi-criteria decision making technique. *Journal of Applied Sciences and Environmental Management*, *23*(8), 1449. https://doi.org/10.4314/jasem.v23i8.7

12. Pamucar, D., & Ecer, F. (2020). Prioritizing the weights of the evaluation criteria under fuzziness: The fuzzy full consistency method – fucom-f. *Facta Universitatis, Series: Mechanical Engineering*, *18*(3), 419–437. https://doi.org/10.22190/FUME200602034P

13. Pamučar, D., Stević, Ž., & Sremac, S. (2018). A new model for determiningweight coefficients of criteria in MCDM models: Full Consistency Method (FUCOM). *Symmetry*, *10*(9), 1–22. https://doi.org/10.3390/sym10090393

14. Pandey, A., & Pandey, M. (2017). Multicriteria Recommender System for Life Insurance Plans based on Utility Theory. *Indian Journal of Science and Technology*, *10*(14). https://doi.org/10.17485/ijst/2017/v10i14/111376

15. Pattnaik, C. R., Mohanty, S. N., Mohanty, S., Chatterjee, J. M., Jana, B., & García-Díaz, V. (2021). A fuzzy multi-criteria decision-making method for purchasing life insurance in india. *Bulletin of Electrical Engineering and Informatics*, *10*(1), 344–356. https://doi.org/10.11591/eei.v10i1.2275

16. Rani, A., Taneja, K., & Taneja, H. (2021a). Life Insurance-Based Recommendation System for Effective Information Computing. *International Journal of Information Retrieval Research*, *11*(2), 1–14. https://doi.org/10.4018/ijirr.2021040101

17. Rani, A., Taneja, K., & Taneja, H. (2021b). Multi Criteria Decision Making (MCDM) based preference elicitation framework for life insurance recommendation system. *Turkish Journal of Computer and Mathematics Education (TURCOMAT)*, *12*(2), 1848–1858. https://doi.org/10.17762/turcomat.v12i2.1523

18. Singh, M., & Pant, M. (2021). A review of selected weighing methods in MCDM with a case study. *International Journal of System Assurance Engineering and Management*, *12*(1), 126–144. https://doi.org/10.1007/s13198-020-01033-3

19. Trung, D. D., Thien, N. V., & Nguyen, N. T. (2021). Application of topsis method in multi-objective optimization of the grinding process using segmented grinding wheel. *Tribology in Industry*, *43*(1), 12–22. https://doi.org/10.24874/ti.998.11.20.12

Note: All the figures and tables in this chapter were made by the author.

Computational Intelligence and Mathematical Applications – Dr. Devendra Prasad et al. (eds)
© 2024 Taylor & Francis Group, London, ISBN 978-1-032-87721-1

Technological Marvels: Pioneering PCOS Detection through the Lens of Machine Learning and Deep Learning

Himanshi Jain[1]

Department of Computer Science and Engineering, Chitkara University, Punjab, India

Shweta Jain[2], Anju Bhandari Gandhi[3]

Department of Artificial Intelligence and Machine Learning, Panipat Institute of Engineering and Technology, Samalkha, Panipat, India

Rishit Aggarwal[4]

Department of Computer Science and Engineering, Graphic Era University, Dehradun, India

S.C. Gupta[5]

Department of Computer Science and Engineering, Panipat Institute of Engineering and Technology, Samalkha, Panipat, India

Abstract: Polycystic ovary syndrome (PCOS), a widespread endocrine disorder and the leading cause of worldwide infertility due to anovulatory insufficiency in women is traditionally diagnosed through ultrasound scans to detect ovarian cysts. However, manual identification is error-prone. This research proposes an advanced machine learning approach using Convolutional Neural Networks (CNNs) with transfer learning, stacking ensemble models, and state-of-the-art techniques to classify PCOS in ultrasound images. The method significantly improves accuracy and reduces training time, with the most successful configuration incorporating a pre-trained VGGNet16 CNN and XGBoost ensemble achieving a remarkable 99.89% accuracy.

Keywords: Machine learning (ML), Bioengineering, Polycystic ovary syndrome (PCOS), Ultrasound images, Image classification

1. INTRODUCTION

The WHO reports that PCOS, or polycystic ovarian syndrome, is a usual endocrine scenario that impacts women of reproductive age. PCOS is a common ailment to diagnose and manage; all that needs to be done is careful implementation of certain widely accepted methodologies for diagnosis along with reasonable therapy tactics. [1-5]

The study revealed that 18% of women had PCOS or polycystic ovarian syndrome while 70% of the women with PCOS who were diagnosed had an inadequate diagnosis. The study's conclusions draw attention to the high incidence of PCOS in society as a whole and the troubling problem of underdiagnosis, suggesting that better education, awareness, and healthcare practices regarding PCOS in that specific population may be necessary. [1-3]

[1]himanshi2157.be21@chitkara.edu.in, [2]jainshweta290204@gmail.com, [3]anjugandhi.cse@piet.co.in, [4]rishitaggarwal1@gmail.com, [5]hod.cse@piet.co.in

DOI: 10.1201/9781003534112-42

Table 42.1 List of acronyms

Acronym	Definition
WHO	World Health Organization
PCONET	CNN architecture for PCOS
SMOTE	Synthetic Minority Oversampling Based Techniques
MRI	Magnetic Resonance Imaging
XAI	Explainable Artificial Intelligence
XGBoost	eXtreme Gradient Boosting
ENN	Edited Nearest Neighbor

CNN-style deep learning designs can be used for diagnosing and identifying PCOS. With the aid of a CNN-based model called PCONET, it can identify cysts in the ovaries from ovarian ultrasound photographs with an accuracy rate of 98.12%. [4-5]

The following steps are included in the pre-processing and image acquisition that is the main focus of this:

(a) Distinctive Image
(b) Boosting of imaging process (Enhancement)
(c) Noise Alleviation
(d) Image Segregation (OTSU)
(e) Feature Extraction (ORB)

The effectiveness, practicality, and legitimacy in the healthcare industry can all be safeguarded through the implementation of (XAI). It includes:

(a) SMOTE + ENN combined version.
(b) Feature selection techniques (To alleviate data dimensionality & optimal feature set selection)
(c) Usage of Bayesian Optimizers along with cross-validation (For optimizing ML algorithms and resulting in enhanced accuracy.

The investigation's administrators contributions and points out are listed below :-

(a) Study of normal ovary and polycystic ovary
(b) Study of previous work done in this field
(c) Use of basic machine learning model for PCOS detection
(d) Use of deep learning model
(e) Proposed stacking-based architecture

2. RELATED WORK AND LITERATURE REVIEW

2.1 Types of Ovaries

This section explains the normal ovary and polycystic ovary with their characteristics (as shown in Fig. 42.1):

(a) *Normal Ovary*—A normal ovary, as an essential component of the female reproductive system, plays a multifaceted role that includes the intricate orchestration of hormonal secretion as well as the orchestration of follicular development.

(b) *Polycystic Ovary*—Multiple small, underdeveloped follicles, which are often detectable through ultrasound imaging, are the defining characteristic of a polycystic ovary. Usually, this condition is called polycystic ovarian morphology.

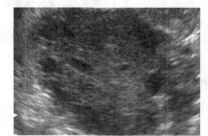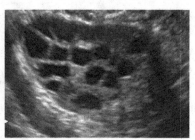

Fig. 42.1 Normal and polycystic ovary

Source: https://www.kaggle.com/datasets/anaghachoudhari/pcos-detection-using-ultrasound-images

2.2 Related Works

Prominent in the accurate diagnosis of Polycystic Ovary Syndrome (PCOS) is the examination of ovarian ultrasound images to detect multiple follicular cysts, a practice requiring the expertise of ultra-sonographers or radiologists. These specialists visually scrutinize ultrasound images for abnormalities in the ovaries, specifically larger cysts, to confirm PCOS.

The development of computer - aided PCOS follicle surveillance systems primarily involves the incorporation of digital image processing methodologies. These techniques are assessed using performance metrics like the (FAR) and (FRR).

In addition to image processing, researchers have integrated machine learning models into PCOS diagnosis, not only finding follicles part but also classifying them as A. PCOS or B. non-PCOS. Deep learning strategies have gained momentum in PCOS detection, using Convolutional Neural Networks (CNNs) to automatically extract features from ultrasound images.

Despite these advances, the segmentation of follicles using digital image processing involves intricate steps, necessitating careful tuning and adding computational complexity. Additionally, the use of deep learning may result in prolonged processing times and high computational power requirements, making it challenging to create user-friendly interfaces for medical professionals and patients. These techniques have shown promise in other healthcare applications, enhancing the performance of machine learning models by combining their predictions. Thus, the study proposes a novel approach using a Convolutional Neural Network (CNN) as a feature extractor & ensemble machine learning (EML) for the correct classification of PCOS and non-PCOS ultrasound images. This approach may offer new insights into improving PCOS diagnosis.

3. METHODOLOGY AND PROPOSED SYSTEM

The methodology used in the study (shown in Fig. 42.2) is divided into four phases:
- Typically used Machine Learning Techniques
- Phase 1Methodology with incorporation of Feature Reduction
- Deep Learning (DL) Technique
- Presented proposed System (Use of Stacking Ensemble Machine Learning (Stack-EML))

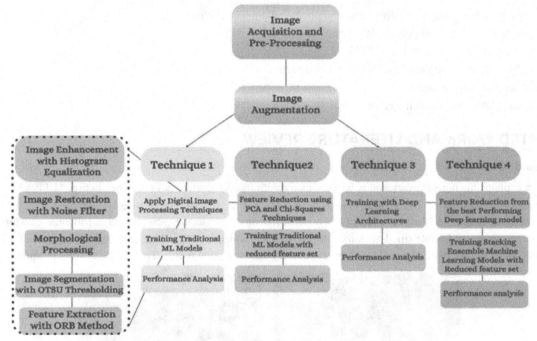

Fig. 42.2 The complete view of research process[6]

Phase 0 (Image acquisition, pre-processing, and augmentation): It includes various pre-processing techniques such as image enhancement, morphological processing, etc).

The implementation of phase0 (shown in Fig. 42.3) and results in Fig. 42.4 is as follows:-

```
In [16]:  import cv2
          import numpy as np
          import matplotlib.pyplot as plt

In [17]:  # Load the image
          image = cv2.imread('ovary_image.jpg', cv2.IMREAD_COLOR)

In [18]:  # Image enhancement (histogram equalization)
          gray_image = cv2.cvtColor(image, cv2.COLOR_BGR2GRAY)
          equalized_image = cv2.equalizeHist(gray_image)
          # Noise reduction (Gaussian blur)
          blurred_image = cv2.GaussianBlur(equalized_image, (5, 5), 0)

In [19]:  # Morphological processing (dilation)
          kernel = np.ones((5, 5), np.uint8)
          dilated_image = cv2.dilate(blurred_image, kernel, iterations=1)

In [20]:  # Image segmentation (OTSU method)
          _, segmented_image = cv2.threshold(dilated_image, 0, 255, cv2.THRESH_BINARY + cv2.THRESH_OTSU)

In [21]:  # Feature extraction (ORB)
          orb = cv2.ORB_create()
          keypoints, descriptors = orb.detectAndCompute(segmented_image, None)
          # Draw the keypoints on the image
          output_image = cv2.drawKeypoints(segmented_image, keypoints, None, color=(0, 255, 0), flags=0)

In [22]:  # Display the results
          plt.figure(figsize=(12, 6))
          plt.subplot(231), plt.imshow(cv2.cvtColor(image, cv2.COLOR_BGR2RGB)), plt.title('Original Image')
          plt.subplot(232), plt.imshow(equalized_image, cmap='gray'), plt.title('Enhanced Image')
          plt.subplot(233), plt.imshow(dilated_image, cmap='gray'), plt.title('Morphological Processing')
          plt.subplot(234), plt.imshow(segmented_image, cmap='gray'), plt.title('Segmented Image (OTSU)')
          plt.subplot(235), plt.imshow(cv2.cvtColor(output_image, cv2.COLOR_BGR2RGB)), plt.title('Feature Extraction (ORB)')
          plt.show()
```

Fig. 42.3 Implementation of image acquisition, pre-processing, and augmentation

Source: Author

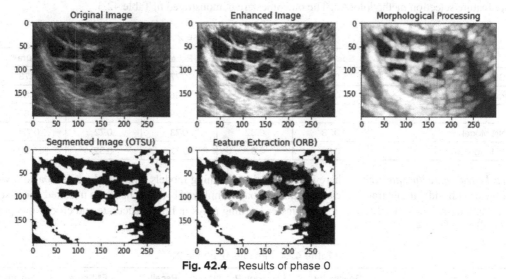

Fig. 42.4 Results of phase 0

Source: Author

Phase 1(Typical Machine Learning Techniques): The set of carefully processed images goes through a critical transformation after the laborious digital image processing workflow—which consists of a series of transformative steps—is finished. These steps include enhancement, noise reduction, morphological processing, and feature extraction. These images are cleverly transformed into a one-dimensional representation by carefully using the 'FLATTEN' function. As is common in machine learning practice, this unidimensional dataset is then divided into stratified divisions, with 70% of the data set aside for training and the remaining 30% for thorough testing. These extracted features form the curated dataset that is produced, which provides the cornerstone for the subsequent implementation of different classical machine learning models. These models, which are based on well-known algorithms, are carefully applied to the data and produce a wide range of accuracy metrics. The results are properly demonstrated in Table 42.2.

Table 42.2   Results of phase 1

Models	Specificity	Precision	Recall	F1-Score	Accuracy
XGBoost Model	0.86	0.85	0.85	0.86	0.86
Gradient Boost Model	0.86	0.856	0.86	0.85	0.855
Random Forest	0.87	0.88	0.88	0.88	0.88
KNN Model	0.84	0.87	0.85	0.85	0.85
SVM Algorithm	0.79	0.81	0.81	0.8	0.81

Phase 2(Typical Machine Learning (ML) Methodology along with best incorporating Feature Reduction): In the previous method, we encountered an issue due to the sheer size of our input image array, measuring 594x50176. This signifies that each row of the image comprises a whopping 50176 features, corresponding to pixel values. Such an overwhelming number of features can lead to overfitting, potentially undermining the predictive capabilities of our models. To mitigate this concern, we turned to feature selection strategies, aimed at curbing the dimensions of our dataset by isolating the most influential features, thereby sidestepping the notorious curse of dimensionality. In pursuit of this objective, the subsequent phase of our study introduced two distinctive feature selection methodologies to our dataset, post-image processing.

The first feature reduction method hinged on the chi-square feature selection strategy, a widely employed technique in machine learning. It assessed the significance of features by scrutinizing their relationships and deviations from expected distributions. The second approach, Principal Component Analysis (PCA), provided an effective means of dimensionality reduction by quantitatively simulating the data. PCA skillfully mapped the authenticated n-dim features to a more manageable k-dimensional set of orthogonal features known as principal components.

These two strategies independently pinpointed and retained the most influential 25000 features out of the initial 50176, each through its distinct methodology. The ensuing datasets, bearing this reduced feature set, set the stage for training traditional machine learning models. The 5 specified models from the previous method were retrained, this time utilizing the datasets divided into 70% for training and 30% for testing, with the added advantage of the reduced feature sets derived from PCA and the Chi-Square feature selection methodologies. The outcomes are demonstrated in Table 42.3.

Table 42.3   Results of phase 2

Models	Specificity	Precision	Recall	F1-Score	Accuracy
XGBoost Model	0.74	0.78	0.77	0.77	0.78
Gradient Boost Model	0.86	0.86	0.86	0.85	0.85
Random Forest	0.89	0.886	0.89	0.88	0.89
KNN Model	0.78	0.82	0.73	0.72	0.778
SVM Algorithm	0.82	0.82	0.82	0.82	0.83

Phase 3 (Deep Learning Technique): Now, there is a use of deep learning models with CNN which is also divided into two parts. Both of these with and without transfer learning, the approach employed by CNN has been nurtured and assessed as well. These outcomes emphasize the crucial function that transfer learning plays in highly accurate and efficacious prediction. The results are shown in Table 42.4.

Table 42.4   Results of phase 3

Models	Specificity	Precision Results	Recall	F1-Score	Accuracy
CNN (without transfer learning)	0.74	0.75	0.78	0.75	0.748
CNN (with VGGNet16)	0.96	0.97	0.96	0.97	0.98

Phase 4 (Proposed System): The proposed technique seamlessly integrates deep learning (DL) & stacking – ensembled based machine learning style (Stack-EML) to facilitate the classification of PCOS or non-PCOS criteria. The foundation of this approach lies in the extraction of pivotal features from the input dataset, a task deftly executed by the deep learning(DL) technique-based feature extraction segment. Subsequently, these essential features are harnessed for classification through the incorporation of a stacking ensemble machine-learning (Stack-EML). The "Flatten layer" plays a pivotal role by generating a

singular feature vector that accumulates the outputs from preceding layers. Consequently, after subjecting the deep learning modelling to the image dataset(input), the outstanding performances emanating from the "flattened layer" materializes as a 1-d array. Notably, the beauty of CNN architecture lies in its intrinsic ability to autonomously discern significant attributes from input images without necessitating human intervention or elaborate image processing.

The subsequent phase of the approach involves the training of stacking ensemble machine learning (Stack-EML)models using the reduced-featureset. This strategy introduces several advantages, including the analysis of heterogeneous weak classifiers in parallel and the aggregation of base classifiers via a meta-learner, thereby fortifying predictive capabilities and reducing variance.

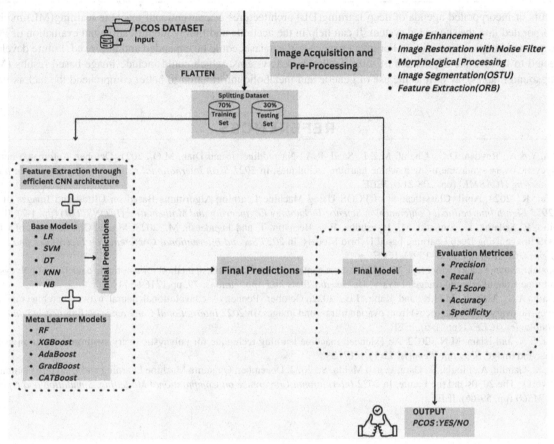

Fig. 42.5 Proposed architecture

Source: Author

In the presented part: the stacked ensemble model (shown in Fig. 42.5), the dataset with reduced features initially feeds into the base learners (BL). At this juncture, 5 continuously recognized conventional ML classifiers, namely Logistic Regression (LR), Gaussian Naive Bayes (GNB), Support Vector Machine (SVM), Decision Tree (DT), & K-Nearest Neighbour (KNN) function as the weak learners (or base classifiers) at level 0 within the stacked model. These paramount models are independently constructed by means of their associated prediction algorithm design. The projections generated via the level 0 models then relocate on to the next level 1, where the highest-level foresight will be generated by a single classification model, additionally referred to as a meta-learner. Level 1 retains the base models compared to Level 0 but launches five distinct classifier varieties to accommodate meta-learners. The aforementioned approach addresses numerous architectural configurations in order to identify the best performer. After checking the predictions extracted from the 5 base models, the meta-learner undergoes training on higher part of the level 0 models to produce the final result. As a consequence, a more extensive stacked ensemble machine learning classification system is demonstrated, featuring the combination of one boosting or bagging classifier acting as the meta-learner and five conventional classifiers serving as base models. The overarching objective is to discriminate between A. PCOS and B. non-PCOS patient ovary ultrasound images effectively. The results are shown in Table 42.5.

Table 42.5 Results of phase 4

Base Models	Meta-learner Models	Specificity	Precision	Recall	F1-Score	Accuracy
Decision Tree	AdaBoost	0.99	0.99	0.98	0.99	0.9959
Support Vector machines	XGBoost	0.99	0.99	1.00	0.99	0.9989

4. CONCLUSIONS

The benefits of incorporated agenda of deep learning(DL) architectures & conventional machine learning(ML) models have been incorporated into the suggested system. It can help in the accurate and timely identification and evaluation of PCOS for both patients and hospitals and clinics. For greater success, the dataset could be expanded and improved. Future developments are expected to involve the application of multi-feature-based systems, which could include image-based results (MRI, CT scan, ultrasound images), as well as the use of genetic and metabolic information to better comprehend the factors impacting PCOS.

REFERENCES

1. Adla, Y.A.A., Raydan, D.G., Charaf, M.Z.J., Saad, R.A., Nasreddine, J. and Diab, M.O., 2021, October. Automated detection of polycystic ovary syndrome using machine learning techniques. In *2021 Sixth international conference on advances in biomedical engineering (ICABME)* (pp. 208-212). IEEE.
2. Khilar, R., 2023, April. Classification of PCOS Using Machine Learning Algorithms Based on Ultrasound Images of Ovaries. In *2023 Eighth International Conference on Science Technology Engineering and Mathematics (ICONSTEM)* (pp. 1-7). IEEE.
3. Chitra, P., Srilatha, K., Sumathi, M., Jayasudha, F.V., Bernatin, T. and Jagadeesh, M., 2023, March. Classification of Ultrasound PCOS Image using Deep Learning based Hybrid Models. In *2023 Second International Conference on Electronics and Renewable Systems (ICEARS)* (pp. 1389-1394). IEEE.
4. Gopalakrishnan, C. and Iyapparaja, M., 2020. Active contour with modified Otsu method for automatic detection of polycystic ovary syndrome from ultrasound image of ovary. *Multimedia Tools and Applications*, 79, pp.17169-17192.
5. Hosain, A.S., Mehedi, M.H.K. and Kabir, I.E., 2022, October. Pconet: A convolutional neural network architecture to detect polycystic ovary syndrome (pcos) from ovarian ultrasound images. In *2022 International Conference on Engineering and Emerging Technologies (ICEET)* (pp. 1-6). IEEE.
6. Suha, S.A. and Islam, M.N., 2022. An extended machine learning technique for polycystic ovary syndrome detection using ovary ultrasound image. *Scientific Reports*, 12(1), p.17123.
7. Jain, S., Gandhi, A., Singla, S., Garg, L. and Mehla, S., 2022, December. Quantum Machine Learning and Quantum Communication Networks: The 2030s and the Future. In *2022 International Conference on Computational Modelling, Simulation and Optimization (ICCMSO)* (pp. 59-66). IEEE.

Computational Intelligence and Mathematical Applications – Dr. Devendra Prasad et al. (eds)
© 2024 Taylor & Francis Group, London, ISBN 978-1-032-87721-1

43 | Transforming Visits: Empowering Guests with QR-Based E-Ticketing

Deepti Dhingra[1]
Assistant Professor, Department of Computer Science and Engineering, PIET, Haryana

Mayank Jindal[2], Rahul Tomer[3], Jatin Rathee[4]
Department of Computer Science and Engineering, PIET, Haryana

Abstract: The use of QR codes in monument ticketing has revolutionized the way visitors access historical sites. In this paper, we have implemented an QR based E-Ticketing system to improve the efficiency of the ticketing process. Using our system, users can easily purchase tickets online and receive a unique QR code on their mobile device. This code can be scanned at the monument entrance, providing contactless access that is both efficient and safe. Our system will also eliminate the use of traditional paper-based ticket and long waiting queues. Once log into the system, the registered user can get the monument information and purchase the ticket. The booking module will generate QR code that can be downloaded and used further as ticket. The Dashboard module, allows the user to access the recent visits, cancel ticket, and re-download the ticket. Our Ticketing system includes essential features such as registration, login, monument information, booking, dashboard, downloadable QR codes, FAQs, and feedback to ensure a convenient and user-friendly experience for the users, making it faster and more convenient for everyone involved.

Keywords: QR code, E-ticketing, Monuments of india

1. INTRODUCTION

Monuments and historical sites are popular tourist destinations, but the traditional ticketing process can be a hassle for both visitors and staff. Digital ticketing solutions, such as QR codes, have gained popularity in recent years due to their convenience and security. QR codes are two- dimensional barcodes that can be scanned using a smartphone, providing a fast and reliable way to purchase and validate tickets proposed by Soon, T. J. [1] This paper focuses on the use of QR code-based e-ticketing systems for monuments, exploring the benefits and challenges of this technology. We will discuss how QR code ticketing can improve the visitor experience by making the ticketing process faster and more convenient, while also enhancing security. Additionally, we will examine potential challenges such as the need for reliable internet connectivity and staff training to ensure a successful implementation. Overall, this paper will highlight the advantages of QR code-based e-ticketing systems for monuments, and how they can provide a modern and efficient solution to traditional ticketing processes. This research aims to present a system proposal for the development of a QR Code-based e-ticketing for monuments. The proposed system will utilize a QR code-based payment method to enhance the ease of ticket purchasing for the public offering a fast and convenient way to purchase and validate tickets, while also enhancing security [2].

[1]deeptimona@gmail.com, [2]mayankjindal40@gmail.com, [3]rahultomer9722@gmail.com, [4]jatinrathee3703@gmail.com

DOI: 10.1201/9781003534112-43

2. LITERATURE SURVEY

In recent years, much work has been done related to e-ticketing systems. Shan et al. [3] proposed a railway ticketing application to maintain the grid-oriented passenger service record using a face recognition platform. A. S. Putra. [4] In 1998, the use of the Internet for booking travel tickets gained popularity, and it now accounts for over two percent of the tourism market. Experts predict that this trend will continue, with an annual growth rate of 7.5%. In the past few years, Bhonsle. [5] introduced an innovative concept aimed at simplifying the user experience. According to this idea, users would only need to verify their identity when booking tickets, thus enhancing the system's friendliness towards guest users. The proposed application will offer a user-friendly interface for booking tickets. Waghmare. [6] proposed that technology is progressing rapidly. Take, for instance, the railway sector, where the introduction of e-ticketing marked a significant advancement. In this system, passengers navigate the official government website to reserve long-distance tickets, which are subsequently printed and presented to the ticket checker when required. The aim of developing an e-ticketing system with a QR code based on IoT is to modernize the ticket-selling operations of a public transportation company. This involves streamlining the purchasing process, ultimately saving time for customers and minimizing the occurrence of human errors [7].

In the past few years, the rapid evolution of information technology has brought about significant changes in various business sectors, including public transportation and museums, where electronic tickets are increasingly being offered as an alternative to traditional paper-based tickets. The primary advantages for stakeholders lie in the hassle-free and paperless nature of electronic tickets, notably in mitigating the risks associated with ticket loss. Furthermore, the overarching goal is to encourage greater adoption of electronic ticketing systems [8].

3. PROPOSED MODEL

An innovative electronic ticketing system designed for monuments, incorporating QR code technology, presents an efficient and user-friendly solution for managing visitor access. The accompanying application boasts a comprehensive array of features, including user registration, login functionality, monument information display, booking capabilities, a user dashboard, downloadable QR codes, a frequently asked questions (FAQs) section, and a feedback mechanism. This holistic approach ensures a seamless and convenient experience for users engaging with the system.

Illustrated in Fig. 43.1 is the proposed model, delineating an online ticket purchasing system that offers a diverse range of payment options. Upon completing a purchase, each ticket generates a unique QR code, affording users the flexibility to either display it on their mobile devices or opt for a printed copy. The system's operational efficiency is further highlighted at the entrance, where staff members utilize dedicated scanners to authenticate the QR codes, subsequently granting access to ticket holders.

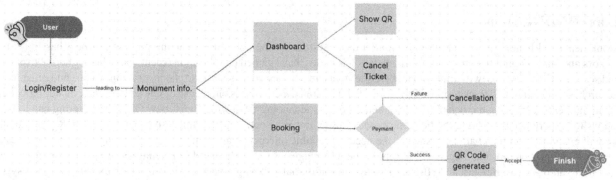

Fig. 43.1 Block diagram of the proposed model

Beyond mere ticket authentication, the system incorporates real-time tracking functionalities to monitor visitor numbers dynamically. In addition, robust security measures are implemented to thwart fraudulent activities, safeguarding the integrity of the ticketing process. Moreover, the system is equipped with reporting and analytics capabilities, providing valuable insights into visitor trends and enabling administrators to make data-informed decisions.

In essence, this e-ticketing system not only streamlines the ticket purchasing process but also enhances overall monument management through its user-centric features and advanced technological components.

As per the proposed model specifications, users are required to initiate their interaction with our system by either logging in or registering. Subsequently, they gain access to a user-friendly dashboard where they can find downloadable QR code-based electronic tickets for their past transactions. Within the dashboard, users also have the option to select a particular monument of interest, view relevant information, and proceed to book their tickets by making the necessary payment as determined by the respective monument authority. Upon successful payment, a QR code is generated, serving as the electronic ticket that will be scanned and verified by the monument authority during the visit.

4. FEATURES OF THE SYSTEM

Registration and login: Users will need to register with the app using their email or phone number and verify their identity. This ensures that only verified users can access the portal and book tickets. Once users are registered, they can log in to the app to access their dashboard and other features.

Monument info: The app will also provide users with information about different monuments, such as their location, opening hours, ticket prices, and other relevant details. Users can browse through the monuments and select the one they want to visit. This feature helps users make informed decisions about which monuments to visit and plan their trip accordingly.

Booking: The booking process on the app is designed to be simple and efficient. Users can select the monument they want to visit, choose the date and time of their visit, and purchase tickets. The app will display the ticket price and payment options. After making the payment, the app will generate a QR code on the screen that serves as the e-ticket for the user's visit. This QR code can be scanned at the monument to gain entry.

Dashboard: The dashboard feature of our system provides users with a convenient way to manage their ticket information. Users can access their upcoming visits, purchase history, and other relevant information on their dashboard. This feature ensures that users can easily keep track of their monument visits and manage their tickets.

Downloadable QR code: Users can download the QR code for their visit from the app, save it on their device, and access it offline if necessary. They can also share the QR code with others who will be visiting the monument with them. This feature provides users with the flexibility to access their e-tickets offline and share them with others.

The FAQs section of the portal is designed to help users find quick answers to common queries related to using the app and visiting the monuments. This feature provides users with the necessary information to plan their trip and use the app effectively.

Validation: To validate the QR code, the monument staff or security personnel can use a QR code scanner to read the code and match it against the information in the database to confirm that it is a valid ticket. If the QR code is valid, the user can enter the monument, and if it is not, they will be denied entry.

Apart from scanning the QR code, the staff can also visually inspect the ticket for any signs of tampering or forgery. They can check the ticket's hologram or security features to ensure that it is genuine.

5. RESULTS AND EXPERIMENTAL SETUP

The above Table 43.1 represents the technical requirements for the system to be met for its smooth working.

Table 43.1 Technical requirements

S. No.	Technical Requirements
1.	Operating System: (Windows 7 or later) or Linux (Ubuntu) or MacOS
2.	Google Chrome Browser
3.	CPU: 2 GHZ (estimated)
4.	RAM: 2 GB (estimated)
5.	Disk space: 500 MB (estimated)

5.1 Dashboard

This module enables users to access a dashboard where they can review their past bookings and view their user profile, encompassing essential details. The profile includes general information for QR code-based e-ticketing at monuments, providing a streamlined experience. The system enhances convenience by offering a comprehensive overview of previous reservations and user details. Figure 43.2 represents the dashboard for general details of the user.

5.2 List/Info

This module enables users to explore a comprehensive list of monument sites by selecting a city. It encompasses brief historical narratives, ticket pricing, operating hours, and photographs of the monument. The module is specifically designed to facilitate a streamlined and efficient QR code-based e-ticketing system for monuments, enhancing the overall experience for visitors and simplifying the ticketing process. Figure 43.3 represents the monument's info.

Fig. 43.2   Dashboard

Fig. 43.3   Monument's info.

5.3 Payment

This module provides users with a convenient and secure means to purchase tickets. The system supports a diverse range of payment methods, including Credit/Debit cards, UPI, and QR codes, allowing users to choose the option that best suits their preferences. Figure 43.4 represents the payment gateway.

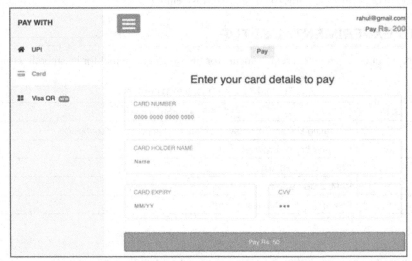

Fig. 43.4   Payment

6. FUTURE SCOPE OF THE PROPOSAL

Integration with Wearable Technology: The e-ticketing system can be integrated with wearable devices such as smartwatches or bracelets in the future. This would allow visitors to access their tickets easily without the need for a phone or printed ticket, thus improving the system's convenience and efficiency.

Integration with Augmented Reality (AR): The e-ticketing system can integrate AR technology, which would provide an immersive and interactive tour of the monument. This integration can enhance the visitor experience and offer a unique and memorable way to learn about the monument's history and significance.

Integration with Digital Payment Systems: As digital payment systems continue to advance, the e-*ticketing system can integrate with new payment options, providing visitors with more choices and* flexibility when purchasing tickets.

7. CONCLUSION

A QR code-based e-ticketing system offers various advantages compared to traditional paper-based ticketing systems, particularly for monuments and other tourist attractions. By allowing visitors to purchase tickets online, the system reduces the need for queuing and paper transactions. Apart from providing convenience to visitors, a QR code-based e-ticketing system also improves the security and accuracy of ticketing. Overall, implementing a QR code-based e-ticketing system for monuments is a modern and efficient solution that aligns with the current trend toward digitization and automation in the tourism industry. This system can significantly improve the visitor experience, reduce operational costs, and increase revenue.

REFERENCES

1. N. K. Dewi and A. S. Putra, "Decision Support System for Head of Warehouse Selection Recommendation Using Analytic Hierarchy Process (AHP) Method," Prosiding International Conference of Universitas Pekalongan, pp. 1- 12, 2021.
2. Soon, T. J. (2008). QR code. synthesis journal, 2008, 59-78.
3. Shan, X., Zhang, Z., Ning, F., Li, S., & Dai, L. (2023). Application and realization of key technologies in China railway e-ticketing system. Railway Sciences, 2(1), 140-156.
4. A. S. Putra, "Konsep Kota Pintar Dalam Penerapan Sistem Pembayaran Menggunakan Kode QR Pada Pemesanan Tiket Elektronik," TEKINFO Jurnal Ilmiah Teknik Informatika, vol. 21, pp. 1-15, 2020.
5. Vayadande, K., Raut, A., Bhonsle, R., Pungliya, V., Purohit, A., & Ingale, V. (2023). Secured Ticketless Booking System to Monuments and Museums Using Cryptography. In Intelligent Communication Technologies and Virtual Mobile Networks(pp. 373-385). Singapore: Springer Nature Singapore.
6. Waghmare, A., Pansambal, S., Pavate, A., & Kumawat, D. (2018). QR code based Railway e-Ticket. IOSR J Eng (IOSR JEN).
7. Ong, Yee Lin, and Noryusliza Abdullah. "Design and Development of Ferry E-Ticketing System with QR Code based on IoT Technology." Applied Information Technology And Computer Science 4.1 (2023): 308-326.
8. Krishnapal Songara, Kushagra Verma, Lakshit Gupta, Madhur Dubey, Prof. Ambrish Srivastav , ticketless entry system to monuments/heritage sites, International Research Journal of Modernization in Engineering Technology and Science (IRJMETS), Volume 4, No.11, November 2022.

Note: All the figures and table in this chapter were made by the author.

Computational Intelligence and Mathematical Applications – Dr. Devendra Prasad et al. (eds)
© 2024 Taylor & Francis Group, London, ISBN 978-1-032-87721-1

Prioritizing the Barriers of EVs adaptation in India using Fuzzy TOPSIS Methodology

Pardeep Kumar

Research Scholar, SRMIST, University Ghaziabad

Sunil Dhull[1], Anju Gandhi[2], Stuti Mehla[3]

Panipat Institute of Engineering & Technology

Abstract: The purpose of the research is to comprehend how customers generally see electric vehicles and the barriers preventing their widespread acceptance. The barriers of EVs are identified from the literature survey. Fuzzy- TOPSIS methodology is applied to rank the identified barriers of EVs adaptation in Indian context. The results show that "high cost of EVs is the main barrier followed by "high battery replacing cost" after the end of useful life of the electrical vehicle's battery. Very long battery charging time in the house hold charging stations is another barrier along with the lack of battery charging infrastructure in the country. If the policymakers and industries try to resolve these barriers, this will help in boosting the sales and adaptation of electrical vehicles in India. Adaptation of EVs will help in mitigating the pollution in country like Indian and will reduce the CO_2 emission drastically, especially in the metro cities.

Keywords: EVs, Fuzzy-TOPSIS, Adaptation, Barriers, Policymakers

1. INTRODUCTION

According to a recent study by Autocar Professions 2022, India sold 999,949 electric vehicles in 2022 a rather small number internal combustion engines driven vehicles. This shows that determining the barriers preventing EV adoption in the Indian market must be done quickly. The aim of this study is to examine and recognize the unique barriers of EVs that either directly or indirectly impede India's adoption of electric vehicles. The present research has two main objectives: i) to identify the barrier to EV adoption in Indian context. To do this, a thorough analysis of earlier research is conducted. Second objective is to rank these barriers according to the importance's level of these barriers. To rank these barriers of EVs, Fuzzy-TOPSIS methodology is used. The study's findings offer an extensive understanding of the key barriers that limits the acceptance of EVs on Indian roads. The methodology employed in the study additionally serves to understand the high-priority barriers that need to be removed on the priority that will help in improving the acceptance of EVs in country like India. Based on the study's findings, manufacturers and public policymakers may develop their next set of policies to improve their sales and ultimately gaining the confidence of the Indian customers in EVs.

2. LITERATURE REVIEW

The dynamics of the EV markets for different geographic locations have been examined and reviewed by researchers worldwide, who have also presented research implications to support policy initiatives and identified potential implementation barriers (Yang, 2022; Cui et al., 2022). The inadequate facilities for charging will add the additional strain on the national grid. The

[1]sunildhull12@rediffmail.com, [2]anjugandhi.cse@piet.co.in, [3]stutimehla.cse@piet.co.in

DOI: 10.1201/9781003534112-44

effectiveness and safety of batteries in high temperatures environment and the suitability of EVs for use in desert regions, Alotaibi et al. (2021). Biresselioglu et al. (2018) the lack of charging infrastructure, Expensive to buy, longer charging period, greater power demands for EVs, and scarcity of accessibility of battery raw materials were found to be impediments to the widespread adoption of EVs. The research acknowledged the following characteristics as predictors of electric car adoption: financial obstacles, poor performance of EVs, poor charging infrastructure, and environmental preservation. Asadi et al. (2022) found in their research that the most important essentials driving the acceptance of electric vehicles (EVs) in country like Malaysian. Kannan et al. (2022) studied the barriers of EVs and indicated that the very expensive to by an electric vehicle and the absence for charging stations are few main barriers. Based on a thorough assessment of the literature covering a range of geographic areas and insights gathered from Indian EV company representatives, we have identified thirteen EVs barriers. Table 44.1 provides an extensive literature overview of EVs barriers.

Table 44.1 List of barriers of EV adaptation

S. No	Drivers	Authors
B-1	Expensive to buy	Murugan & Marisamynathan (2022), Tarei et al. (2022), Smith et al. (2017, Khazaei (2019), Munshi et al. (2022)
B-2	Uncertain travelling distance	An et al. (2023), Sriram et al. (2022), Smith et al. (2017), Kannan et al. (2022), Noel et al., 2020
B-3	Long battery charging time	Pourgholamali et al. (2023), An et al. (2023), Sriram et al. (2022), Tarei et al. (2022), Khazaei (2019), Smith et al. (2017), Murugan & Marisamynathan (2022)
B-4	vehicle as it is work in progress technology	Murugan &Marisamynathan (2022), Tarei et al. (2022), Smith et al. (2017)
B-5	High maintenance Cost	Murugan & Marisamynathan (2022), Tarei et al. (2022),
B-6	Lacking Repair and maintenance Centers	Tarei et al. (2022), (Wang et al., 2017), (Lieven et al., 2011
B-7	Excessive battery changing expenditure	Sriram et al. (2022), Murugan & Marisamynathan (2022), Smith et al. (2017)
B-8	Low confidence in Electric vehicles	Tarei et al. (2022), (Habla et al., 2020), Smith et al. (2017)
B-9	Uncertain Resale Value	Pourgholamali et al. (2023), An et al. (2023), Khazaei (2019), Setiawan et al. (2022), Murugan & Marisamynathan (2022),
B-10	Safety	Murugan & Marisamynathan (2022), Tarei et al. (2022)
B-11	Limited Choice of EVs	Tarei et al. (2022), Smith et al. (2017),
B-12	Insufficient public charging stations	Tarei et al. (2022), Khazaei (2019), Goel et al. (2021), Patyal et al. (2021)
B-13	Poor Knowledge	Kannan et al. (2022), Setiawan et al. (2022)

3. METHODOLOGY

The barriers of EV adaptation were selected from the literature survey and shown in Table 44.1. These barriers were further discussed with the professionals working in the field of automobile sector. The three perspectives were used to rank barriers of EV adaptation in the present research. This is the reason the experts were taken from automobile manufacturing, experts in automobile marketing and public opinion in the relevant field. The suggestions given by these experts were assimilated in the questionnaire. The professionals from academia and industries employed in the domain of automobile sector were further revised this questionnaire for ease of understanding besides clarity. The questionnaire was sent to the experts who have more than ten years of experience. In each criterion three decision makers were requested to complete the survey. The questionnaire was sent on email and requested to fill the questionnaire and after 15 days a reminder email was sent and after one-month personal meetings were requested and responses were collected.

TOPSIS methodology is most commonly applied methodology for prioritizing the real-world problems. According to Yu (2002), TOPSIS methodology has one major limitation that it cannot work very efficiently in vague and uncertain decision-making environment. This vague and uncertain environment can be easily handled by means of fuzzy system along with TOPSIS. This encourages the decision-makers to incorporate immeasurable and inadequate data in the decision-making. Frequently, the vague reply is stated by the decision-maker rather than the exact value Lee (2006). As a result, it is very difficult to measure this qualitative data. The Fuzzy-TOPSIS procedure was first used by Bellman and Zadesh in 1970 to solve real-life decision-making problems. Gumus (2009), recommended that the fuzzy TOPSIS methodology is very good technique in comparison to traditional TOPSIS methodology.

Step I: The criteria and alternative ratings specified by decision makers

The decision makers were taken from industry/ manufacturers, Sales and Marketing of EVs and public. The criteria and alternatives rating were specified by three experts in each perspective for thirteen barriers shown in Table 44.3 and Table 44.4. The linguistic terms are changed into fuzzy number using a conversion scale using Table 44.2.

Step II: Average fuzzy weights of criteria (Barriers)

The linguistic terms were combined and aggregated fuzzy score was calculated

Table 44.2 Criteria and alternative rating in linguistic terms (fuzzy rating)

Criteria weight in linguistic terms	[VL] Very low	[L] Low	[M] Medium	[H] High	[VH] Very high
Membership function	[1, 1, 3]	[1, 3, 5]	[3, 5, 7]	[5, 7, 9]	[7, 9, 9]
Linguistic term used for alternative	[NI] Not important	[LI] Less important	[FI] Fairly important	[I] Important	[VI] Very important
Membership function	[1, 1, 3]	[1, 3, 5]	[3, 5, 7]	[5, 7, 9]	[7, 9, 9]

Step III: Computation (fuzzy decision matrix)

The linguistic weights provided by the experts/DMs were changed into fuzzy rating. The resulting matrix was used to make fuzzy decision-matrix for alternatives ($\tilde{A}$).

Table 44.3 Linguistic assessment of the criteria

Criteria	DM-1	DM-2	DM-3	DM-4	DM-5	DM-6	DM-7	DM-8	DM-9
Automobile manufacturers C-1	H	H	H	H	H	VH	M	H	VH
Sales & marketing of authorized dealers C-2	H	M	H	H	VH	H	M	H	H
Public perspective [C-3]	VH	VH	VH	VH	H	VH	VH	H	VH

Table 44.4 Linguistic assessment of alternatives (Barriers)

S. No.	Barriers	Automobile manufacturers C-1			Sales & marketing of authorised dealers C-2			Public perspective [C-3]		
BR-1	Expensive to buy	VI	VI	VI	VI	I	VI	VI	VI	VI
BR-2	Excessive battery changing expenditure	VI	I	VI	VI	VI	I	VI	VI	VI
BR-3	Uncertain travelling distance	I	I	VI	I	VI	I	I	VI	VI
BR-4	Long battery charging time	I	VI	I	I	I	VI	I	VI	I
BR-5	Low confidence in Electric	FI	I	FI	I	FI	FI	I	I	FI
BR-6	Vehicle as it is work in progress technology	FI	FI	I	FI	FI	I	I	I	VI
BR-7	Insufficient public charging stations	I	I	I	VI	VI	VI	VI	VI	VI
BR-8	Lacking repair and maintenance centres	VI	I	I	VI	VI	I	I	I	VI
BR-9	Poor knowledge	FI	I	FI	FI	FI	I	I	FI	FI
BR-10	Limited choice of EVs	FI	FI	FI	FI	LI	FI	I	I	FI
BR-11	High maintenance cost	I	I	I	I	FI	I	FI	FI	I
BR-12	Uncertain resale value	VI	I	VI	I	I	VI	I	I	I
BR-13	Safety	I	I	I	FI	I	FI	I	I	FI

Step IV: Aggregate fuzzy evaluations for alternative

If fuzzy weights and evaluations for the alternatives of the mth decision makers are specified by $\tilde{x}_{ijm} = (a_{ijm}, b_{ijm}, c_{ijm})$, then the value shown can represent the fuzzy rating ($\tilde{x}_{ij}$) of the alternative with respect to each criterion. The value of a_{ij} was taken to be minimum, the value of b_{ij} is taken as the mean value and value of c_{ij} is taken as the maximum value.

Step V: Normalized fuzzy decision matrix

The aggregate fuzzy weights are normalized by taking all criteria's and alternatives using linear conversion scale (same scale).

Step VI: Normalized Weight Matrix Calculation

The normalized weights can be calculated by multiplying $(\tilde{s}_{ij})$ with $(\tilde{w}_j)$. These weights are normalized fuzzy decision matrix with weights of the estimated criteria.

Step VII: Fuzzy Positive Ideal Solution (FPIS) and Fuzzy Negative Ideal Solution (FNIS)

The TOPSIS technique is based on the theory (for benefit criteria) that the selected alternative should have shortest geometric distance from FPIS and for cost criteria it should have longest geometric distance from the FNIS (Bellman and Zadesh, 1970).

Step VIII: Distance from FPIS and FNIS of individual alternative

The distances (d_i^*, d_i^-) of individual weighted alternative i = 1, 2, 3, ..., x from the FPIS to FNIS are determined. Vertex technique was adopted to get the geometric distance from FPIS and FNIS.

Step IX: Ranking of the alternatives/barriers

The barrier with maximum CCi value is placed at top position in table, whereas the barrier having lowest CCi value is placed at the lowest position in the table. The alternative which is ranked one has least geometric distance from FPIS and maximum from FNIS. Table 44.5 & 44.6 shows prioritizing of barriers of EVs.

Table 44.5 Linguistic assessment of alternatives (Barriers of EV)

Barrier	Automobile Manufacturers C-1	Sales & Marketing of Authorized Dealers C-2	Public Perspective [C-3]
BR-1	0.654	0.600	0.665
BR-2	0.587	0.600	0.665
BR-3	0.571	0.586	0.591
BR-4	0.571	0.586	0.575
BR-5	0.485	0.514	0.499
BR-6	0.485	0.514	0.575
BR-7	0.654	0.572	0.665
BR-8	0.571	0.600	0.575
BR-9	0.485	0.514	0.483
BR-10	0.396	0.403	0.499
BR-11	0.554	0.527	0.483
BR-12	0.587	0.586	0.557
BR-13	0.554	0.514	0.499

Table 44.6 Aggregated closeness coefficients for alternative (Barriers of EVs)

Barrier	d*i	di⁻	CCi
BR-1	9.38	16.55	0.638
BR-2	10.11	16.22	0.616
BR-3	11.13	15.55	0.583
BR-4	11.24	15.37	0.578
BR-5	14.07	14.03	0.499
BR-6	13.13	14.41	0.523
BR-7	9.61	16.27	0.629
BR-8	11.13	15.51	0.582
BR-9	14.22	13.89	0.494
BR-10	15.21	11.64	0.434
BR-11	13.19	14.34	0.521
BR-12	11.27	15.38	0.577
BR-13	13.17	14.37	0.522

The Fig. 44.1 shows the closeness coefficient values of barriers of EVs in India. Higher the value of the closeness co-efficient, more important is the barriers and the barrier with minimum value is the least important barrier. In the figure given below barriers with highest value is 0.638 (Expensive to buy) and has maximum impact. The barrier with minimum value "Limited Choice of EVs" (0.434) has least impact on EV adaptation in India.

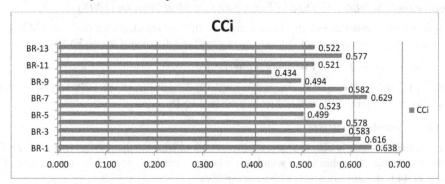

Fig. 44.1 Shows the average CCi values of barriers of EVs

4. CONCLUSION

Due to high initial cost of EVs in India is the first and foremost barrier. After that "Excessive battery changing expenditure" is the second most important barriers. The running cost of the EVs is less but after the end of useful life the battery changing cost is very high. The people think that is more than the fuel prize during that tenure. Very long charging time is ranked third and again most important barriers in EVs adaptation in India. A domestic house hold charger developed by the car manufacturers will take around eight hours to charge the EVs in households the time to charge the vehicle is very high. Lack of charging infrastructure is fourth and important barrier in EV adaptation in country like India and it will take more time to setup more charging stations especially away from metro cities. Fifth barrier is the "Limited Choice of EVs" are available in developing country like India. The people do not have much options or varieties in lower or middle range prize. If the policy makers focus on these barriers, the sale of EVs in India will rise very rapidly. This will help in reducing the pollution level in metro cities in India.

REFERENCES

1. Alotaibi, S.; Omer, S.; Aljarallah, S.; Su, Y. Potential implementation of EVs–Features, Challenges and User perspective. In Proceedings of the 2021 IEEE Green Energy and Smart Systems Conference (IGESSC), Long Beach, CA, USA, 1–2 November 2021; pp. 1–10.
2. Asadi, R., Moradi, M., & Hasanzadeh, M. N. (2022). Reliability Evaluation of Power System based on Demand Response Program in the Presence of the Electric Vehicles. Research and Technology in the Electrical Industry, 1(1), 85-94.
3. Bellman, R. E., & Zadeh, L. A. (1970). Decision-making in a fuzzy environment. Management science, 17(4), B-141.
4. Biresselioglu, M. E., Kaplan, M. D., & Yilmaz, B. K. (2018). Electric mobility in Europe: A comprehensive review of motivators and barriers in decision making processes. Transportation Research Part A: Policy and Practice, 109, 1-13.
5. Dhull, S., & Narwal, M. (2016). Drivers and barriers in green supply chain management adaptation: A state-of-art review. Uncertain Supply Chain Management, 4(1), 61-76.
6. Kannan, R., Rajasekaran, S., Stallon, S. D., & Anand, R. (2023). Improved indirect instantaneous torque control-based torque sharing function approach of SRM drives in EVs using hybrid technique. ISA transactions.
7. Khazaei, H. (2019). The influence of personal innovativeness and price value on intention to use of electric vehicles in Malaysia. European Online Journal of Natural and Social Sciences, 8(3), pp-483.
8. Lee, K. H. (2006). First course on fuzzy theory and applications (Vol. 27). Springer Science & Business Media.
9. Munshi, T., Dhar, S., & Painuly, J. (2022). Understanding barriers to electric vehicle adoption for personal mobility: A case study of middle-income in-service residents in Hyderabad city, India. Energy Policy, 167, 112956.
10. Murugan, M., & Marisamynathan, S. (2022). Analysis of barriers to adopt electric vehicles in India using fuzzy DEMATEL and Relative importance Index approaches. Case Studies on Transport Policy, 10(2), 795-810.
11. Noel, L., de Rubens, G. Z., Kester, J., & Sovacool, B. K. (2020). Understanding the socio-technical nexus of Nordic electric vehicle (EV) barriers: A qualitative discussion of range, price, charging and knowledge. Energy Policy, 138, 111292.
12. Cui, Y., Liu, J., Cong, B., Han, X., & Yin, S. (2022). Characterization and assessment of fire evolution process of electric vehicles placed in parallel. Process Safety and Environmental Protection, 166, 524-534.

13. Setiawan, A. D., Zahari, T. N., Purba, F. J., Moeis, A. O., & Hidayatno, A. (2022). Investigating policies on increasing the adoption of electric vehicles in Indonesia. Journal of Cleaner Production, 380, 135097.

14. Smith, B., Olaru, D., Jabeen, F., & Greaves, S. (2017). Electric vehicles adoption: Environmental enthusiast bias in discrete choice models. Transportation Research Part D: Transport and Environment, 51, 290-303.

15. Sreeram, K., Surendran, S., & Preetha, P. K. (2022). A Review on Single-Phase Integrated Battery Chargers for Electric Vehicles. Information and Communication Technology for Competitive Strategies (ICTCS 2021) Intelligent Strategies for ICT, 751-765.

16. Yang, C. (2022). Running battery electric vehicles with extended range: Coupling cost and energy analysis. Applied Energy, 306, 118116.

17. Yu, C. S. (2002). A GP-AHP method for solving group decision-making fuzzy AHP problems. Computers & Operations Research, 29(14), 1969-2001.

18. An, Y., Gao, Y., Wu, N., Zhu, J., Li, H., & Yang, J. (2023). Optimal scheduling of electric vehicle charging operations considering real-time traffic condition and travel distance. Expert Systems with Applications, 213, 118941.

19. Pourgholamali, M., Homem de Almeida Correia, G., Tarighati Tabesh, M., Esmaeilzadeh Seilabi, S., Miralinaghi, M., & Labi, S. (2023). Robust Design of Electric Charging Infrastructure Locations under Travel Demand Uncertainty and Driving Range Heterogeneity. Journal of Infrastructure Systems, 29(2), 04023016.

Note: All the tables and figure in this chapter were made by the author.

Computational Intelligence and Mathematical Applications – Dr. Devendra Prasad et al. (eds)
© 2024 Taylor & Francis Group, London, ISBN 978-1-032-87721-1

45 | Arduino Based Vehicle Accident Alert System Using GPS

Shanu Khare[1], Navpreet Kaur Walia[2]

Computer Science & Engineering Chandigarh University, India

Abstract: In India, individuals believe street accidents to be the most frequent thing to occur, and call it is an appalling occasion or destiny instead of responding towards it. In India, the total deaths because of road accidents is around 1,50,000 per year which is equal to 400 accidents per day. There are a few causes related to street mishaps, a couple of them incorporate over speeding, drunken driving, red light hopping, abstaining from driving wellbeing estimates, for example, caps and safety belts, utilizing cell phones, sounding, absence of traffic sense, and so on which generally happen on the expressways, and the roadways which are far away from the city. The fundamental objective of this exertion is serving the individuals/relations who are presented to occurrences in the spots that are confined or a long way from the private localities, and to watch the coincidences happening period and home, and pardon is the standing of influenced individuals. Subsequently, here we propose. This system is capable enough to detect the accidents in vehicles and send the alert to nearby rescue centres and hospitals about the location and whereabouts of the accident.

Keywords: Sensor, Arduino microcontroller, GPS module, Bluetooth module, Alert message

1. INTRODUCTION

The developing technology has additionally expanded the traffic risks and the street mishaps. Because of the absence of best crisis offices accessible in our nation the lives of the individuals are under high hazard. Accident alert system significant point is to save life of individuals involves in accidents. This is improved security frameworks for vehicles. GPS are profoundly valuable in the present period. Hardware is fitted on to the vehicle in such a way, that it is not noticeable to anybody so nobody can see it. Therefore, it is utilized as an undercover unit which ceaselessly or by any hinder to the framework, sends the area information to the checking unit. This Accident alert system ,which is in vehicle, it detects the Accident and the area of the accident happened and it sends GPS directions to the predefined portable, PC or different gadgets which would be included. The primary objective of this exertion is serving the persons who are presented to incidents in the places and roads that are confined or a long way from the residential neighborhood, and to observe the accident fashionable period and place, and what is the status of the persons. In this manner, here we propose. This system is capable enough to detect the accidents in vehicles and send the alert to nearby rescue centers and hospitals about the location and whereabouts of the accident. It sends a text message which contains the venue, time and date of the accident which is very accurate. In case there is no need to send the signals, there is a physical system in which one can turn off the sending of signal so that wrong signal is not sent to the nearby rescue groups. Whenever the accident happens, the sensors that are placed senses it, and sends the message to the micro-controller. It then transmits the the alert message through the systems to the nearby rescue centers like Police stations, hospitals etc. The signal is sent through the various technologies present in the system like GPS and sensors with the help of which the exact location of the event is known and the appropriate help is sent to the needed location.

[1]shanukhare0@gmail.com, [2]navpreet.walia12@gmail.com

DOI: 10.1201/9781003534112-45

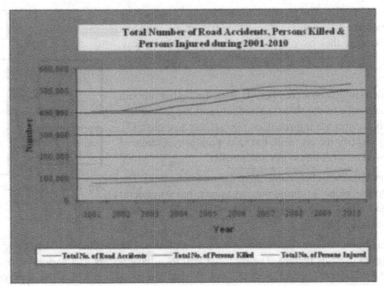

Fig. 45.1 Entire amount of roads coincidences, examination of govt.of india from 2001 to 2010[2]

"According to World Health Statistics 2008 reported in the Report on Road Safety, RTIs were the 9th leading cause of death in 2004, and if current trends continue, they will replace diabetes and HIV/AIDS as the 5th leading cause of death by 2030. [2]. In today's India, car accidents are one of the top causes of death. In 2005, nearly 1 lakh people died in 4.4 lakh road accidents. In 2008, 1.2 lakh people died in accidents, and in 2020, the death toll is expected to be over l.36 lakh for 5 lakh accidents [2]. As a result, the United Nations has declared 2011-20 as the Decade of Action on Road Safety, and has urged all member countries to develop a decadal action plan for execution in their own countries by 2020, in order to stabilize and reverse the rising trend in road accidents. Furthermore, a survey shows that every minute an injured crash victim waits for emergency medical help can make a significant difference in their survival rate; for example, analysis shows that reducing accident reaction time by one minute correlates to a 6lives saved.. an efficient approach for reducing traffic fatalities, therefore, is to reduce the time between when an accident occurs and when first responders, such as medical personnel, arrive. To determine whether an accident has happened, automatic collision notification systems use sensors integrated in the vehicle. These devices rapidly dispatch emergency medical staff to catastrophic incidents +- IEEE Advancing Technology for Humanity 1238. Eliminating the time between accident occurrence and first responder dispatch reduces fatalities by 6.

2. LITERATURE REVIEW

2.1 Proposed System

The suggested framework includes a new approach to road accident treatment that is based on the use of microcontrollers, haptic components, GPS modules, Bluetooth modules, and programming. A microcontroller (MC) is a little computer that consists of a processor, memory, and programmable data/yield peripherals on a single integrated circuit. The Arduino Mega Microcontroller has been chosen as the central component of this framework. Arduino is an open- source gadget prototyping platform that uses easily configurable hardware and code. It is recommended for artisans, architects, professionals, and anybody interested in creating intuitive objects or situations. Similarly, the tactile components, which are a basic and small unit utilised to view a few marvels, are the second important piece of the proposed framework. A detecting unit, a preparation unit, a handset unit, and a power unit are the four basic components of each sensor hub. The use of the Global Positioning System (GPS), which is a satellite-based route framework made up of a system of 24 satellites deployed into orbit by the United States Department of Defense, is the third important component of the framework. In addition, one of the most important aspects of this project is the use of Bluetooth technology to convey accident-related data.

2.2 Problem Formulation

Vehicle accidents on roads is one of the main issues that we face in everyday lives i.e., typically on the highways and the easy way to comply with the conference paper formatting requirements is to use this document as a template and simply type your text into it. outskirts of the city. The problem we are trying to solve here is critical and needs immediate attention because it involves many innocent lives. Many people died because they have an accident in a disused highway, where an ambulance cannot reach the specific spot on the time, which results in human loss.

Therefore, the main problem that is observed here is with the higher amount of deaths which is a consequence of the vehicle accidents and the reason it is caused is because the people who are injured does not get any response [3]. To rapidly and accurately identify the accident, the said work uses a range of sensors and modules, as well as an Arduino microcontroller.

Because of this system, several researchers came up in the matching field. The concept has been given to the globe on numerous occasions, but its general use has been limited due to a lack of execution, a lack of a suitable location for the system, or other obstacles [4].

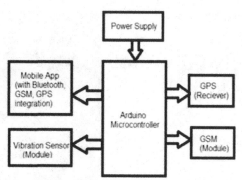

Fig. 45.2 A vehicle accident alert systems block diagram

Lexus Enform [6] is a brand-new feature that comes standard on all new Lexus vehicles. The driver must have a smart phone and be logged into the app to use this feature. The motorist will then have access to a variety of options, including GPS and phone directions. [6] "Safety Connect is a feature of technology that detects an accident or the deployment of the airbag system using a force sensor on the vehicle's rear end and sends an automated message to the response service centre via the smart phone [6]. The Lexus Enform system requires a lot of upkeep; after the second year, it costs around dollar 260 each year to keep it working." [7].

Similarly, "OnStar Corporation, a GM subsidiary, began providing accident notification services in the United States before expanding to Ecuador and Venezuela under the name Chevystar [8]. Automatic Crash Response, Stolen Vehicle, Tracking, Turn-by-Turn Navigation, and Roadside Assistance are among the features [8]. In online critiques of the system, overpricing, poor service, and a lack of assistance have all been mentioned, underlining its inefficiencies [9]. ASAD, on the other hand, only offers one type of detection and alerting service. The device obtains the driver's GPS location from the network provider and sends a text message to the driver's smartphone." Because the user just pays once, there are usually no additional costs.

The digitally notifying system combines V2V and V2I communications as a result of which it is capable of predicting early accidents in vehicles, is capable of notifying hospitals nearby within a very short span of time to reduce casualties, and the reporting of pertinent accident data [10]. It uses a vehicle's GPS link to send information to a service centre via an Internet connection supplied by roadside equipment, which responds quickly [10].

Using the smartphone's built-in sensors, the author of [11] describes a method for automatically identifying and notifying an accident. "This technique [11] assumes the presence of a mobile wireless telecommunications system. The phone is then mounted on a stand that runs parallel to the automobile longitudinally and transversally [11]. The driver does not need to interfere because the cell phone will deliver a signal [11]. The problem with this technology is that it can cause the system to slow down if the user touches the phone. Because of the reduced acceleration, the driver will be unable to use his smart phone while driving for whatever reason. The suggested solution has an advantage over this strategy in that the hardware installation ensures correct sensor alignment rather than depending on built-in hardware devices in autos. When deciding whether or not to reduce the system's errors, the system evaluates both force and velocity."

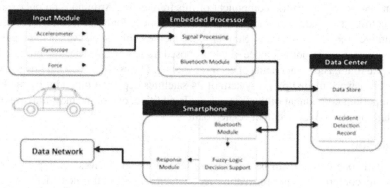

Fig. 45.3 Architecture for automated smart accident detection

According to the writers of [12]," the system had a GPS device, an entrenched wireless modem, bang devices, and an entrenched CPU. The accident device detects the collision, the GPS device finds the automobile, and the information is sent to a response centre via a wireless modem and embedded CPU [12]. The obtainability of GM On star as a in height-price option for clienteles is mentioned in the article, prompting the establishment of a low-cost option for users [12]. There are certain positions where GPS is fit in the vehicle. Using the position of these, the system communicates the information to the central department via text message. It does not use wireless means of its own, instead uses phone's internet connection to do so."

3. PROJECT OVERVIEW/SPECIFICATIONS

As previously stated, the suggested framework is implemented using a number of programming bundles and equipment modules. For all intents and purposes, we have utilized the accompanying equipment modules alongside Arduino Uno microcontroller and C, C++ language for framework coding process: [13]

1. *Tremor device:* Correspondingly, we have utilized the advanced tremor sensor. The recurrence for the instrument, when associated with Arduino Uno microcontroller, ranges after 31 Hzto 65535 Hz. Additionally, the vibration sensor designed by setting an alert for the stun affectability to enlist a stun. It the whole thing by trembling the device spasmodically, resulting in a "1" yield (as a pointer of trembling or development).

2. *GPS Component:* NEO6MV2 module as a GPS chip for this venture. It requires four wires associations which is very simple to consolidate by Arduino Uno MCU [14]: GND stick, the power stick, the receive jot (RX) associated with empower GPS module to get advanced sign from Arduino presentation that the GPS needs to record another area, lastly the spreader stick(TX) that empowers GPS unit to communicate the area arranges (possibility and longitude) to Arduino finished computerized I/O pins. This association has a inconsequential concern where the GPS module utilizes a 3.0 V rationale. Including two resistors as a simple voltage divider notion, on the other hand, can explain this issue. The power development in the circuit was controlled by the resistors.

3. *GSM Module:* We need to utilize the SIM900A GSM imperfection. It tends to be effectively associated through Arduino utilizing 4 wires associations: the GND stick, the voltage stick, the spreader stick(TX), lastly the recipient stick. To be able to access the SIM card and transmit SMS messages, the GSM module needs be set up with a useful correspondence recurrence band (or even accomplish more correspondence functionalities) [15].

3.1 Hardware Specification

At the equipment development stage, the total framework engineering of the proposed framework included:

1. *Vibration Sensor:* A vibration transducer, for example, is a device that merges a laser to transform vibrations into its electrical identical, like voltage. It's also known as a vibration transducer.

2. *GSM Component:* The Worldwide Organization for Moveable Infrastructures (GSMC) is a computerized portable communication framework extensively rummage-sale in Europe and other shares of the world.

3. *GPS:* The GPS (Worldwide Putting Classification) is a wireless triangulation organization that consents terrestrial, marine, and aerial customers to determine their precise position, speed, and period 24 hours a day, in any climate condition, wherever on the earth. It will detect the accident's location.

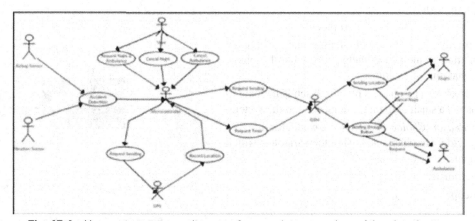

Fig. 45.4 Usage circumstance diagrams for complete procedure of the classification

4. *Arduino Microcontroller:* MCU (short for microcontroller unit) is a miniature processer (SoC) with a processor, reminiscence, and programmable information/harvest peripherals on a solitary combined tour.

5. *Power supply:* Is a type of electronic expedient that provisions electronic control to an electrical consignment. A power supply's primary function is to charge one form of electrical vitality.

6. *Accelerometer:* This 3-axial component collects information about the car's current acceleration along three orthogonal dimensions. It is also used in fuzzy logic component for decision support in order to get the velocity of the vehicle[16].

7. *Gyroscope:* The vehicle's spin and tilt are detected by the gyroscope, "which scans the data in degrees per second. This rate of rotation is used to evaluate if the vehicle has flipped or rotated to one side or the other."

8. *Force Sensor:* It is located at four sides of the car (each on one side with tyres) to get the force of the accident and it relative impact.

3.2 Software Specification

On a software level, the following tools and programmes were employed to assist us in effectively building sketches and successfully completing the future effort:

1. *Microsoft Word 2013:* It is a Microsoft expression computer database that permits users to category utilizing useful gears such as a spell and parsing checking, introducing benches and diagrams, and saving documents. To assist us in creating this report, we used Microsoft Word software as well as tools to develop the system architecture and system flow chart [17].

2. *Forum. Arduino Website:* The Arduino website is a really useful resource. It gave us with a wealth of information that aided us in finishing our system. It aided us in discovering problems and learning about them from other people's experiments and experiences on the internet. We also learned how to obtain the libraries wanted to track the code, as well as in what way to usage these public library in the code with the system's hardware elements and parts. We also learned how to read the various hardware bits and pieces. It also gave us with extra ciphers for other works that are relevant to our enterprise.

3. *Arduino IDE:* An exposed foundation Arduino Software (IDE) aids in the creation of codes and their ordering, as well as the transfer of these programmes to the Arduino panel. It can be used on slightly podium and functioning organization, including [18]

Gaps, Mac OS, and Linux. The atmosphere that is made up is being programmed in C language.

4. *Auto-desk Simulant Connected Website:* Autodesk and Tours-io is a influential connected tool that ropes the majority of Arduino hardware mechanisms, including the "Uno, Mega, Nano", and additional popular panels. It has a number of tools that help with fast draught prototyping and correcting. It aided us in the design of our microelectronic tours and the connection of hardware-replicated mechanisms [19].

3.3 Smartphone

The smartphone application serves as both a decision-making tool and a means of requesting assistance. It constitutes of following components:

1. Bluetooth module
2. Uncertain reason choice support module
3. An app that is capable of sharing the emergency information to the nearby help centres like hospitals, police stations, fire stations

1. *Bluetooth Module:* The phone's bluetooth sends the data to the stations via a small chip that is embedded in the system.

2. *Uncertain Reason Choice Provision:* It evaluates whether or not an accident has happened using the variables; which serve as input:

   1. Accelerometer
   2. Gyroscope
   3. Speed
   4. Force [20]

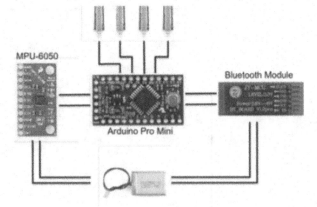

Fig. 45.5 Implementation architecture

"There are three possible ranges for each variable: low, medium, and high. When analyzing a detection circumstance, the decision support now has 81 alternative input combinations to consider."

3. *Response Module:* The sensors are capable enough to send the emergency information to the nearby help centers such as hospitals, police stations etc. It is also capable of sending the exact location of the place which also constitutes of date and time of the coincidence. The manuscript mails container be utilized to double-check the storage data [21].

3.4 Data Center

The observations of acceleration, rotation, force and speed are processed. These observations are then stored in Data Store. Also, it stores the contingent data which can be further used by experts in reports of the accident". For each user, it lists the following information:

1. Accelerometer readings
2. Gyroscope readings
3. Force readings
4. Velocity of the automobile
5. Position of the accident
6. The occasion of the accident [22]

4. RESULT

The suggested study aims to conduct research that will lead to the development of a method for saving the lives of those engaged in terrible road accidents on the outskirts of town, far from basic medical services. The outcome of this research paper will be to minimize the chances that occur every single day and to reduce the lives lost due to carelessness. These include accidents through over- speeding, drink and drive, not following the traffic lights, accidents because of low visibility, accidents caused due to vehicle part failures.

The intended goal will be reached by breaking down the task into the following goals:

1. Accident Alert
2. Detection of Intensity of accident
3. Coordinate location sharing with control room and hospital

5. CONCLUSIONS

New communication technologies incorporated into the automotive sector have the potential to provide better help to those injured in traffic accidents by lowering emergency response times and boosting the amount of information available to rescuers immediately before they begin the rescue. This system is highly optimized and intelligent to the extent that it automatically sends location of the coincidence to the nearest infirmary through GPS and GSM. As a consequence, the rescuers will not have to waste time looking for the accident site. A lots of millions of innocent lives would be saved with the help of this Research paper implementation.

REFERENCES

1. "Statistics - Suraya Foundation." Suraya Foundation. N.p., n.d. Web. 03 June 2013.
2. Kawach, Nadim. "Abu Dhabi Has a Car Crash Every 4.5 Minutes." Emirates 24/7. N.p., 25 Oct. 2010. Web. 04 June 2013.
3. "UAE Accident Report for 2011." Khaleej Times. N.p., 22 Apr. 2012.
4. Olarte, Olivia. "Road Accidents Top Emergency Ambulance Calls in Abu Dhabi." Khaleej Times. Khaleej Times Online, 13 Oct. 2011.
5. Evanco, William (1996) The impact of rapid incident detection on freeway accident fatalities. Mitretek Systems, Inc., WN96W0000071.
6. "Enform." Lexus, Toyota. N.p., n.d. Web. 15 June 2014.
7. Johnston, Casey. "The Pursuit of Connection: The 120K Lexus App and Car Combo." Ars Technica. N.p., n.d. Web. 15 June 201. [8] "Auto Security — Car Safety — Navigation System — OnStar." OnStar. N.p., n.d. Web. 15 June 2014.
8. "OnStar Reviews." Consumer Affairs. N.p., n.d. Web. 15 June 2014
9. J. Breckling, Ed., The Analysis of Directional Time Series: Applications to Wind Speed and Direction, ser. Lecture Notes in Statistics. Berlin, Germany: Springer, 1989, vol. 61.

10. Fogue, Manuel. et al., "Prototyping an automatic notification scheme for traffic accidents in vehicular networks," Wireless Days (WD), 2011 IFIP , vol., no., pp.1,5, 10-12 Oct. 2011

11. Montague, Albert, PH.D. Mobile Wireless Phone with Impact Sensor, Detects Vehicle Accidents/thefts, Transmits Medical Exigencyautomatically Notifies Authorities. Patent US2005/0037730 A1. 17 Feb. 2005. Print.

12. Gabler, Hampton. "Development of an Automated Crash Notification System: An Undergraduate Research Experience." Cite Seer X Beta. N.p., 18 Oct. 2000. Web.

13. Bhumakar, S. P. "Intelligent Car System for Accident Prevention Using ARM-7."International Journal of Emerging Technology and Advanced Engineering. N.p., 4 Apr. 2012.

14. Article by Youth Ki Awaaz "3 Lives Are Lost Every 10 Minutes In India Because We Don't Pay Attention To This Issue" [15] https://www.irjet.net/archives/V5/i3/IRJET-V5I357.pdf http://ethesis.nitrkl.ac.in/5742/1/E-92.pdf.

15. Yellamma Pachipala, Tumma srinivas Rao, G Siva, Nageswara Rao and D Baburao, "An IoT Based Automatic Accident Detection and Tracking System for Emergency Services", Jour of Adv Research in Dynamical Control Systems, vol. 12, no. 02, pp. 111117, 2020

16. Elie Nasr, Elie Kfoury and David Khoury, "An IoT Approach to Vehicle Accident Detection Reporting and Navigation", IEEE Explore, 2016.

17. Madhu Mitha Gowshika and Jayashree, "Vehicle Accident Detection System By Using GSM And GPS", IRJET, 2019

18. Himanshu Arora, Samyak Jain and Sanket Anand, Real Time Safety Alert System for Car published in the year 2019 in IEEE.

19. Rajvardhan Rishi, Sofiya Yede, Keshav Kunal and Nutan V. Bansode, Automatic Messaging Model for tracking vehicles and Accident Detection published in IEEE in the year 2020.

20. Yellamma Pachipala, Kaza Lalitha, Putta Anupama and Kodali Lakshmi Bhavani, "Object Tracking System utilizing IoT", International Journal of Innovative Technology and Exploring Engineering (IJITEE), vol. 9, no. 3, pp. 1357-1362, January 2020.

21. S. Mohana, "Internet of Things based Accident Detection System", IEEE Explore, 2019. [25] Pachipala Yellamma, P. Dileep Kumar, K. Sai Pradeep Reddy, L. Sri Harsha and N Jagadeesh, "Probability of Data Leakage in Cloud Computing", International Journal of Advanced Science and Technology, vol. 29, no. 6, pp. 3444-3450, 2020

22. M. Wegmuller, J. P. von der Weid, P. Oberson, and N. Gisin, "High resolution fiber distributed measurements with coherent OFDR," in Proc. ECOC'00, 2000, paper 11.3.4, p. 109.

Note: All the figures (except Fig. 45.1) in this chapter were made by the author.

Computational Intelligence and Mathematical Applications – Dr. Devendra Prasad et al. (eds)
© *2024 Taylor & Francis Group, London, ISBN 978-1-032-87721-1*

46 | Comparative Analysis of Heart Failure Detection Using Quantum Enhanced Machine Learning

Avni[1], Arunima Jaiswal[2], Amritaya Ray[3], Swati Gola[4]
Department of Computer Science and Engineering, Indira Gandhi Delhi Technical University for Women, Kashmere Gate, Delhi

Abstract: This paper explores the use of quantum-based machine learning algorithms for predicting heart disease. Traditional methods face challenges in processing complex healthcare data efficiently. To address this, quantum computing techniques, including dimensionality reduction and feature optimization, are applied. Quantum machine learning models, such as Variational Quantum Classifier (VQC), Quantum Support Vector Classification (QSVC) and Quantum Neural Networks (QNN), leverage these enriched datasets. This convergence of quantum computing and healthcare has the potential to transform medical care, cut costs, and save lives, presenting a significant advancement in predictive medicine.

Keywords: Machine learning, Quantum computing, Quantum machine learning, Feature selection, Heart disease, Variational quantum classifier, Quantum support vector classifier, Quantum neural networks

1. INTRODUCTION

Heart disease is a global health challenge, responsible for a substantial number of fatalities and health complications. The ability to predict and prevent heart disease is a critical pursuit in healthcare, as it holds the potential to save lives and reduce the burden on healthcare systems (WHO website). In this context, machine learning has surfaced as a valuable asset, aiding in the development of predictive models. However, as healthcare datasets become increasingly intricate and voluminous, there is a growing need for more robust and efficient techniques to handle the complexity of these data.

In this paper, we are introducing the integration of quantum-based machine learning algorithms into the realm of heart disease prediction. This endeavor is fueled by the urgent need to address the ongoing global healthcare challenge posed by cardiovascular diseases, which continue to exert a substantial toll on human health and well-being. Existing machine learning methods have certainly made significant strides in risk assessment; however, they often grapple with the formidable computational complexities inherent in healthcare datasets.

2. DATASET DESCRIPTION

The UCI Cleveland Heart Disease dataset is a widely used dataset for heart disease prediction and classification. It contains 1025 instances and 14 attributes, including both categorical and continuous variables. Some of the dataset attributes are (UCI Cleveland website) Sex, Age, Fasting Blood Sugar, Serum Cholesterol (chol), Chest Pain Type (CP) and many others.

This dataset is of great utility in creating and validating models for forecasting the probability of heart disease, relying on a patient's characteristics. It has been widely employed in healthcare-focused projects involving machine learning and data analysis.

---

[1]avni065btcse20@igdtuw.ac.in, [2]arunimajaiswal@igdtuw.ac.in, [3]amritaya014btcse20@igdtuw.ac.in, [4]swati025btcse20@igdtuw.ac.in

DOI: 10.1201/9781003534112-46

3. LITERATURE REVIEW

The proposed quantum K-means clustering method by Kavitha et al. (2023) was used to detect heart disease using a quantum circuit approach that achieved an accuracy of 96.4%, which is higher than the accuracy achieved by the classical K-means clustering method (94%).

The proposed quantum machine learning model by Alsubai et al. (2023) achieved an accuracy of 98% in classifying input data with both categories of affected and not affected.

The accuracy of the proposed system for heart failure detection by Kumar et al. (2021) was evaluated using several metrics. The Quantum Random Forest (QRF) algorithm achieved the highest accuracy of 91.97%, followed by the Quantum K-Nearest Neighbor (QKNN) algorithm with an accuracy of 90.33%.

A hybrid quantum classification method proposed by Heidari et al. (2022) deals with early heart rate detection. Their HQNN achieves a notable accuracy of 96.43% in 10-fold cross-validation. In the 70/30 train/test split ratio, HQRF stands out with accuracy of 94.31% for the test set. For the Statlog dataset, HQNN achieves the highest accuracy of 97.78% in 10-fold cross-validation. In the 70/30 train/test split ratio, HQRF achieves the accuracy of 90.52%.

A paper by Karthick et al. (2022) suggests that employing machine learning (ML) is a viable approach to mitigate and comprehend symptoms of heart disease. The achieved accuracies for the models Support Vector Machine (SVM), Gaussian Naive Bayes, Logistic Regression, LightGBM, XGBoost, and Random Forest, are 80.32%, 78.68%, 80.32%, 77.04%, 73.77%, and 88.5%, respectively.

4. METHODOLOGY

Our principal methodology of implementation is outlined as follows:

First we applied dimensionality reduction techniques like PCA, LDA, Projection Pursuit, and SVD.

Then we proceeded to Quantum Feature Optimization: This quantum feature optimization phase embodies the application of quantum principles to enhance the efficiency of information processing based on the underlying principles of Grover's Algorithm.

Quantum Models on Enhanced Datasets: Next, our focus converges on the application of quantum models to datasets fortified through previous stages. QSVC comes into play, representing a quantum-amplified extension of the classical SVM, supported by the Qiskit library. Additionally, VQC, implemented with the available functionality of Pennylane and Grover's algorithm, are introduced.

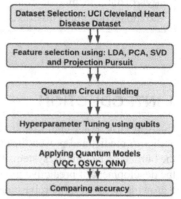

Fig. 46.1 Quantum model building workflow

4.1 Dimensionality Reduction

Dimensionality reduction involves methods to decrease the quantity of input variables within training data. In cases where data possesses high dimensionality, it is frequently beneficial to reduce this dimensionality by mapping the data into a lower-dimensional subspace that encapsulates the core characteristics of the data. (Javier et al., 2022)

Principal Component Analysis (PCA): It is a feature extraction technique that transforms a higher-dimensional feature space into a lower-dimensional one. During this reduction in dimensions, PCA aims to preserve as much of the original dataset's information as possible, while minimizing the correlation between the resulting Principal Components. (Jolliffe et al., 2016)

Linear Discriminant Analysis (LDA): It is an algorithm utilized for both multi-class classification, where it predicts class labels, and dimensionality reduction. It generates a projection of a training dataset aimed at optimally segregating examples based on their assigned class labels. (Van Otten et al., 2023)

Projection Pursuit: It is an iterative method employed for dimensionality reduction with a focus on preserving noteworthy data features. (GeeksforGeeks website)

Singular Value Decomposition (SVD): It is a mathematical technique used to break down a P x Q matrix, given as M and composed of mathematical values, into three constituent matrices. (Mahendru, 2023)

4.2 Quantum Feature Enhancement

Qubits: Quantum bits, or simply "qubits," are the fundamental units of quantum information in quantum computing and quantum information theory. They are analogous to classical bits but differ in several key ways due to the principles of quantum mechanics (Rietsche et al., 2022). Quantum qubits have four key properties which are Superposition, Entanglement, Quantum Interference and Measurement.

Grover's Algorithm: Grover's algorithm is primarily designed for searching an unsorted database of N elements in $O(\sqrt{N})$ time, whereas classical search algorithms require $O(N)$ time. In machine learning applications that involve searching through large datasets. Grover's algorithm can be used to accelerate certain subroutines within quantum machine learning algorithms, making them more optimized (Nikhade et al., 2022).

Quantum circuit: Quantum circuits are a fundamental concept in quantum computing, and they serve as the quantum counterpart of classical digital circuits (Ju et al., 2007). Key components of quantum circuits include: Qubits, Quantum Gates, Quantum Measurements, Quantum Algorithms. The process of quantum algorithm starts by generating a parameterized quantum circuit designed to meet the needs of the dataset. These parameters are updated iteratively to improve results.

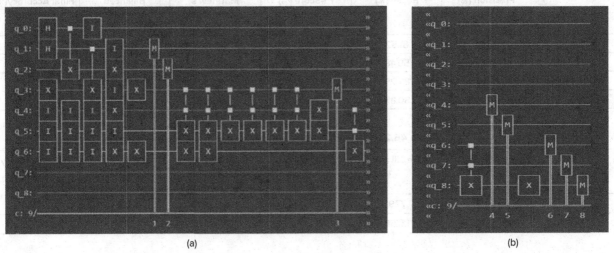

Fig. 46.2   Quantum circuits generated by our qubit hyperparameter tuning code

4.3 Quantum Machine Learning Models

The Variational Quantum Classifier (VQC): It is a machine learning algorithm powered by quantum computing that performs classification tasks. It employs a hybrid quantum-classical approach, utilizing a variational quantum circuit with adjustable parameters to process quantum-encoded data. (Qiskit website)

Quantum Support Vector Classifier (QSVC): It is a machine learning algorithm powered by quantum computing, that adapts the classical Support Vector Machine (SVM) for classification tasks. It harnesses quantum computing's unique characteristics, such as superposition and quantum interference, to potentially provide computational advantages in solving complex classification problems (Suzuki et al., 2023).

Quantum Neural Network (QNN): These represent a burgeoning frontier in the intersection of quantum computing and deep learning, offering a novel paradigm for processing and extracting features from quantum data. In this novel approach, quantum states are employed to encode input data, and unitary transformations. Leveraging the principles of superposition and entanglement in quantum systems, QNNs surpass classical counterparts in terms of computational capacity for specific tasks. (QNN-Qiskit website)

5. RESULT AND DISCUSSION

In this paper, our aim is to showcase the usage of different Quantum machine learning models using various preprocessing techniques which are used for dimensionality reduction of the datasets. Specifically, we focus on the classical encoding and comparing Quantum machine learning models such as VQC, QSVC and QNN on our dataset and optimize its result by reducing the dimensionality of the dataset using various methods such as PCA, LDA, SVD and Projection Pursuit.

Given the limited number of qubits available on current quantum computers, researchers are exploring alternative strategies. Our experiments involve encoding data into two dimensions for two qubits, where SVD fetched the most accurate results.

We employed three quantum machine learning models VQC, QSVC and QNN. After evaluating their performance according to the following tables, our conclusion is that the VQC outperformed QSVC and QNN. It is worth highlighting that we conducted 150 iterations of VQC for each dimensional reduction algorithm. An important observation relates to the execution time: VQC managed to complete 150 iterations within a relatively short span of 1.5-2 hours, while QSVC proved to be considerably time-consuming, requiring over 5 hours for execution. The implementation of QNN was less time consuming than others, but fetched us a comparatively low accuracy of 81%. This observation underscores VQC's superiority as a refined quantum-enhanced machine learning model, both in terms of accuracy and efficiency in computation and execution time.

Table 46.1 Comprehensive metrics of VQC model

	Precision (%)	Recall (%)	F1-Score (%)	Matthew's Correlation Coefficient (%)	Balanced Accuracy (%)	Final Accuracy(%)
LDA	86.21	81.52	83.8	64.49	82.43	82.32
PCA	81.18	59.48	68.66	23.81	63.07	61.59
SVD	91.86	82.29	86.81	71.03	86.0	85.37
PP	90.48	80.85	85.39	68.7	84.71	84.12

Table 46.2 Comprehensive metrics of QSVC model

	Precision (%)	Recall (%)	F1-Score (%)	Matthew's Correlation Coefficient (%)	Balanced Accuracy (%)	Final Accuracy (%)
LDA	72.7	77.9	75.03	46.28	72.84	72.84
PP	55.10	55.68	55	5.88	52.94	52.73

Table 46.3 Accuracy achieved by QNN model

Model	Accuracy (%)
QNN	81

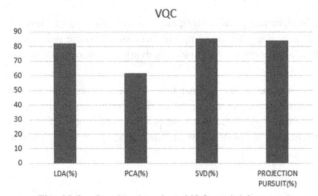

Fig. 46.3 Graphical analysis VQC model Accuracy

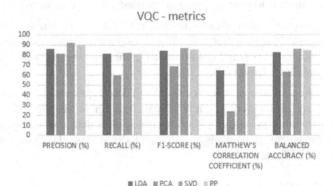

Fig. 46.4 Graphical analysis VQC model metrics

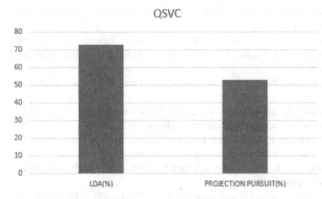

Fig. 46.5 Graphical analysis of QSVC model accuracy

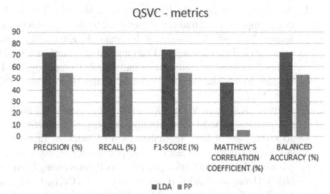

Fig. 46.6 Graphical analysis of QSVC model metrics

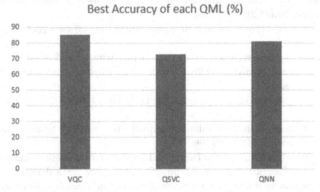

Fig. 46.7 Graphical analysis of QML model Accuracies

6. CONCLUSIONS

To conclude this study, we arrive at insights of major significance in the realm of quantum machine learning. The Variational Quantum Classifier (VQC) has established itself as the most potent quantum machine learning model. Its consistent outperformance of Quantum Support Vector Classification (QSVC) and Quantum Neural Networks (QNN) on multiple fronts, including accuracy and computational efficiency, underscores its supremacy in prediction-based models.

Our exploration also underscores the pivotal role of dimensionality reduction in the context of quantum machine learning. Singular Value Decomposition (SVD) emerges as the standout among the dimension reduction techniques, consistently surpassing Linear Discriminant Analysis (LDA), Projection Pursuit and Principal Component Analysis (PCA). This finding underscores the integral role of SVD in optimizing data representation for quantum processing.

VQC's speed in completing 150 iterations within mere hours is in stark contrast to QSVC's substantial time and computational resource requirements. This substantial disparity in execution time substantiates the practicality and applicability of VQC for real-world use cases. QNN evaluated 15-39 epochs within an efficient time window; the calculated accuracy and relevant metrics were still less than those of VQC. However, QNN serves as a better performing alternative to QSVC.

REFERENCES

1. World Health Organization: WHO. (2019, June 11). Cardiovascular diseases. https://www.who.int/health-topics/cardiovascular-diseases#tab=tab_1
2. UCI Machine Learning Repository. (n.d.-b). https://archive.ics.uci.edu/dataset/45/heart+disease
3. Kavitha, S.S., Kaulgud, N. Quantum K-means clustering method for detecting heart disease using quantum circuit approach. Soft Comput 27, 13255–13268 (2023).

4. Alsubai, S., Alqahtani, A., Binbusayyis, A., Sha, M., Gumaei, A., & Wang, S. (2023). Heart failure detection using instance quantum circuit approach and traditional predictive analysis. Mathematics, 11(6).

5. Kumar, Y., Koul, A., Sisodia, P. S., Shafi, J., Kavita, K., Gheisari, M., & Davoodi, M. B. (2021). Heart failure detection using Quantum-Enhanced Machine Learning and traditional machine learning techniques for the internet of artificially intelligent medical things. Wireless Communications and Mobile Computing, 2021, 1–16.

6. Heidari, Hanif, and Gerhard Hellstern. "Early Heart Disease Prediction Using Hybrid Quantum Classification." Cornell University, Aug. 2022.

7. Karthick, K., Aruna, S., Samikannu, R., Kuppusamy, R., Teekaraman, Y., & Thelkar, A. R. (2022). Implementation of a heart disease risk prediction model using machine learning. Computational and Mathematical Methods in Medicine, 2022, 1–14.

8. Alotaibi, Fahd Saleh. "Implementation of Machine Learning Model to Predict Heart Failure Disease." International Journal of Advanced Computer Science and Applications, vol. 10, no. 6, Science and Information Organization, Jan. 2019.

9. Mancilla, Javier, and Christophe Père. "A Preprocessing Perspective for Quantum Machine Learning Classification Advantage Using NISQ Algorithms." arXiv (Cornell University), Cornell University, Aug. 2022

10. Jolliffe, I. T., & Cadima, J. (2016). Principal component analysis: a review and recent developments. Philosophical Transactions of the Royal Society A, 374(2065), 20150202

11. Van Otten, N. (2023, November 7). Linear Discriminant Analysis Made Simple & How To Tutorial In Python. Spot Intelligence. https://spotintelligence.com/2023/11/07/linear-discriminant-analysis-lda/#:~:text=Linear%20Discriminant%20Analysis%20(LDA)%20is,between%20classes%20in%20a%20dataset.

12. GeeksforGeeks. (2022, December 30). Projection pursuit using Python. https://www.geeksforgeeks.org/projection-pursuit-using-python/

13. Mahendru, K. (2023, November 7). Master Dimensionality Reduction with these 5 Must-Know Applications of Singular Value Decomposition (SVD) in Data Science. Analytics Vidhya. https://www.analyticsvidhya.com/blog/2019/08/5-applications-singular-value-decomposition-svd-data-science/

14. Rietsche, R., Dremel, C., Bosch, S., Steinacker, L., Meckel, M., & Leimeister, J. M. (2022). Quantum computing. Electronic Markets, 32(4), 2525–2536.

15. Nikhade, A. (2022, January 6). Grover's search algorithm | simplified - Towards Data Science. Medium. https://towardsdatascience.com/grovers-search-algorithm-simplified-4d4266bae29e

16. Y. -L. Ju, I. -M. Tsai and S. -Y. Kuo, "Quantum Circuit Design and Analysis for Database Search Applications," in IEEE Transactions on Circuits and Systems I: Regular Papers, vol. 54, no. 11, pp. 2552-2563, Nov. 2007, doi:10.1109/TCSI.2007.907845.

17. VQC — Qiskit 0.32.1 documentation. (n.d.). https://qiskit.org/documentation/stable/0.32/stubs/qiskit.aqua.algorithms.VQC.html

18. Suzuki, T., Hasebe, T., & Miyazaki, T. (2023). Quantum support vector machines for classification and regression on a trapped-ion quantum computer. arXiv (Cornell University). https://doi.org/10.48550/arxiv.2307.02091

19. Quantum Neural Networks - QISKIT Machine Learning 0.7.0. (n.d.). https://qiskit.org/ecosystem/machine-learning/tutorials/01_neural_networks.html

Note: All the figures and tables in this chapter were made by the author.

Computational Intelligence and Mathematical Applications – Dr. Devendra Prasad et al. (eds)
© 2024 Taylor & Francis Group, London, ISBN 978-1-032-87721-1

Empirical Analysis of Depression Detection Using Bat and Firefly Algorithm

Kritika Shrivastava[1], Arunima Jaiswal[2]
Computer Science & Engineering, Indira Gandhi Delhi Technical University For Women, Delhi, India

Suresh Kumar[3]
Computer Science & Engineering, Galgotias College of Engineering & Technology, Greater Noida, India

Nitin Sachdeva[4]
Computer Science & Engineering, Amity University, Noida, Uttar Pradesh, India

Abstract: Depression, a widespread and exhausting mental health condition, imposes substantial burdens on both individual sufferers and society. Early and precise diagnosis of depression is essential to enable timely intervention and effective therapeutic measures. Recent years have witnessed the emergence of machine learning techniques as valuable tools for facilitating early depression detection. Diverse machine learning classifiers, such as Random Forest, k-nearest Neighbours, Decision Tree, and Gradient Boosting, are employed to construct predictive models. After the initial classifier assessment, this paper introduces the Firefly & Bat Swarm Optimization technique. They are deployed to fine-tune the hyperparameters of the machine learning classifiers to maximize predictive accuracy, thereby enhancing the overall performance of depression detection models.

Keywords: Firefly, BAT, Machine learning classifiers, KNN, Decision tree, Random forest, Gradient boosting

1. INTRODUCTION

A complex mental health condition called depression is defined by enduring sadness and hopelessness as well as a lack of interest in or enjoyment from routine activities. It may impact a person's feelings, ideas, and actions, among other areas of their life [1]. Many studies have added to our knowledge of depression's causes, risk factors, and treatment options. Depression is a primary global health concern [2]. Developing successful prevention and intervention strategies requires a multifaceted understanding of depression that takes into account biological, genetic, environmental, and psychological factors. [3]

1.1 Machine Learning used in Detecting Depression

Because machine learning can analyse and interpret large amounts of data and extract patterns and insights that traditional methods may find difficult to extract, ML is used to detect depression [4]. Text, speech, and physiological signals are just a few data sources machine learning algorithms can examine to find minute patterns that might be signs of depression [5].

1.1 Optimization Techniques used in Detecting Depression

Optimization techniques extract meaningful patterns and features from various data sources. By reducing false positives and negatives, these methods seek to raise the overall dependability of depression detection models. To improve the performance of depression detection models, recent research has investigated the integration of ML, feature selection, and ensemble methods. [4]

[1]kritika010mtcse22@igdtuw.ac.in, [2]arunimajaiswal@igdtuw.ac.in, [3]suresh.kumar@galgotiacollege.edu, [4]nsachdeva@amity.edu

DOI: 10.1201/9781003534112-47

In this paper, we have detected depression using the depression anxiety dataset. Various Machine learning classifiers have been used and their accuracies have been compared. Then, we optimised the accuracies using BAT and Firefly optimization techniques.

2. RELATED WORK

The field of depression detection has drawn a lot of attention lately to the association between machine learning and mental health. Several research works have investigated how well machine learning classifiers can identify depression-related patterns from various data sources. Most notably, Wang et al. [6] showed how natural language processing methods applied to social media data could be used to identify depression symptoms early on. Continuing this work, Al-Mosaiwi and Johnstone [7] examined how self-reported emotional states could be used to test machine learning algorithms, highlighting the significance of feature selection in enhancing classification accuracy. Additionally, Gkotsis et al. [8] investigated how to combine textual and visual cues with multimodal data to create more reliable depression detection models.

Research attempts that have recently been conducted have made significant progress in exploring depression detection through optimisation techniques. Notably, Liang et al. [9] improved the efficacy of machine learning models for depression detection using electroencephalogram (EEG) signals by utilizing Particle Swarm Optimisation (PSO), a metaheuristic optimisation algorithm. Their research showed how effective PSO is at fine-tuning the classifiers' parameters, leading to increased precision in identifying depressive states. Furthermore, Zhang et al. [10] looked into how Genetic Algorithms could optimize feature selection for physiological signal-based depression detection.

Recently, studies investigating optimisation techniques for depression detection have ventured into the novel field of Bio-Inspired Algorithms. In particular, there has been increased interest in using the Bat Algorithm for depression detection. Wu et al. [11], for example, looked into how well the Bat Algorithm performed when it came to fine-tuning the parameters of a support vector machine (SVM) classifier to differentiate between depressive states and electroencephalogram (EEG) signals. The study showed that the Bat Algorithm greatly enhanced the classification performance by adjusting the SVM parameters. This innovative research demonstrates how feature selection and parameter optimisation for depression detection can be made simpler by using optimisation techniques inspired by nature. In a noteworthy study, Yang and Wu [12] used the Firefly Algorithm to optimize a deep neural network's parameters for better depression detection accuracy using electroencephalogram (EEG) signals. Their investigation revealed that the Firefly Algorithm successfully traversed the complicated parameter space, improving the deep neural network's ability to discriminate between depressive states.

3. METHODOLOGY

The implementation part of this paper is done in the following steps:

1. *Data Collection:* We have used the depression anxiety dataset [13], a public dataset available at Kaggle. This dataset contains a total of 19 columns. Some of them are as follows:

    A. BMI contains these values: Class I Obesity, Normal, Overweight, Underweight.

    B. PHQ score- The Patient Health Questionnaire assesses the severity of depression.

    C. Depression severity contains None, Mild, Moderate, Severe, and Minimal values.

    D. Sleepiness- contains these values: False, True

2. *Data Pre-processing:* First, we will address issues related to missing data and outliers. Then, we will segregate the dataset into training and testing sets to facilitate model evaluation.

3. *Feature Optimization:* The most relevant and useful features must be chosen and refined to increase the model's accuracy. This process is known as feature optimization. [14] In the dataset used in this research, we selected the depression diagnosis column as our target variable.

4. *Model Training:* We train the following machine learning classifiers on the pre-processed dataset: K-Nearest Neighbours (KNN), Decision Tree, Random Forest, and Gradient Boosting.

5. *Classification:* We then classify the patients as depressed or not depressed.

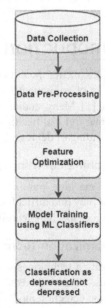

Fig. 47.1 Proposed methodology

In this paper, we have used the following machine-learning classifiers in detecting depression:

A. K-Nearest Neighbours (KNN)—KNN functions because data points sharing similarities often belong to the same class. It computes the distance between a data point and its k nearest neighbours, with 'k' being a user-defined parameter. The choice of 'k' plays a pivotal role as it influences the algorithm's sensitivity to data outliers and noise. [15]

B. Decision Tree—A decision tree is constructed to make binary decisions, such as determining whether a patient is suffering from depression or not. The model continues to branch the data until it reaches a predefined threshold. Decision Trees offer the advantage of interpretability, enabling the identification of influential features in depression detection. [16]

C. Random Forest—Random Forest is an ensemble learning approach that combines numerous decision trees to enhance predictive accuracy and minimize overfitting. Each tree within the forest is trained on a random subset of the data and a random subset of the features. This approach mitigates the variance associated with an individual decision tree and improves the model's ability to generalize to new patient data. [16]

D. Gradient Boosting—It generates decision trees sequentially, with each tree attempting to rectify the errors of its predecessor. Gradient Boosting can detect depression by establishing a robust predictive model by iteratively refining the mistakes of the prior models. [17]

1. Optimisation techniques are frequently utilised in depression detection to improve the effectiveness and reliability of diagnostic models by increasing their accuracy.

2. *Firefly Algorithm*: The FSO algorithm, inspired by the blinking behaviour of fireflies, is a nature-inspired optimization technique.

3. *Feature Selection:* The FSO algorithm is utilized to identify the most salient features for depression detection. The feature subset derived through FSO is subsequently employed as input for machine learning models. [18]

4. *Hyperparameter Tuning:* FSO is manipulated for fine-tuning hyperparameters of ML algorithms intended for depression detection. This optimization aims to identify the optimal combination of hyperparameters, thus enhancing the overall model performance.

   Experimental Results: The FSO algorithm is employed for feature selection and hyperparameter optimization in depression detection models.

5. *Application to Depression Detection:* The feature subsets obtained through FSO-based feature selection and the optimized hyperparameters are subsequently applied to the training and testing of machine learning models for depression detection. [18]

   (a) *BAT Algorithm*: BSO is a bio-inspired optimization algorithm grounded in the principles of echolocation, a behaviour exhibited by bats. [11]

   (b) *Feature Selection:* Depression detection is characterized by considering various attributes, encompassing physiological data, behavioural patterns, and self-reported questionnaires.

   (c) *Parameter Tuning:* The efficacy of machine learning models employed in depression detection often hinges upon the judicious tuning of hyperparameters.

   (d) *Ensemble Learning:* The amalgamation of multiple depression detection models into an ensemble framework is a strategy BSO may facilitate. [11]

3. RESULTS

This section compares the accuracies of various ML classifiers like K-Nearest Neighbors, Random Forest, Decision Tree, and Gradient Boosting. Then, we optimised the accuracies of these classifiers using two swarm optimization techniques: BAT and Firefly. Then, we compared the optimised accuracy of these ML classifiers.

Figure 47.2 is a bar graph of different Machine Learning Classifiers. The accuracy of the decision tree is the highest at 80.5%.

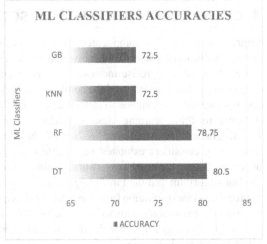

Fig. 47.2 Accuracies comparison of ML classifiers

Figure 47.3 shows a bar graph of different ML Classifiers like K-Nearest Neighbors, Random Forest, Decision Tree, and Gradient Boosting with optimized accuracy. BAT-optimized decision tree accuracy was the highest at 85.4%.

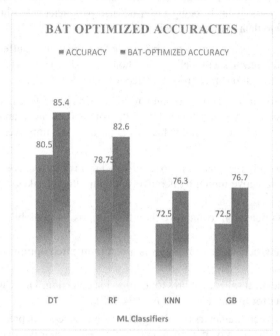

Fig. 47.3 ML classifiers BAT optimized accuracies

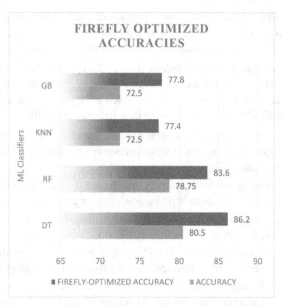

Fig. 47.4 Firefly optimized accuracies

Figure 47.4 shows a bar graph of different ML Classifiers like K-Nearest Neighbors, Random Forest, Decision Tree, and Gradient Boosting with optimized accuracy. Firefly-optimized decision tree accuracy was the highest at 86.2%.

Figure 47.5 shows a bar graph of different ML Classifiers like K-Nearest Neighbors, Random Forest, Decision Tree, and Gradient Boosting with Bat and Firefly optimized accuracies. As we can see from the graph, Firefly-optimized accuracy on the decision tree classifier is the highest at 86.2%.

4. CONCLUSION AND FUTURE SCOPE

Depression is a highly complicated mental health disorder that exerts a substantial societal and individual burden, necessitating more advanced and precise methods for early identification. In this paper, we have investigated the combination of ML classifiers and optimization algorithms to detect depression. We used four different machine learning classifiers, decision trees, random forest, gradient boosting, and K-Nearest Neighbors for this task. These classifiers exhibited favourable outcomes regarding accuracy. Then, we used two swarm optimization techniques, the Bat Algorithm and the Firefly algorithm. These optimization

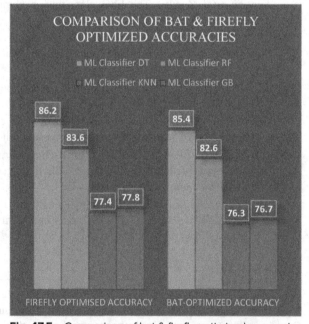

Fig. 47.5 Comparison of bat & firefly optimized accuracies

approaches draw inspiration from the natural behaviours of bats and fireflies. A marked performance enhancement was achieved by applying these algorithms to fine-tune our ML models' parameters. The combination of machine learning classifiers with Bat and Firefly algorithms for the detection of depression captures a powerful and transformative strategy in the continuous effort to address depression.

As technology advances and datasets expand in size and complexity, the potential for more precise, timely, and efficacious mental health assessment becomes increasingly within reach. This research marks a critical advancement toward that aspiration, promoting the development of solutions that possess the capacity to revolutionize mental healthcare and, ultimately, the lives of those with depression.

REFERENCES

1. Krishnan, The molecular neurobiology of depression. Nature, 455(7215), 894–902.
2. De Choudhury, Mining Twitter Data for Health Research: A Comparison Between Depression and Breast Cancer. Proceedings of the SIGCHI Conference on Human Factors in Computing Systems
3. Guntuku, Detecting depression and mental illness on social media: An integrative review. Current Opinion in Behavioral Sciences, 18, 43-49.
4. Fatima, Feature selection and classification for EEG-based depression detection. Computers in Biology and Medicine, 111, 103346.
5. Subasi, A survey on depression detection with EEG signals. Computer Methods and Programs in Biomedicine, 171, 1-13.
6. Wang, Social media as a sensor of depression in individuals. *Proceedings of the 26th CHI Conference on Human Factors in Computing Systems*.
7. Al-Mosaiwi, In an absolute state: Elevated use of absolutist words is a marker specific to anxiety, depression, and suicidal ideation. *Clinical Psychological Science*.
8. Gkotsis, The language of mental health problems in social media. In *Proceedings of the 11th International Conference on Language Resources and Evaluation (LREC 2018)*.
9. Liang, Depression detection based on feature optimization algorithm of EEG signals. IEEE Access.
10. Zhang, A novel feature selection method using genetic algorithms for depression detection. Journal of Medical Systems.
11. Wu, An improved bat algorithm for parameter optimization of SVM and its application in EEG feature selection for depression detection. Frontiers in Neuroscience.
12. Yang, Firefly algorithm: Recent advances and applications. In Studies in Computational Intelligence (Vol. 769). Springer.
13. https://www.kaggle.com/datasets/shahzadahmad0402/depression-and-anxiety-data
14. Kumar, Piyush, et al. "Feature based depression detection from twitter data using machine learning techniques." Journal of Scientific Research 66.2 (2022): 220-228.
15. Smys, "Analysis of deep learning techniques for early detection of depression on social media network-a comparative study," Journal of Trends in Computer Science and Smart Technology, vol. 3, pp. 24--39, 2021.
16. Z. Xu, "Sub phenotyping depression using machine learning and electronic health records," Learning Health Systems, vol. 4, p. e10241, 2020.
17. Priya, "Predicting anxiety, depression and stress in modern life using machine learning algorithms," Procedia Computer Science, vol. 167, pp. 1258--1267, 2020.
18. Yang, X.-S. "Firefly Algorithms for Multimodal Optimization." In: Stochastic Algorithms: Foundations and Applications. Springer.

Note: All the figures in this chapter were made by the author.

Computational Intelligence and Mathematical Applications – Dr. Devendra Prasad et al. (eds)
© 2024 Taylor & Francis Group, London, ISBN 978-1-032-87721-1

48 | Comparative Analysis of Optimisation Techniques for Depression Detection

Kritika Shrivastava[1], Arunima Jaiswal[2]

Computer Science & Engineering, Indira Gandhi Delhi Technical University for Women, Delhi, India

Javed Miya[3]

Computer Science & Engineering, Galgotias College of Engineering & Technology, Greater Noida, India

Nitin Sachdeva[4]

Computer Science & Engineering, Amity University, Noida, Uttar Pradesh, India

Abstract: Depression, a widespread mental health disorder, exerts an intense impact on a significant global population. Accurate and timely diagnosis of depression is of crucial importance for effective treatment and intervention. Recent years have witnessed the emergence of machine learning techniques as valuable tools for facilitating early depression detection. This research paper outlines a comprehensive methodology for the classification of depression utilising machine learning classifiers, complemented by an in-depth exploration of the enhancement of classification outcomes via the application of three distinct optimisation techniques: Grey Wolf Optimization, Particle Swarm Optimization, and Cuckoo Search. Various Machine learning classifiers, including K-Nearest Neighbours, Decision Tree, Random Forest and Gradient Boosting, are employed. The initial classifier models are carefully trained and rigorously evaluated to establish a baseline performance. Empirical findings underscore the efficacy of the proposed methodology. The ML classifiers exhibit substantial performance enhancements following optimisation techniques, manifesting in heightened accuracy.

Keywords: Cuckoo, Particle swarm, Grey wolf, Machine learning classifiers, KNN, Decision tree, Random forest, Gradient boosting

1. INTRODUCTION

Depression is typified by facing sadness and hopelessness as well as a lack of interest in or enjoyment from routine activities. It profoundly affects a person's thoughts, feelings, and physical health in addition to the typical ups and downs of life. Typical symptoms include changes in appetite and sleep patterns, exhaustion, trouble focusing, and a feeling of worthlessness. [1] Numerous things, such as a genetic predisposition, chemical imbalances in the brain, trauma, or extended stress, can cause depression. Psychotherapy, medicine, and lifestyle modifications are frequently used in conjunction for treatment. [2]

Figure 48.1 is a flowchart about common symptoms of depression that most people experience. Some people may experience some of these symptoms, while others may experience all of them. The intensity of these symptoms may vary from person to person.

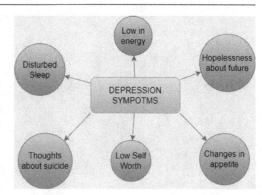

Fig. 48.1 Common symptoms of depression

[1]kritika010mtcse22@igdtuw.ac.in, [2]arunimajaiswal@igdtuw.ac.in, [3]javed.miya@galgotiacollege.edu, [4]nsachdeva@amity.edu

DOI: 10.1201/9781003534112-48

Since depression is a complex mental health condition with many contributing factors, machine learning is used to detect depression because it can analyse large amounts of diverse data and identify patterns that may be indicative of depressive symptoms [3]. Because ML algorithms can process and interpret data from multiple sources, including text, speech, and physiological signals, they can reveal hidden relationships. [4] By optimizing and improving the performance of machine learning models utilized in the diagnostic process, optimisation techniques are essential in the detection of depression. These methods aim to maximize predictive accuracy and reduce errors by optimizing the model's parameters. [4]

In this paper we have detected depression using EHR [Electronic Health Record] dataset. Various Machine learning classifiers have been used like KNN, Decision Tree, Random Forest, Gradient Boosting and their accuracies have been compared. Then we have optimised the accuracies of these ML Classifiers using three swarm optimization techniques: Particle Swarm Optimization [PSO], Cuckoo Swarm Optimization (CSO), and Grey Wolf Swarm Optimization (GWO). Then we have compared the optimised accuracies of these classifiers.

2. RELATED WORK

In this section first we have discussed previous work related to Machine learning in detecting depression. Then, we have discussed previous work related to optimization techniques.

Machine learning for depression detection has attracted a lot of interest lately as scientists work to improve the precision and effectiveness of testing techniques. Numerous investigations have looked at different strategies for solving this issue using different features and datasets. For example, a model utilizing speech and language patterns was proposed by Smith et al. to detect depressive symptoms in audio recordings [5]. In a different direction, Zhang et al. concentrated on creating a strong classifier for depression detection using electroencephalogram (EEG) signals [6]. Furthermore, Wang et al. work delves into the field of social media analysis and shows how valuable features can be extracted from user-generated content to aid in the early detection of depression [7]. Even though these studies offer insightful information, efforts to improve techniques and investigate new modalities continue in order to raise the overall effectiveness of machine learning-based depression detection systems.

Recent research initiatives have achieved noteworthy advancements in the study of depression detection via optimisation methodologies. Khurram Jawad [8] predicted depression on the Reddit dataset using various ML and DL models like KNN, SVR, decision trees, ResNet, VGG, and LeNet. He then applied the Cuckoo Search optimization technique to improve the testing accuracy by 98.7%. Yan Guo et al. [9] presented an EEG-based approach for detecting mild depression. They proposed a novel multi-objective particle swarm optimization (MOPSO) for the detection of depression. Their model's objectives include minimizing the number of misclassifications, maximizing the external distance, and minimizing the internal distance.

3. METHODOLOGY

In this section, we have discussed the methodology used during the implementation part of this paper. Each step is discussed in detail and explains the machine learning classifiers used. Then we have also explained the three swarm optimization techniques used in this paper: Particle Swarm Optimization [PSO], Cuckoo Swarm Optimization (CSO), and Grey Wolf Swarm Optimization (GWO).

Figure 48.2 is the flowchart of our methodology while implementing this research study. Optimization Techniques include Cuckoo Search Optimization, Particle Swarm Optimization and Grey Wolf Optimization. We have used four Machine Learning Classifiers: K-Nearest Neighbors [KNN], Decision Tree, Random Forest and Gradient Boosting.

The implementation part of this paper is done in the following steps:

1. *Data Collection:* We have used the EHR dataset [Electronic Health Record], a public dataset available at Kaggle. This dataset contains columns like Patients Temperature, Pulse, Blood Pressure, etc. [10]

2. *Data Pre-processing:* First, we will address issues related to missing data and outliers. Then, we will segregate the dataset into training and testing sets to facilitate model evaluation. The target variable of the EHR dataset is the SCORE Column, which contains four different values- 0,1,2,3. The target variable has multiclass

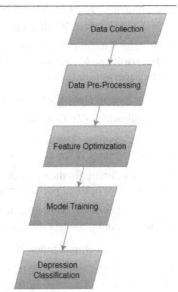

Fig. 48.2 Proposed methodology

labels, and we have to pre-process it to convert it into a binary classification task which will contain binary tags (e.g., 0 & 1 where 0 represents non-depressed and 1 illustrates depressed). To achieve this, it's necessary to establish a binary classification threshold. E.g., we can choose 1 as our threshold value. We will create a fresh binary target column called "binary label" based on the specified threshold. Any values exceeding the threshold will be assigned a value of 1, while values below or equal to the threshold will be designated as 0. This approach effectively transforms our dataset into a binary classification challenge.

3. *Feature Optimization:* In order to increase the model's accuracy, the most relevant and useful features must be chosen and refined. This process is known as feature optimisation. The objective is to prioritise and identify the most significant features in order to enhance depression detection algorithms' performance. [8] In the dataset used in this research, we selected the SCORE column as our target variable. In this research we have optimised the accuracies of these ML Classifiers using three swarm optimization techniques: Particle Swarm Optimization [PSO], Cuckoo Swarm Optimization (CSO), and Grey Wolf Swarm Optimization (GWO).

4. *Model Training:* We train the following machine learning classifiers on the pre-processed dataset: K-Nearest Neighbours (KNN), Decision Tree, Random Forest, Gradient Boosting. These classifiers can be divided into various groups according to their fundamental assumptions and methods of instruction. Gradient Boosting and Random Forest are ensemble techniques, whereas Decision Tree is an independent model. An instance-based technique that does not require building explicit models during training is K-Nearest Neighbours.

Decision Tree:

Type: Single tree-based model.

Characteristics: Makes decisions based on a series of feature tests.

Example: CART (Classification and Regression Trees), ID3.

Random Forest:

Type: Ensemble method.

Characteristics: Builds multiple decision trees and combines their predictions.

Example: Random Forest.

Gradient Boosting:

Type: Ensemble method.

Characteristics: Builds trees one after the other, fixing mistakes in each one.

Examples: XGBoost, AdaBoost, Light-GBM.

K-Nearest Neighbours (KNN):

Type: Instance-based method.

Characteristics: Identifies classes according to the feature space's k nearest neighbours' majority class.

Example: K-Nearest Neighbours.

5. *Classification:* We then classify the patients as depressed or not depressed.

Optimisation techniques are frequently utilised in the context of depression detection to improve the effectiveness and reliability of diagnostic models by increasing their efficiency and accuracy. The explanation of each of these optimization techniques used in this paper is as follows:

Particle Swarm Optimization (PSO)

The application of PSO in the context of depression detection unfolds as follows:

Feature Selection: The identification of salient features or variables associated with depression constitutes a fundamental preliminary stage in the construction of an effective detection model. [9]

Parameter Tuning: In the realm of machine learning, numerous algorithms and models feature hyperparameters that necessitate calibration for optimal performance. [9]

Ensemble Learning: The utilization of PSO extends to the optimization of ensemble learning strategies, whereby multiple machine learning models are combined to enhance overall predictive performance in the realm of depression detection. [9]

Cuckoo Swarm Optimization (CSO)

Cuckoo Swarm Optimization (CSO) stands as a nature-inspired optimization algorithm informed by the reproductive behavior of cuckoo birds. This algorithm has been formulated to leverage the concept of Levy flights, facilitating its application in the resolution of optimization problems. CSO is classified as a metaheuristic approach, adept at discerning the global optima within a multi-dimensional search space by virtue of its bio-inspired foundation. The fundamental underpinnings of CSO may be encapsulated in the ensuing manner:

Grey Wolf Swarm Optimization (GWO)

Grey Wolf Optimizer (GWO) is an optimization algorithm grounded in the emulation of the collective behavior exhibited by grey wolves in a pack. This algorithm was introduced in 2014 and serves as a nature-inspired metaheuristic for addressing intricate optimization problems. The operational behavior of GWO adheres to mathematical formulations that simulate the predatory and social interactions inherent to grey wolves in their natural habitat. The algorithm's principal objective is the discovery of the global optimum or proximate optimal solutions for complex optimization problems.

4. RESULTS

In this section first we have compared the accuracies on various machine learning classifiers like K-Nearest Neighbors, Random Forest, Decision Tree, and Gradient Boosting. Then we have optimised the accuracies of these classifiers using three swarm optimization techniques: Particle Swarm Optimization [PSO], Cuckoo Swarm Optimization (CSO), Grey Wolf Swarm Optimization (GWO). Then we compared the optimised accuracy of these ML classifiers.

Figure 48.3 is a bar graph of different Machine Learning Classifiers like K-Nearest Neighbors, Random Forest, Decision Tree, and Gradient Boosting with their Accuracies. The accuracy of the decision tree classifier is the highest at 91.23%.

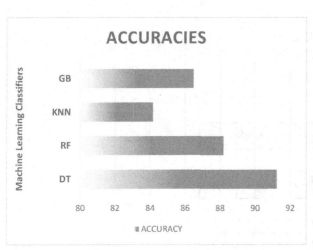

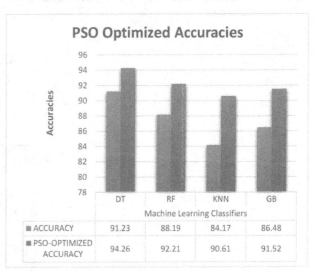

Fig. 48.3 Accuracies obtained by applying ML classifiers

Fig. 48.4 Accuracies obtained by applying PSO optimization technique

Figure 48.4 shows a bar graph of different Machine Learning Classifiers like K-Nearest Neighbors, Random Forest, Decision Tree, and Gradient Boosting with optimised accuracies. PSO-optimized decision tree accuracy was the highest at 94.26%.

Figure 48.5 shows a bar graph of different ML Classifiers like K-Nearest Neighbors, Random Forest, Decision Tree, and Gradient Boosting with GWO optimised accuracies. GWO-optimized accuracy on the decision tree classifier is the highest at 95.52%.

Figure 48.6 shows a bar graph of different ML Classifiers like K-Nearest Neighbors, Random Forest, Decision Tree, and Gradient Boosting with Cuckoo optimised accuracies. Cuckoo-optimized accuracy on the decision tree classifier is the highest at 94.77%.

Figure 48.7 shows a bar graph of different ML Classifiers with optimised accuracies with swarm optimization techniques GWO, PSO and Cuckoo Optimization techniques. GWO-optimized accuracy on the decision tree classifier is the highest at 95.52%.

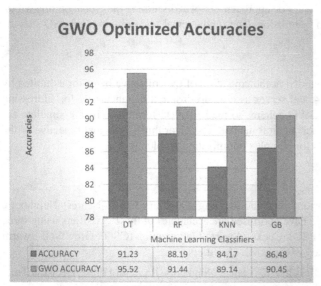

Fig. 48.5 Accuracies obtained by applying GWO optimization technique

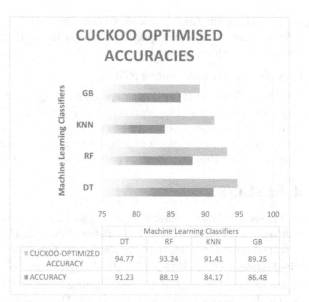

Fig. 48.6 Accuracies obtained by applying cuckoo optimization technique

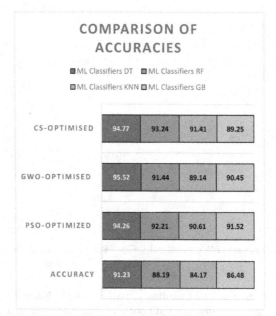

Fig. 48.7 Comparison of optimization techniques

5. CONCLUSION AND FUTURE SCOPE

In this research paper we thoroughly examined the critical task of using machine learning (ML) classifiers to identify depression in individuals. After this stage, we used three different optimisation methods, which are Grey Wolf Optimisation (GWO), Particle Swarm Optimisation (PSO), and Cuckoo Search (CS), to try to improve the results that the classifiers produced. Electronic Health Records (EHR) served as our source of data for this study, and it has provided insightful information about the possibilities of data-driven approaches for the early detection of depression.

The empirical results of this study clearly demonstrate how effective the machine learning classifiers we used—Decision Tree, Random Forest, Gradient Boosting, and K-Nearest Neighbours [KNN]—are. These classifiers performed exceptionally well in identifying subtle patterns and unique characteristics that point to depression. As such, this study sheds light on these classifiers' ability to support quick and early therapeutic interventions. Moreover, the integration of GWO, PSO, and CS optimization techniques has decisively played a pivotal role in augmenting the classifiers' overall performance. By judiciously fine-tuning model parameters and hyperparameters, these optimisation algorithms have granted the classifiers with the ability to achieve superior accuracy.

Future research and development of the suggested methodology may include real-time data stream integration, more optimisation algorithms, and the creation of user-friendly applications for widespread use in clinical settings. This will promote a more proactive and approachable approach to the diagnosis and treatment of depression.

REFERENCES

1. S. Mamidisetti and A. M. Reddy, "A Stacking-based Ensemble Framework for Automatic Depression Detection using Audio Signals," International Journal of Advanced Computer Science and Applications, vol. 14, no. 7, 2023.
2. Schuch, Felipe Barreto, and Brendon Stubbs. "The role of exercise in preventing and treating depression." Current sports medicine reports 18.8 (2019): 299-304.
3. Shobhika, "Prediction and comparison of psychological health during COVID-19 among Indian population and Rajyoga meditators using machine learning algorithms," Procedia computer science, vol. 218, pp. 697--705, 2023.
4. Mahmud, "Machine learning approaches for predicting suicidal behaviors among university students in Bangladesh during the COVID-19 pandemic: A cross-sectional study," Medicine, vol. 102, p. e34285, 2023.
5. Smith, A., Jones, B., & Johnson, C. (2019). "Depression Detection from Audio Recordings Using Machine Learning." Journal of Machine Learning in Mental Health, 5(2), 78-92.
6. Zhang, L., Wang, Y., & Li, Y. (2020). "EEG-Based Depression Detection Using Machine Learning Algorithms." International Conference on Bioinformatics and Biomedical Engineering, 215-224.
7. Wang, J., Li, C., & Zhang, L. (2018). "Detecting Depression from Social Media: A Text Mining Approach." Journal of Medical Internet Research, 20(4), e448.
8. K. Jawad, "Novel Cuckoo Search-Based Metaheuristic Approach for Deep Learning Prediction of Depression," Applied Sciences, vol. 13, p. 5322, 2023.
9. Guo, Yan, Haolan Zhang, and Chaoyi Pang. "EEG-based mild depression detection using multi-objective particle swarm optimization." 2017 29th Chinese Control And Decision Conference (CCDC). IEEE, 2017.
10. https://www.kaggle.com/datasets/hansaniuma/patient-health-scores-for-ehr-data/data
11. Kumar, Piyush, et al. "Feature based depression detection from twitter data using machine learning techniques." Journal of Scientific Research 66.2 (2022): 220-228.
12. Smys, "Analysis of deep learning techniques for early detection of depression on social media network-a comparative study," Journal of Trends in Computer Science and Smart Technology, vol. 3, pp. 24--39, 2021.
13. Z. Xu, "Subphenotyping depression using machine learning and electronic health records," Learning Health Systems, vol. 4, p. e10241, 2020.
14. Priya, "Predicting anxiety, depression and stress in modern life using machine learning algorithms," Procedia Computer Science, vol. 167, pp. 1258--1267, 2020.
15. Prasadu, "Mental Illness and trauma prediction based on machine learning," International Research Journal of Modernization in Engineering Technology and Science, vol. 5, no. 4, 2023.

Note: All the figures in this chapter were made by the author.

Computational Intelligence and Mathematical Applications – Dr. Devendra Prasad et al. (eds)
© 2024 Taylor & Francis Group, London, ISBN 978-1-032-87721-1

49 | The Impact of Augmented Reality: Applications in Education

Khushi Rao[1]

Indira Gandhi Delhi Technical University For Women, Kashmere Gate

Daniel Kr Brahma[2], Jagriti Das[3]

Central Institute of Technology Kokrajhar, Assam

Alongbar Wary[4], and Gaurav Indra[5]

Indira Gandhi Delhi Technical University For Women, Kashmere Gate

Abstract: In today's ever-changing educational scene, the use of technology has spurred the rise of smart education systems that cater to varied learning demands. Among these breakthroughs, augmented reality (AR) and virtual reality (VR) have had a significant impact on the acceptance and change of ed education. However, the hardware requirements and accompanying limits frequently pose barriers to wider adoption. This study takes a different approach, focusing on an innovative method that utilizes current infrastructure, reducing the hardware requirement often associated with AR and VR. We give a thorough examination of our technique, proving its efficacy through statistical validation. Notably, our research reveals that using our product improves children's concentration levels significantly. Our contribution improves the usefulness of AR and VR technologies in education by resolving hardware restrictions and generating real enhancements, marking a significant milestone in the continuous search for optimizing learning experiences.

Keywords: Unity 3D, Vuforia engine, Augmented reality, Education generative AI

1. INTRODUCTION

In recent years, Immersive technologies, including augmented reality and virtual reality, have had a significant influence on education. By blending physical and digital realms, it offers immersive experiences that enhance engagement and information retention. These technologies, driven by sophisticated algorithms, collectively provide dynamic educational environments by redefining traditional learning approaches. These developments promote inclusion, accessibility, and international cooperation in addition to improving educational quality. Examples of these innovations include AR-enhanced smart classrooms and immersive VR simulations. Our research program is at the forefront of the application of augmented reality, using it to create a seamless learning environment for students by integrating interactive simulations and 3D models into their actual surroundings. The advent of augmented reality (AR) on smartphones is a significant development that provides students with a cheap and easily available way to interact with realistic 3D representations. This can boost student excitement and reduce educational gaps across socioeconomic backgrounds. Its user-friendly design and ability to function independently of advanced smart classroom infrastructures are in line with our mission to improve education via the use of accessible, immersive, and empowering technology[1,2].

---
[1]khushi115bteceai21@igdtuw.ac.in, [2]u20cse1009@cit.ac.in, [3]u20cse1062@cit.ac.in, [4]alongbarwary@igdtuw.ac.in, [5]gaurav.indra.in@ieee.org

DOI: 10.1201/9781003534112-49

Our research addresses the essential issue of seamlessly integrating AR models into the realm of education. The most important problem we want to address is how to create inclusive and engaging learning experiences for all kids, regardless of their background or circumstances. Our hypothesis proposes that strategically merging AR on generally accessible platforms such as smartphones might transform interactive learning methodologies.

The underlying aim is a future in which learning is not just immersive but also intrinsically inclusive and inspiring for students of diverse backgrounds.

The paper is presented in various sections: the first section presents the introduction, the next is the Literature Survey, the third section highlights the methodology, and finally, the results and discussion are presented.

2. LITERATURE SURVEY

It has been brought to light through extensive research over the past decades that technology can be implemented and harnessed to facilitate and enhance the way of learning for learners, known as Technology Enhanced Learning(TEL). TEL is employed to offer versatility in the learning approach. Mobile learning leverages technology by prioritizing mobility and personalization in education[3]. E-comics engage elementary students by offering captivating illustrations for effective educational use [4].

The cornerstone upon which the bulk of standardization initiatives progress is precisely stated vocabularies with terminology, their related definitions, and authoritative references. Exact terminology required ontologies typically play a crucial role in the field of digital infrastructure [5]. Saidin et al. [1] identified that technology's role in education is known to engage students, yet its improper im-implementation can lead to passive learning. This integration immerses students in subjects like Medicine, Chemistry, and History, with advantages over traditional methods. However, while the potential is clear, AR's limitations also necessitate further exploration. Similar research by Hidayat et al. [6] states that the educational sector is witnessing a surge in AR adoption, appealing to stakeholders aiming to enhance learning quality. However, a gap in the literature remains regarding the direct application of AR in elementary education.

3. METHODOLOGY

3.1 Tool Selection

To achieve our objectives, we strategically leveraged two powerful tools: Unity 3D and Vuforia Engine. The cross-platform development engine Unity 3D as shown in Fig. 49.1 is strong and flexible. It is often utilized for creating interactive 2D and 3D programs, such as video games, simulations, educational software, etc. In our project, we opted for Unity 3D due to its versatility across platforms, its intuitive user interface, and its powerful 3D graphics tools. On the other hand, as shown in Fig. 49.2, the Vuforia Engine was carefully chosen as the foundation of our augmented reality software because of its extraordinary powers in creating and specifying picture targets.

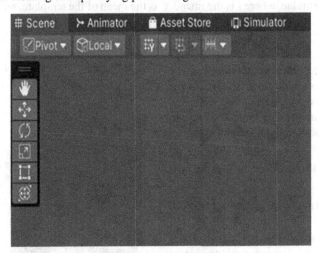

Fig. 49.1 Unity 3D Interface

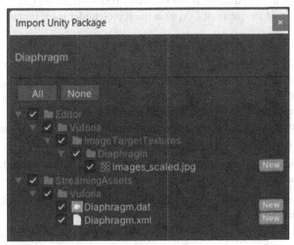

Fig. 49.2 Importing vuforia in unity

3.2 Image Target Creation

The development of image targets was one of the most important steps we took for the advancement of our AR project. In order to offer the accuracy and dependability we aimed for, from our AR application, this step was extremely essential. After careful consideration, we opted to work with the Vuforia Engine to create image targets, mainly because of its remarkable accuracy and extensive feature set.

We effectively detected images and configured them as image targets using the Vuforia Engine, as shown in Fig. 49.3. These image targets effectively represented real objects or images that would cause our AR software to do certain behaviors. By employing Vuforia Engine's image target creation capabilities, we harnessed some mathematical algorithms and techniques to detect and configure images as image targets.

Fig. 49.3 Creating image target using vuforia engine

Feature Detection and Extraction: Techniques like Harris Corner Detection or Scale-Invariant Feature Transform (SIFT) are employed to detect and extract distinctive features from an image. These features are represented by keypoint coordinates (x, y) and associated descriptors. Harris Corner Response Function is as mentioned, where $\lambda 1$ and $\lambda 2$ are the eigen-values of the image(s) gradient matrix [7].

$$R = \lambda 1 \lambda 2 - k(\lambda 1 + \lambda 2) \tag{1}$$

Homography Calculation: Homography is used to map points from one image to another. It's essential for aligning the augmented content with the image target [8].

$$s*[x' : y' : 1] = H * [x : y : 1] \tag{2}$$

Here, in Eqn. (5), (x, y) are the coordinates of the homography matrix (H) in the source image(s), and (x', y') are the transformed coordinates.

Image Matching: Image targets are often compared with stored reference images using techniques like Normalized Cross-Correlation (NCC) or Mean Squared Error (MSE) for similarity scoring [9].

$$y(u,v) = \frac{\sum [f(x,y) - \overline{f}_{u,v}][t(x-u, y-v) - \overline{t}]}{\sum_{x,y}\left[f(x, y - \overline{f}_{u,v})\right]^2 \sum_{u,v}[t(x-u, y-v) - \overline{t}]^2} \tag{3}$$

The above equation is Normalized Cross-Correlation (NCC) formulae, where f is the image, t⁻ is the mean of the template, and f⁻u,v is the mean of f (x, y) in the region under the template.

3.3 3D Image Rendering

A three-dimensional virtual scene or model is converted into a two-dimensional image using a complex method known as three-dimensional image rendering that is suitable for printing or projecting on a computer screen. In our educational project, we used 3D rendering technology to completely transform the learning process. Our primary objective was to effortlessly incorporate AR into educational settings, and one of the fundamental techniques we used was the overlaying of 3D models on 2D images.

This procedure involved building incredibly detailed 3D models that matched the instructional material we sought to impart as shown in Fig. 49.4.

Let's now go through some essential ideas and mathematical formulas in order to accomplish this:

i) *World Space:* In a 3D scene, each item is positioned using its (x, y, and z) coordinates in the world space. This is the basis of the scene, where all of the elements are independent of one another and the viewpoint of the camera.

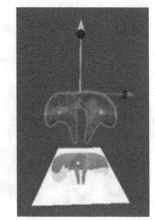

Fig. 49.4 3D image rendering

ii) *View Space (Camera Space):* A new coordinate system known as view space or camera space is introduced by the camera. It appears that the camera has replaced the original origin and that everything is defined in relation to it. Translation and Rotation are required to go from view space to world space:

$$\text{ViewMatrix} = \text{Translation} * \text{Rotation} \tag{4}$$

where Translation is the Camera Position and Rotation is the Camera Rotation.

Here, as mentioned in Eqn. (8), the 'Translation' matrix moves objects to be in the camera's reference frame, and the 'Rotation' matrix aligns the objects with the camera's orientation.

Project Transformation: The view space coordinates are transformed through a projection process to give the appearance of depth and perspective. When an object is projected in perspective, it gets smaller the further it is from the camera. The Projection Matrix can be described as the product of one's perspective with the surrounding.

$$\text{ProjectionMatrix} = \text{Perspective} * \text{Other} \tag{5}$$

4. RESULTS & DISCUSSION

The outcomes of the study demonstrated the effective incorporation of the AR technology into the educational application. The engaging and immersive quality of the application was useful in making learning more interesting and effective. A new era of dynamic and interactive learning experiences was brought about by this revolutionary feature. The captivating and immersive features of the program not only drew users in but also helped to improve the overall efficacy of the learning process.

In the above Fig. 49.5, the demo image of the application is being shown. It shows that the 3D image is projected within the image target created and the brief description of the object is also shown on the top of the projected 3D image. When testing it with 135 primary school students it was observed that the students were much more interested as compared to the traditional way of teaching which proves that the use of the application in Educational Institution is worth it.

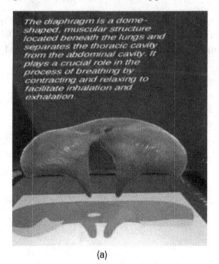
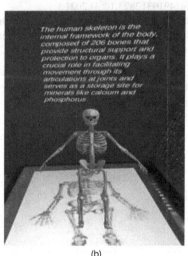

(a) (b)

Fig. 49.5 Sample images taken from application (a) Diaphragm (b) Human skeleton

In the above Table 49.1 it can be observed that the students of each age group had a greater average test score after using the AR application and the difference of about 10% can be observed between the test score before and after using the AR application.

Table 49.1 Analysis of the Data

Age group	Gender	No. of participants	Avg. test score before using AR application	Avg. test score after using AR application
8–11	Male	40	6.5	7.44
8–11	Female	34	7.11	7.93
12–14	Male	30	6.33	7.25
12–14	Female	29	7.24	7.71

5. CONCLUSION AND FUTURE SCOPE

In conclusion, the combination of AR and VR has drastically changed education, ushering in a period of immersive and interactive learning paradigms. The potential of these advances in technology to transform education is apparent, but hardware requirements have posed significant barriers to widespread implementation. Our technical input has resulted in the introduction of an innovative solution that makes use of current infrastructure while avoiding hardware restrictions. We are positive that with the use of this AR enhanced learning technique the grasping power of the students can be increased by 4-5% as it becomes much easier to understand and learn. Evolving technologies will most certainly reduce hardware limits, allowing for even greater usage. Notably, our work will include a regional language aspect, making the application more accessible to a wider people. By allowing students to learn in a language they are acquainted with, we bridge understanding barriers and assure inclusivity.

As we traverse the educational landscape, our contribution represents an important step forward in improving the accessibility and effectiveness of AR and VR technology. As these technologies progress, we envisage a future in which learning transcends boundaries, providing engaging and individualized experiences for learners all across the world.

REFERENCES

1. Saidin, Nor F. et al.:A Review of Research on Augmented Reality in Education: Advantages and Applications. International Education Studies 8.3 (2015)
2. Akçayır, Murat et al.: Advantages and Challenges Associated With Augmented Reality for Education: A Systematic Review of the Literature. Educational Research Review 20,1-11(2017)
3. Zhu, Zhi-Ting et al.: A research framework of smart education. Smart learning environments3,1-11(2016)
4. Uskov, Vladimir L. et al.: Smart education and smart e-learning. Springer, Singapore (2021)
5. Hoel, Tore et al.: Standards for smart education–towards a development framework. Smart Learning Environments5(1),1-25(2018)
6. Hidayat H., Sukmawarti S. et al.: The application of augmented reality in elementary school education. Smart Learning Environments pp.e14910312823- e14910312823.10(3),2021
7. Sánchez, Javier et al.:An Analysis and Implementation of the Harris Corner De- tector.. Image Processing On Line (2018)
8. Chum, Ondřej et al.:The Geometric Error for Homographies. Computer Vision and Image Understandings 97(1),86-102(2005)
9. Yoo, Jae-Chern et al.:Fast Normalized Cross-Correlation. Circuits, systems and signal processing 28,819-843(2009)

Note: All the figures and table in this chapter were made by the author.

Computational Intelligence and Mathematical Applications – Dr. Devendra Prasad et al. (eds)
© 2024 Taylor & Francis Group, London, ISBN 978-1-032-87721-1

Revolutionizing Attendance Management: Deep Learning and HOG-Based Smart System for Educational Institutions

Sunesh[1]

Maharaja Surajmal Institute of Technology, Janakpuri, New Delhi

Mamta[2]

PIET, Samalkha Haryana

Priyanka[3]

Maharaja Surajmal Institute of Technology, Janakpuri, New Delhi

Aakanksha[4]

PIET, Samalkha Haryana

Abstract: The manual attendance system relies on paper-based methods to keep track of attendance, following a traditional approach. Although this system is simple, cost-effective, and easy to implement, it suffers from several drawbacks. These include time-consuming processes, the potential for errors, and the inability to analyze data in real-time. To address these challenges, this paper proposes a smart attendance system based on deep learning, combined with the histogram of oriented gradient (HOG) technique. The proposed model aims to overcome the aforementioned issues by offering improved accuracy, streamlined automation, and real-time data analysis capabilities. The model utilizes HOG to extract facial features and employs a Support Vector Machine (SVM) classifier for face identification. Furthermore, the proposed system provides notifications to administrators and parents via email or SMS, keeping them informed about attendance status. By leveraging advanced technologies and techniques, this smart attendance system enhances the efficiency and effectiveness of attendance management in educational institutions.

Keywords: Deep learning, Face detection, Report generation, Support vector machine (SVM), Histogram of oriented gradient

1. INTRODUCTION

The objective of the face recognition system is to streamline and expedite the manual attendance process. By implementing an automatic attendance system, faculty members can efficiently manage attendance, save time, and leverage data-driven insights to enhance teaching and learning experiences. The system captures students' images and utilizes a facial recognition model to record attendance. An algorithm compares the captured images with a comprehensive master database containing students' details.

The implementation of Automatic attendance system involves three phases. The first phase focuses on capturing images of students present in the classroom, ensuring that all students are accounted for. In the second phase, the system identifies faces through a dynamic process where the webcam captures images and compares them with pre-fed data for recognition. The third

[1]mamta1.tarar@gmail.com, [2]erhsmalik@gmail.com, [3]priyankanandal@msit.in, [4]er.aakanksha@yahoo.co.in

DOI: 10.1201/9781003534112-50

phase involves generating a list of attendees and absentees and notifying parents via email and SMS when their children are absent. Additionally, the administrator receives notifications regarding attendance details. The face recognition system utilizes a technique to compare and validate real humans from captured images. Machines require training using an SVM classifier to efficiently detect and recognize faces. The process begins by converting the image to grayscale to identify faces within it. The system analyzes each pixel and its surrounding area to determine the extent to which pixels are darker compared to adjacent ones. This information is used to create gradients, which are combined from multiple faces to form a comprehensive pattern for comparison with the target image. This process is known as the histogram of gradients. In this proposed model, optimized attendance surveillance provides the administrator with a list of attendees and absentees, significantly saving time and improving efficiency.

2. RELATED WORK

The related work section describes various methodologies and techniques used in facial recognition systems for automated attendance management. These systems employ a combination of technologies such as convolutional neural networks (CNN), PCA, Eigenface values, YOLO V3 algorithm, local binary patterns (LBP), Haar cascades, Gabor filters, SVM, generative adversarial networks (GAN), and deep learning.

S. Sawhney proposed a methodology for leveraging facial recognition technology in conjunction with convolutional neural networks, PCA, and Eigenface values to construct an automated attendance management system for students in a class (CNN) [1]. An unified framework has been developed on Microsoft Azure's face API in conjunction with YOLO V3 (You Only Look Once) algorithm to facilitate face detection within a database (face database) [2]. YOLO V3 count the students in a picture, differentiate between recognized and unrecognized faces, and display the results in spreadsheet. A web-based face-recognition student attendance system using KNN is utilized to categorizes the student faces, and deep learning is used to build facial embedding, and convolutional neural networks (CNN) are used to detect faces in photos [3].

The face recognition portion is completed using the local binary pattern (LBP) approach, the face detection portion is completed using the viola-jones algorithm method [4].

The combination of face recognition technology with the local binary patterns histograms (LBPH) algorithm for face detection with Haar feature-based cascades, and distance-based clustering is used[9]. The suggested solution automatically logs everyone's attendance in a classroom setting and provide a spreadsheet to user for output outlining of attendance. A novel approach is introduced using the Local Binary Pattern (LBP) algorithm in conjunction with advanced image processing techniques such as Contrast Adjustment, Bilateral Filter, Histogram Equalization, and Image Blending[10]. The approach emphasizes the distinctive characteristics of several attributes, like the face, eyes, and nose of individuals. Different real-time scenarios are taken into consideration in order to assess the effectiveness of various face recognition systems[11]. It offers advice on how to handle tricky situations like spoofing and avoiding student proxy.

Three convolutional neural networks that were trained on our data are used to implement transfer learning. In terms of high accuracy of prediction and reasonable training time, the all three networks give a good performance.

In order to detect students, face a model has been constructed to categorize all persons face from a recorded image, using a set of rules, or the LBP method [18].

The recommended framework is utilized for modules encompassing face detection, mask detection, face recognition, and generation of attendance reports [26]. The haar cascade classifier is used to recognize faces and facemasks. The eigenface-mams and local binary pattern histogram were two methods for face recognition that were researched[27] they used a database, which contains the student's name, picture, roll number, and attendance time, will be linked to the system. The three steps in this application are the key ones. It will first capture pictures. Next, compare them to the current photographs that are kept in the main database. Thirdly, it will mark missing the remaining pupils from that class and automatically present all the matched photographs on a spreadsheet.

Author describe a system that makes use of a variety of facial recognition models, including VGG-FACE, Facenet, Openface, and DeepFace, in order to identify the person with a higher degree of accuracy[28]. A practical perspective is proposed on technical advancements, making it easier to employ facial recognition technology to supplement information available online, to verify faces, and to suggest potential uses for a variety of applications [29]. In order to accomplish the desired outcomes, AttendXN and a hierarchical feedforward network employ ResNet as the technique. A system is designed that utilise a Multi-Task Cascaded Convolutional Neural Network (MTCNN) [30]. In order to extract the facial landmark from the image, The

Face Alignment Network (FAN) is also applied to the image. FaceNet is used to extract data from the face. The retrieved face embedding from the Face Detection System is then categorized.

3. PROPOSED ATTENDANCE SURVEILLANCE MODEL

This section presents the proposed model mainly composed four distinct modules: Optimized Face Identification, marking attendance, report generation and alert generation. Proposed model is depicted in Fig. 50.1.

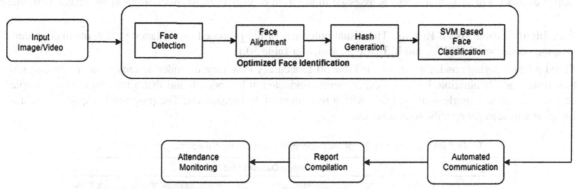

Fig. 50.1 Proposed attendance surveillance model

3.1 Optimized Face Identification

This module process involves detection of faces, aligning them, encoding their features, and classifying them using a trained SVM classifier. Machine learning-based face identification involves several key steps in the process:

1. **Face detection:** This initial step focuses on identifying faces within an image. In this, the concept of Histogram of Oriented Gradient (HOG) is utilized to convert the image into black and white, analyze pixel patterns, and establish gradients.
2. **Face alignment:** Faces in the captured image may appear tilted or positioned at different angles. To address this, estimation process locates 68 specific points on the face. These identified points provide the basic structure of face and enable geometric operations to bring face into a standardized orientation.
3. **Hash Generation:** In order to compare images, they need to be transformed into a numerical representation. In furtherance, deep convolutional networks are trained to encode images and generation of a unique hash for each face
4. **SVM based Face classification:** The final step involves using a classifier Support Vector Machine (SVM), to classify the images based on their encoded features. The classifier is trained to analyze measurements from a new testing image and identify the person with the closest match.

3.2 Attendance Monitoring

In this Module, Python dictionaries are utilized to update the status of a candidate on a specific date based on the minimum face distance between an unknown image and known images. During this step, the face differences between known and unknown images are compared, and the minimum face distance of the unknown image is returned. Python dictionaries are then employed to update the candidate's status on the given date. The time and status (whether present or absent) are stored in the dictionaries. The program is capable of handling input of unknown images from both the webcam and a directory.

3.3 Report Compilation

The result compilation module generates the attendance reports based on custom parameters. The administrator inputs custom parameters to generate a report within a specific time frame. By utilizing file handling traits in Python, the proposed module effectively read and write CSV file, enabling the generation of the report based on the provided parameters.

3.4 Automated Communication

In this module, the proposed model sends notifications to its stakeholders through email and SMS alerts. Email notifications are used to inform absentee guardians about their wards' absences, while admins receive emails with an attached "report.csv" containing overall customized records. SMS alerts are also sent to absentee guardians using Fast2SMS services, which provide bulk SMS, quick SMS, and DLT services.

4. RESULT AND DISCUSSIONS

The proposed attendance surveillance model was thoroughly evaluated through a series of experiments on Python. The results of these experiments demonstrated the effectiveness of the proposed model, and the best outcomes were achieved. The performance evaluation of proposed model is assessed in the form of visual results, Accuracy and automated communication results.

1. **Face Identification Visual Results:** The visual outcomes of the proposed surveillance model, obtained under distinct imaging conditions (IC1, IC2, and IC3), are presented in Table 50.1.

   Based on the imaging conditions shown in Table 50.1, accuracy may vary, considering factors such as image resolution, face pose, and illumination. Based on experiments conducted, it has been found that a maximum of 10 people can be accommodated in a single well-lit image with a resolution of 16 megapixels. The program is capable of scanning one image at a time as per specific requirements.

Table 50.1 Visual results of face identification under three imaging conditions

2. **Accuracy Results and Analysis:** Based on the distinct imaging conditions as presented in face identification visual results, the accuracy of proposed model is calculated and depicted in Table 50.2.

Table 50.2 Accuracy results of proposed attendance surveillance model

Image	Total Person present in image	Details present in dataset	No Face Detection	Correctly Identified	Incorrect identification	Not Identified	Accuracy %
IC1	13	9	0	9	0	0	100
IC2	10	10	2	6	1	1	60
IC3	16	16	2	9	0	4	15.25

However, there are some limitations and drawbacks to consider. Firstly, the liveliness of an image cannot be confirmed at present. Secondly, the recognition accuracy is not 100%, which means there might be instances of misidentification. Moreover, the system struggles with blurred images and faces wearing spectacles, affecting its performance. Despite these drawbacks, the system provides the option to generate custom attendance reports in CSV format for administrators. It is important to consider these limitations when evaluating the economic viability and practicality of implementing the system.

3. **Automated Communication Results:** The automated communication is performed in two forms: email and SMS. Figure 50.2 illustrates the outcomes of automated communication. Figure 50.2 consists of two parts: Part a displays the results of email communication, while Part b exhibits the outcomes of SMS communication.

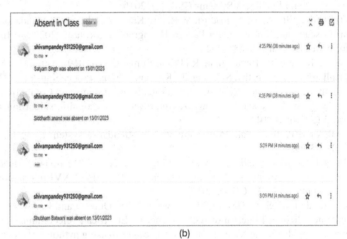

(a)   (b)

Fig. 50.2 Automated communication results (a) E-mail results, (b) SMS result

5. CONCLUSION AND FUTURE SCOPE

This paper introduces a proposed Attendance Surveillance model with Deep Learning and HOG-Based Automation for attendance monitoring. Some of finding of present paper are given below:

(a) The proposed model for attendance monitoring encompasses four distinct modules that work together to achieve efficient and automated attendance management.

(b) The Optimized Face Identification module utilizes machine learning techniques to detect and align faces, generate unique hashes, and classify faces using an SVM classifier. The Attendance Monitoring module utilizes Python dictionaries to update candidate statuses based on face distance calculations and stores time and attendance information.

(c) The Report Compilation module generates attendance reports based on custom parameters, leveraging Python's file handling capabilities. Lastly, the Automated Communication module sends notifications to stakeholders through email and SMS alerts, ensuring timely and customized communication.

(d) However, it is important to address certain limitations such as image quality, camera distance, and the inability to differentiate real faces from photographs. Resolving these challenges is crucial before implementing the model on a commercial scale. With further improvements, this model holds great potential to save time and enhance attendance management processes in various institutions.

REFERENCES

1. S. Sawhney, K. Kacker, S. Jain, S. N. Singh and R. Garg, "Real-Time Smart AttendanceSystem using Face Recognition Techniques," 2019 9th International Conference on Cloud Computing, Data Science & Engineering (Confluence), 2019

2. Khan, S., Akram, A. & Usman, N. Real Time Automatic Attendance System for Face Recognition Using Face API and OpenCV. *Wireless Pers Commun* **113**, 469–480 (2020).

3. Sutabri, Tata, Ade Kurniawan Pamungkur, and Raymond Erz Saragih. "Automatic attendance system for university student using face recognition based on deep learning." *International Journal of Machine Learning and Computing* 9.5 (2019): 668- 674.

4. Shah, K., Bhandare, D., Bhirud, S. (2021). Face Recognition-Based Automated Attendance System. In: Gupta, D., Khanna, A., Bhattacharyya, S., Hassanien, A.E., Anand, S., Jaiswal, A. (eds) International Conference on Innovative Computing and Communications. Advances in Intelligent Systems and Computing, vol 1165. Springer, Singapore.

5. S. Dev and T. Patnaik, "Student Attendance System using Face Recognition," 2020 International Conference on Smart Electronics and Communication (ICOSEC), 2020

6. Z. -h. Lin and Y. -z. Li, "Design and Implementation of Classroom Attendance System Based on Video Face Recognition," 2019 International Conference on Intelligent Transportation, Big Data & Smart City (ICITBS), 2019

7. E. Winarno, I. Husni Al Amin, H. Februariyanti, P. W. Adi, W. Hadikurniawati and M.T. Anwar, "Attendance System Based on Face Recognition System Using CNN-PCA Method and Real-time Camera," 2019 International Seminar on Research of Information Technology and Intelligent Systems (ISRITI), 2019

8. B. Tej Chinimilli, A. T., A. Kotturi, V. Reddy Kaipu and J. Varma Mandapati, "Face Recognition based Attendance System using Haar Cascade and Local Binary Pattern Histogram Algorithm," 2020 4th International Conference on Trends in Electronics and Informatics (ICOEI)(48184), 2020

9. Agarwal, R., Jain, R., Regunathan, R., Pavan Kumar, C.S. (2019). Automatic Attendance System Using Face Recognition Technique. In: Kulkarni, A., Satapathy, S., Kang, T., Kashan, A. (eds) Proceedings of the 2nd International Conference on Data Engineering and Communication Technology. Advances in Intelligent Systems and Computing, vol 828. Springer, Singapore

10. Bah, Serign Modou, and Fang Ming. "An improved face recognition algorithm and its application in attendance management system." *Array* 5 (2020): 100014.

11. Chakraborty, Partha, et al. "Automatic student attendance system using face recognition." *Int. J. Eng. Adv. Technol.(IJEAT)* 9.3 (2020): 93-99.

12. L. Fung-Lung, M. Nycander-Barúa and P. Shiguihara-Juárez, "An Image Acquisition Method for Face Recognition and Implementation of an Automatic Attendance System for Events," 2019 IEEE XXVI International Conference on Electronics, Electrical Engineering and Computing (INTERCON), 2019

13. I. Q. Mundial, M. S. Ul Hassan, M. I. Tiwana, W. S. Qureshi and E. Alanazi, "Towards Facial Recognition Problem in COVID-19 Pandemic," 2020 4rd International Conference on Electrical, Telecommunication and Computer Engineering (ELTICOM), 2020

14. Suresh, V., et al. "Facial recognition attendance system using python and OpenCv." *Quest Journals- Journal of Software Engineering and Simulation* 5.2 (2019): 18-29.

15. K. M. M. Uddin, A. Chakraborty, M. A. Hadi, M. A. Uddin and S. K. Dey, "Artificial Intelligence Based Real-Time Attendance System Using Face Recognition," 2021 5thInternational Conference on Electrical Engineering and Information CommunicationTechnology (ICEEICT), 2021

16. K. Sanath, M. K, M. Rajan, V. Balamurugan and M. E. Harikumar, "RFID and Face Recognition based Smart Attendance System," 2021 5th International Conference on Computing Methodologies and Communication (ICCMC), 2021

17. Alhanaee, Khawla, et al. "Face Recognition Smart Attendance System using Deep Transfer Learning."*Procedia Computer Science* 192 (2021): 4093-4102.

18. K. Preethi and S. Vodithala, "Automated Smart Attendance System Using Face Recognition," 2021 5th International Conference on Intelligent Computing and Control Systems (ICICCS), 2021

19. Naufal, Ghani Rizky, et al. "Deep learning-based face recognition system for attendance system." *ICIC Express Lett. Part B Appl* 12 (2021):

20. Jadhav, Sharad R., Bhushan U. Joshi, and Aakash K. Jadhav. "Attendance System Using Face Recognition for Academic Education." *Computer Networks and Inventive Communication Technologies*. Springer, Singapore, 2021. 431-436.

21. Wati, Vera, et al. "Security of facial biometric authentication for attendance system." *Multimedia Tools and Applications* 80.15 (2021): 23625-23646.

22. AB, Evanjalin, and Reshma RS. "Face Recognition System Attendance System using Raspberry Pi." *Irish Interdisciplinary Journal of Science & Research (IIJSR) Vol* 5 (2021): 84-91

23. Hung, Bui Thanh, and Nguyen Nhi Khang. "Student attendance system using face recognition." *Proceedings of Integrated Intelligence Enable Networks and Computing*. Springer, Singapore, 2021. 967- 977.

24. Y. Chen and X. Li, "Research and Development of Attendance Management System Based on Face Recognition and RFID Technology," 2021 IEEE International Conference on Information Communication and Software Engineering (ICICSE), 2021

25. L. Agarwal, M. Mukim, H. Sharma, A. Bhandari and A. Mishra, "Face Recognition Based Smart and Robust Attendance Monitoring using Deep CNN," 2021 8th International Conference on Computing for Sustainable Global Development (INDIACom), 2021, pp. 699-704.

26. Suhaimin, Mohd Suhairi Md, et al. "Real-time mask detection and face recognition using eigenfaces and local binary pattern histogram for attendance system." *Bulletin of Electrical Engineering and Informatics* 10.2 (2021): 1105-1113.

27. K. Mridha and N. T. Yousef, "Study and Analysis of Implementing a Smart Attendance Management System Based on Face Recognition Tecqnique using OpenCV and Machine Learning," 2021 10th IEEE International Conference on Communication Systems and Network Technologies (CSNT), 2021, pp. 654- 659

28. V. A, R. R. Krishna and R. V. U, "Facial Recognition System for Automatic AttendanceTracking Using

29. an Ensemble of Deep-Learning Techniques," 2021 12th International Conference on Computing Communication and Networking Technologies (ICCCNT), 2021, pp. 1-6

30. A. S. Rafika, Sudaryono, M. Hardini, A. Y. Ardianto and D. Supriyanti, "Face Recognition based Artificial Intelligence With AttendX Technology for Student Attendance," 2022 International Conference on Science and Technology (ICOSTECH), 2022, pp. 1-7

31. A. Chowanda, J. Moniaga, J. C. Bahagiono and J. Sentosa Chandra, "Machine Learning Face Recognition Model for Employee Tracking and Attendance System," 2022 International Conference on Information Management and Technology (ICIMTech)

Note: All the figures and tables in this chapter were made by the author.